CHEMISTRY RESEARCH AND APPLICATIONS

ANALYTICAL CHEMISTRY OF CADMIUM: SAMPLE PRE-TREATMENT AND DETERMINATION METHODS

CHEMISTRY RESEARCH AND APPLICATIONS

Applied Electrochemistry
Vijay G. Singh (Editor)
2009. ISBN: 978-1-60876-208-8

Handbook on Mass Spectrometry: Instrumentation, Data and Analysis, and Applications
J. K. Lang (Editor)
2009. ISBN: 978-1-60741-580-0

Handbook of Inorganic Chemistry Research
Desiree A. Morrison (Editor)
2010. ISBN: 978-1-61668-010-7

Handbook of Inorganic Chemistry Research
Desiree A. Morrison (Editor)
2010. ISBN: 978-1-61668-712-0 (E-book)

Solid State Electrochemistry
Thomas G. Willard (Editor)
2010. ISBN: 978-1-60876-429-7

Mathematical Chemistry
W. I. Hong (Editor)
2010. ISBN: 978-1-60876-894-3

Mathematical Chemistry
W. I. Hong (Editor)
2010. ISBN: 978-1-61668-440-2 (E-book)

Physical Organic Chemistry: New Developments
Karl T. Burley (Editor)
2010. ISBN: 978-1-61668-435-8

Physical Organic Chemistry: New Developments
Karl T. Burley (Editor)
2010. ISBN: 978-1-61668-469-3 (E-book)

Chemical Sensors: Properties, Performance and Applications
Ronald V. Harrison (Editor)
2010. ISBN: 978-1-60741-897-9

Macrocyclic Chemistry: New Research Developments
Dániel W. Fitzpatrick and Henry J. Ulrich (Editors)
2010. ISBN: 978-1-60876-896-7

Electrolysis: Theory, Types and Applications
Shing Kuai and Ji Meng (Editors)
2010. ISBN: 978-1-60876-619-2

Energetic Materials: Chemistry, Hazards and Environmental Aspects
Jake R. Howell and Timothy E. Fletcher (Editors)
2010. ISBN: 978-1-60876-267-5

Chemical Crystallography
Bryan L. Connelly (Editor)
2010. ISBN: 978-1-60876-281-1

Chemical Crystallography
Bryan L. Connelly (Editor)
2010. ISBN: 978-1-61668-513-3 (E-book)

Heterocyclic Compounds: Synthesis, Properties and Applications
Kristian Nylund and Peder Johansson (Editors)
2010. ISBN: 978-1-60876-368-9

Influence of the Solvents on Some Radical Reactions
Gennady E. Zaikov, Roman G. Makitra, Galina G. Midyana and Liliya I. Bazylyak
2010. ISBN: 978-1-60876-635-2

Dzhemilev Reaction in Organic and Organometallic Synthesis
Vladimir A.D'yakonov
2010. ISBN: 978-1-60876-683-3

Analytical Chemistry of Cadmium: Sample Pre-Treatment and Determination Methods
Antonio Moreda-Piñeiro and Jorge Moreda-Piñeiro
2010. ISBN: 978-1-60876-808-0

Binary Aqueous and CO2 Containing Mixtures and the Krichevskii Parameter
Aziz I. Abdulagatov, Ilmutdin M. Abdulagatov and Gennadii V. Stepanov
2010. ISBN: 978-1-60876-990-2

Advances in Adsorption Technology
Bidyut Baran Saha and Kim Choon Ng (Editors)
2010. ISBN: 978-1-60876-833-2

Electrochemical Oxidation and Corrosion of Metals
Elena P. Grishina and Andrew V. Noskov
2010. ISBN: 978-1-61668-329-0

Electrochemical Oxidation and Corrosion of Metals
Elena P. Grishina and Andrew V. Noskov
2010. ISBN: 978-1-61668-329-0 (E-book)

Modification and Preparation of Membrane in Supercritical Carbon Dioxide
Guang-Ming Qiu, Rui Tian, Yang Qiu and You-Yi Xu
2010. ISBN: 978-1-60876-905-6

Thermostable Polycyanurates: Synthesis, Modification, Structure and Properties
Alexander Fainleib (Editor)
2010. ISBN: 978-1-60876-907-0

Combustion Synthesis of Advanced Materials
B. B. Khina
2010. ISBN: 978-1-60876-977-3

Structure and Properties of Particulate-Filled Polymer Composites: The Fractal Analysis
G. V. Kozlov, Y. G. Yanovskii and G. E. Zaikov
2010. ISBN: 978-1-60876-999-5

Information Origins of the Chemical Bond
Roman F. Nalewajski
2010. ISBN: 978-1-61668-305-4

Cyclic β-Keto Esters: Synthesis and Reactions
M.A. Metwally and E. G. Sadek
2010. ISBN: 978-1-61668-282-8

Wet Electrochemical Detection of Organic Impurities
F. Manea, C. Radovan, S. Picken and J. Schoonman
2010. ISBN: 978-1-61668-661-1

Wet Electrochemical Detection of Organic Impurities
F. Manea,
C. Radovan,
S. Picken
and J. Schoonman
2010. ISBN: 978-1-61668-491-4
(E-book)

Quantum Frontiers of Atoms and Molecules
Mihai V. Putz (Editor)
2010.ISBN: 978-1-61668-158-6

Molecular Symmetry and Fuzzy Symmetry
Xuezhuang Zhao
2010. ISBN: 978-1-61668-528-7

Molecular Symmetry and Fuzzy Symmetry
Xuezhuang Zhao
2010. ISBN: 978-1-61668-375-7 (E-book)

Molybdenum and Tungsten Cofactor Model Chemistry
Carola Schulzke and Prinson P. Samuel
2010. ISBN: 978-1-61668-750-2

Molybdenum and Tungsten Cofactor Model Chemistry
Carola Schulzke and Prinson P. Samuel
2010. ISBN: 978-1-61668-828-8 (E-book)

Rock Chemistry
Basilio Macías and Fidel Guajardo (Editors)
2010. ISBN: 978-1-60876-563-8

Boron Hydrides, High Potential Hydrogen Storage Materials
Umit B. Demirci and Philippe Miele (Editors)
2010. ISBN: 978-1-61668-362-7

Tetraazacyclotetradecane Species as Models of the Polyazacrown Macrocycles
Ryszard B. Nazarski
2010. ISBN: 978-1-61668-487-7

Tetraazacyclotetradecane Species as Models of the Polyazacrown Macrocycles
Ryszard B. Nazarski
2010. ISBN: 978-1-61668-9001 (E-book)

Chemical Reactions in Gas, Liquid and Solid Phases: Synthesis, Properties and Application
G.E. Zaikov and R.M. Kozlowski (Editors)
2010. ISBN: 978-1-61668-671-0

Chemical Reactions in Gas, Liquid and Solid Phases: Synthesis, Properties and Application
G.E. Zaikov and R.M. Kozlowski (Editors)
2010. ISBN: 978-1-61668-906-3 (E-book)

Ion Transfer at Liquid/Liquid Interfaces
Rodrigo Alejandro Iglesias and Sergio Alberto Dassie
2010. ISBN: 978-1-61668-684-0

Calixarene Complexes with Solvent Molecules
O.V. Surov, M.I. Voronova and A.G. Zakharov
2010. ISBN: 978-1-61668-755-7

Nobel Laureates and Nanotechnologies of Applied Quantum Chemistry
Vladimir A. Babkin and Gennady E. Zaikov
2010. ISBN: 978-1-61668-849-3

Nobel Laureates and Nanotechnologies of Applied Quantum Chemistry
Vladimir A. Babkin and Gennady E. Zaikov
2010. ISBN: 978-1-61728-057-3 (E-book)

CHEMISTRY RESEARCH AND APPLICATIONS

ANALYTICAL CHEMISTRY OF CADMIUM: SAMPLE PRE-TREATMENT AND DETERMINATION METHODS

ANTONIO MOREDA-PIÑEIRO
AND
JORGE MOREDA-PIÑEIRO

Nova Science Publishers, Inc.
New York

LIBRARY OF CONGRESS CATALOGING-IN-PUBLICATION DATA

Moreda-Piñeiro, Antonio.
Analytical chemistry of cadmium : sample pre-treatment and determination methods / Antonio Moreda-Piñeiro, Jorge Moreda-Piñeiro.
p. cm.
Includes index.
ISBN 978-1-60876-808-0 (hardcover)
1. Cadmium--Environmental aspects--Analysis. 2. Environmental chemistry. I. Moreda-Piñeiro, Jorge. II. Title.
TD196.C28M67 2009
628.5'2--dc22

2009048761

Published by Nova Science Publishers, Inc. ✢ New York

CONTENTS

PREFACE

A brief description of sampling strategies, preservation and handling of environmental samples (liquids, solids and air) are commented with particular emphasis on the subsequent cadmium determination.

The main analytical techniques for cadmium determination are briefly described, and they are compared in base on their advantages and drawbacks. Therefore, UV-visible spectrometry, atomic spectrometry techniques, neutron activation analysis, and some electrochemical methods are considered. In addition, the use of automatic analyzers for environmental monitoring are also described.

The main sample pre-treatment methods for liquid and solid environmental samples are also discussed. Well-established methodologies as well as current trends are reviewed for cadmium determination in environmental matrices such as water, soils, sediments, biota, and atmospheric particulate matter.

In addition, cadmium speciation, mainly cadmium fractionation in soil, sediment, sludge and atmospheric particulate matter (BCR method), and isolation and separation of cadmium bound to metallothioneis in biota is also reviewed. Finally, a source of literature on standard and official methods for cadmium determination in environmental materials is also provided.

ABBREVIATIONS AND ACRONYMS

The following abbreviations and acronyms are used in this Chapter:

AAS	atomic absorption spectrometry
AC	activated carbon
ADC	analog to digital converter
AdSV	adsorptive striping voltammentry
AE	anion exchange
AES	atomic emission spectrometry
AFS	atomic fluorescence spectrometry
AMPDAA	2-acetylmercaptophenyldiazoaminoazobenzene
APDC	ammonium pyrrolidin dithiocarbamate
ASE	accelerated solvent extraction
ASV	anodic stripping voltammetry
AT	aminothiophenol
AXAD	Amberlite XAD
BiBDD	bismuth boron doped diamond
BiFE	bismuth film electrode
BIFF-AAS	beam injection flame furnace – atomic absorption spectrometry
Br-PADAP	2-(5-bromo-2-pyridylazo)-5-(diethylamino)-phenol
BTADAP	2-[2-benzothiazolylazo)-5-dimethylaminophenol
BZA	benzoylacetone
CCE	N,N'-bis(2-hydroxy-5-nitrobenzyl)-4,13-diazadibenzo-18-crown-6
CCD	charge coupled device
CDTA	1,2-diaminocyclohexane-N,N,N'N'-tetraacetic acid
CE	capillary electrophoresis
CEC	cationic exchange capacity
CFA	continuous flow analysis
CFME	continuous flow micro-extraction
CID	charge injection device
cLC	capillary liquid chromatography
CMC	critical micellar concentration
CNT	carbon nanotubes
CNAA	cyclic neutron activation analysis

CPE	cloud point extraction
CRM	certified reference material
CS	continuous source
CSV	cathodic stripping voltammetry
CTAB	cetyltrimethyl ammonium bromide
CV-AAS	cold vapor - atomic absorption spectrometry
CV-AFS	cold vapor - atomic fluorescence spectrometry
CV-ICP-OES	cold vapor - inductively coupled plasma - optical emission spectrometry
CV-ICP-MS	cold vapor - inductively coupled plasma - mass spectrometry
CVG	chemical vapor generation
CVG-AAS	chemical vapor generation - atomic absorption spectrometry
CVG-AFS	chemical vapor generation - atomic fluorescence spectrometry
CVG-ICP-OES	chemical vapor generation - inductively coupled plasma - optical emission spectrometry
CVG-ICP-MS	chemical vapor generation - inductively coupled plasma - mass spectrometry
DAAB	diazoaminobenzene
DAD1	N,N'-bis((1R)-1-ethyl-2-hydroxyethyl)ethanediamide
DAD2	N,N'-bis((1S)-1-benzyl-2-hydroxyethyl)-ethanediamide
DAFS	dispersive atomic fluorescence spectrometry
DC	direct current
DDAB	didodecyldimethylammonium-bromide
DDC	diethyldithiocarbamate
DDDE	dibenzylammonium dibenzyldithiocarbamate
DDTP	ammonium O,O-diethyl-dithiophosphate
DEM	2-(diethylamino) ethyl methacrylate
DGNAA	delayed gamma ray neutron activation analysis
DIBK	2,6-dimethyl-heptan-4-one
DIHN	direct injection high efficiency nebulizer
DLLME	dispersive liquid - liquid micro-extraction
DMF	dimethylformamide
DNDAA	2,4-dinitrophenyldiazoaminoazobenzene
DPASV	differential pulse anodic striping voltammentry
DPV	differential pulse voltammentry
DQQH	2,2'diquinolyl ketone 2-quinolylhidrazone
DTC	bis(carboxymethyl)dithiocarbamate
EcHG	electrochemical hydride generation
EDL	electrodeless discharged lamp
EDTA	ethylendiaminotetraacetic acid
EDXRF	energy dispersive X-ray fluorescence spectrometry
EDS	energy dispersive system
ENAA	epithermal neutron activation analysis
EOF	electroosmotic flow
ESI-MS	electrospray ionization - mass spectrometry
ESI-MS-MS	electrospray ionization tandem - mass spectrometry

ETAAS	electrothermal atomic absorption spectrometry
ETV	electrothermal vaporization
FAAS	flame atomic absorption spectrometry
FAES	flame atomic emission spectrometry
FI	flow injection
FIA	flow injection analysis
FIT	flame in-tube atomizer
Freon-TF	1,1,2 trichlorofluoroethane
GAC	green analytical chemistry
GBHA	glyoxal-bis (2-hydroxyanil)
GC	green chemistry
GCE	glassy carbon electrode
GECE	graphite-epoxy composite electrode
HCL	hollow cathode lamp
HDAA	o-hydroxybenzenediazoaminoazobenzene
HDPBA	N1-hydroxy-N1,N2-diphenylbenzamidine
HEDC	bis (2-hydroxyethyl)dithiocarbamate
HEPA	high efficiency particulate
HMDE	hanging mercury drop electrode
HDz	dithizone (diphenylthiocarbazone)
HEP-XRF	high-energy polarized X-ray fluorescence spectrometry
HF	hollow fibre
HF-LPME	hollow fibre – liquid phase micro-extraction
HF-LLLME	hollow fibre – liquid-liquid-liquid micro-extraction
HG-AAS	hydride generation - atomic absorption spectrometry
HG-AFS	hydride generation - atomic fluorescence spectrometry
HG-ICP-OES	hydride generation - inductively coupled plasma - optical emission spectrometry
HG-ICP-MS	hydride generation inductively coupled plasma - mass spectrometry
HHPN	hydraulic high pressure pneumatic nebulizer
HI-HCL	high intensity - hollow cathode lamp
HMA-HMDTC	hexamethyleneammonium hexamethylenedithiocarba- mate
HPLC	high-performance liquid chromatography
HPN	high pneumatic nebulizer
HpDTC	heptyldithiocarbamate
HR-CS AAS	high resolution - continuous source atomic absorption spectrometry
HS-SDME	head space single-drop micro-extraction
H_2DZ	ditizone
IAT	integrated atom trap
IBMK	4-Methyl-2-pentanone
ICP	inductively coupled plasma
ICP-OES	inductively coupled plasma-optical emission spectrome-try
ICP-MS	inductively coupled plasma-mass spectrometry
IDA	isotopic dilution analysis
IE11	N-propylsalicylaldimine
IIP	ionic imprinted polymer

IS	inmunosorbent
ISE	ion selective electrode
ISO	International Organization for Standardization
JIN	jet impact nebulization
JIV	jet impact vaporization
KR	knotted reactor
LEAFS	laser-excited atomic fluorescence spectrometry
LIF	laser-induced atomic fluorescence spectrometry
LLE	liquid - liquid extraction
LLME	liquid – liquid micro-extraction
LLLME	liquid-liquid-liquid micro-extraction
LPME	liquid phase micro-extraction
LPAT	long-path absorption tube
LS	line source
LSASV	linear sweep anodic stripping voltammetry
LSV	linear sweep voltammetry
LTDC	low-temperature directed crystallization
MCA	multichannel analyzer
MDE	mercury drop electrode
MIP	moleculary imprinted polymer
MLP	metallothionein-like protein
MPT	microwave plasma torch
MSIS	multimode sample introduction system
MT	metallothionein
MWCNT	multi-walled carbon nanotubes
NAA	neutron activation analysis
NDAFS	non-dispersive atomic fluorescence spectrometry
NIST	national institute for standardization and testing
NMR	nuclear magnetic resonance spectroscopy
NPV	normal pulse voltammetry
OES	optical emission spectrometry
PAN	1-(2-pyridylazo)-2-naphthol
PAR	4-(2-pyridylazo)resorcinol
PDHP	potassium dihydrogen phosphate
PGNAA	prompt gamma ray neutron activation analysis
PHP	potassium hydrogen phosphate
PHWE	pressurized hot water extraction
PLE	pressurized liquid extraction
PMBP	1-phenyl-3-methyl-4-benzoyl-5-pyrazolone
PN	pneumatic nebulizer
PNIPAAm	poly (N-isopropylacrylamide)
PONPE 7.5	polyethyleneglycolmonop-nonylphenylether
PPT	phenanthraquinone monophenylthiosemicarbazone
PSE	pressurized solvent extraction
PTFE	polytetrafluoroethylene
PUF	polyurethane foam

PVC	polyvinylchloride
PVME	poly(vinyl methyl ether)
PVP	polyvinylpyrrolidinone
QADP	2-[2-(4-Methylquinolyl)azo]-5-diethylaminophenol
QTA	quartz tube atomizer
RF	radiofrequency
RP	reverse-phase
SCA	single channel analyzer
SCE	saturated calomel electrode
SDME	single – drop micro-extraction
SE	size-exclusion
SFA	segmented-flow analysis
SG	functionalized silica gel
SFE	supercritical fluid extraction
SMDE	static mercury drop electrode
SMandT	standard, measurements and testing programme
SPE	solid phase extraction
S-PE	screen-printed electrode
SPME	solid phase microextraction
SR	synchrotron radiation
SR-XRF	synchrotron radiation induced X-ray fluorescence spectrometry
SRTrfCCP	single ring electrode radio frequency capacitavely coupled plasma torch
STAT	slotted tube atom trap
STPF	stabilized temperature platform furnace
STWAT	slotted tube water cooled atom trap
SWASV	squared wavelength anodic striping voltammentry
SWNT	single-walled carbon nanotubes
SWV	squared wavelength voltammentry
TAA	thioacetamide
TAM	2(2-thiazolylazo)-5-dimethylaminophenol
TAN	1-(2-thiazolylazo)-2-naphthol
TDBA	tetradecyldimethylbenzylammonium
THF	tetrahydroforan
THGA	transversal heated graphite atomizer
TMAH	tetramethylammonium hydroxide
TMAOH	tetramethylammonium hydroxide
Tm-APP	tetra(m-aminophenyl)porphyrin
TN	thermospray nebulizer
TOF	time of flight
TOPO	1,1,1-trifluoro-4-mercapto-4-(2-thienyl)but-3-en-2-one
TREN	tris(2-aminoethyl) amine
TRIS	Tris(hydroxymethyl)aminomethane
TSA	thiosalicylic acid
TSC	thermostated spray chamber
TSFF-AAS	thermospray flame furnace – atomic absorption spectrometry

TSP	total suspended particles
TTA	1-(2-thenoyl)-3,3,3-trifluoraceton
TUDS	two-unit desolvatation system
Tween 80	polyethylene glycol sorbitan monooleate
TXRF	total reflection X-ray fluorescence spectrometry
UN	ultrasonic nebulizer
UV	ultraviolet
VIS	visible
WCAT	water cooled atom trap
WDXRF	wavelength dispersive X-ray fluorescence spectrometry
WDS	wavelength dispersive system
XRF	X-ray fluorescence spectrometry
ZnPDC	Zn-piperazinedithiocarbamate
1-BPzDC	1-benzylpiperazinedithiocarbamate
4-VP	4-vinylpyridine
5-Br-PADAP	2-(5-bromo-2-pyridylazo)-5-diethylaminophenol
8-HQ	8-hydroxiquinoline

Chapter I

CADMIUM. ENVIRONMENTAL IMPLICATIONS

Cadmium is used in the industrial fields due to its excellent features. However, this element is one of the most toxic metals, and anthropogenic sources of cadmium are negative to the environment and health. Renal toxicity, pancreatic cancer, or enhanced tumor growth, are some of the reported illness related to environmental, occupational, or dietary exposure to cadmium [1].

Several studies have pointed out that man's cadmium intake, as least for non-smokers, comes principally from the ingestion of foods rather than from inhalation of cadmium in air. However, it must be mentioned that cadmium levels of foods have substantially decreased during the past several decades due to the progressive control of cadmium emissions to the environment [2].

Cadmium levels in the environment vary widely. In ambient air, cadmium is mainly associated with airbone and estimated concentrations ranging from 0.1 to 5 ng m^{-3} in rural areas, from 2 to 15 ng m^{-3} in urban areas, and from 15 to 150 ng m^{-3} in industrialized areas, have been established [3,4]. Differences in cadmium levels between indoor and outdoor air are not reported. Nevertheless, this is true only in non-smoking environments because smoking may substantially affect indoor ambient air for cadmium levels.

Cadmium air concentrations can be higher in certain occupational environments, although occupational exposure standards in such environments have been progressive reduced from 100 to 200 μg m^{-3}, to 2 to 50 μg m^{-3}, with the control requirements to maintain cadmium-in-blood and cadmium-in-urine below certain levels to assure no adverse human health effects from cadmium occupational exposure [5].

The average cadmium contents in the oceans can be lower than 5 ng L^{-1} [3], although other reference values have also been reported; within 5-20 ng L^{-1} [6], around 100 ng L^{-1} [7], or within 10 to 100 ng L^{-1} [4]. These levels can be higher in estuarine waters and around certain coastal areas. Greater variations are quoted for the cadmium contents of rainwater, fresh waters, and surface waters in urban and industrialized areas. Levels from 10 to 4000 ng L^{-1} have been reported in the literature depending on specific location and whether or not total cadmium or dissolved cadmium is measured [3,4].

The average natural abundance of cadmium in the earth's crust has been reported to be from 0.1 to 0.5 mg kg^{-1}, but much higher and much lower values have also been cited depending on a large number of factors. For instance, cadmium levels are lower in soils derived from igneous and metamorphic rocks (tipically from 0.02 to 0.2 mg kg^{-1}), whereas soils from sedimentary rocks offer much higher values (up to 25 mg kg^{-1}).

Concerning foodstuffs, cadmium levels vary widely and cadmium contents may be affected by the agricultural practices utilised in the particular areas such as phosphate fertiliser, sewage sludge and manure application. Leafy vegetables (lettuce and spinach) and certain staples (potatoes and grain) foods content relatively high cadmium values (from 30 to 150 μg kg^{-1}). Meat and fish normally exhibit lower cadmium contents, from 5 to 40 μg kg^{-1}. Some animal tissues such as kidney and liver content extraordinarily high cadmium levels (mainly associated with metallothioneins), and they can be to 1.0 mg kg^{-1} [3].

Chapter II

SAMPLING

In addition to the developement and application of fast sample pre-treatments and accurate analytical methods to determine trace elements, reliable environmental studies can only be assessed after adequate sampling and sample conservation procedures.

II.1. SAMPLING PROGRAMS

The sampling program must ensure that a representative material is sampled, and also it must take account of temporal and spatial variability of such environmental samples. Representative samples are comonly obtained after random (random sampling) or systematic (systematic sampling) selection of the environmental material from different sampling sites, and also, after collecting different samples from each sampling point. Random sampling consists of collecting samples randomly, ensuring that all parts of the system to be sampled can be statistically collected. This sampling startegy allows to find “hot spots” (locations with higher concentrations of pollutants), which can be later evaluated after a systematic sampling strategy. In constrat, systematic sampling consists of dividing the area to be sampled into an imaginary square or triangular grid, and samples are collected at the nodes of the grid, at the center of the spaces defined by a grid, or randomly within the spaces defined by a grid.

In addition to the physical state and location of the different sampling sites and the number of field duplicate samples, the sampling plans also consider the selection and maintenance of the sampling equipment, as well as the use of bottle blanks, field blanks, trip blanks and equipment rinse samples (equipment or rinsate blanks). These strategies are mainly focused to detect contamination sources, such as the effectiveness of the cleanning of sampling devices and sample containers (bottle or material blanks), or contamination from the environment around during the sampling and the sample transport (field blanks and trip blanks). In addition, background samples are also useful for environmental sampling. These are samples of the media similar to the test sample matrix and are taken near to the time and place where the analyte may be present at background level [8]. The volume or the amount of sample to be collected must be also especified in the sampling programme. It must be noticed that there are several guidelines on quality control in sampling for environmental materials, such as the EN ISO European guidelines based on the ISO international guidelines [9], or the American ASTM guidelines.

II.2. Sampling for Water

Because surface waters include different types, such as creeks or rivers, lakes, estuaries, seas, effluents and groundwater (well water), there is no a general procedure for sampling. After the establishment of the location of the sampling sites (both random or systematic sampling), either glass or plastic bottles used in sampling and conservation of samples for trace metals determination must be previously cleanned by soaking in 10% nitric acid for at least 48 hours, and rinsing several times with ultrapure water. Before collecting the water sample, a general practice consists of rinsing the sample container two or three times with the sample to be collected. However, this is not recommended for sampling waters containing high amounts of suspended solids and/or grease because these materials can be adsorbed on the walls of the container and the collected samples can be enriched with them.

II.2.1. Surface Waters

For creeks and rivers, samples are comonly collected from the mid-channel at the surface (surface water) by immersing the sampling container at a fixed location and by retrieving. If it is not possible to take the sample by submerging the container by hand, laboratory forceps or a holder with a sliding sleeve can be used. Similarly, if the sample is to be taken from the mainstream by reaching out from the bank then it may be best to attach the container to a segmented rod made up to the appropriate length. In shallow streams the sample is usually taken from approximately the top third of the water, or from a level between 20 cm and 30 cm below the water surface.

Similarly, surface or depth water can be sampled in lakes at the fixed sampling locations according with a random or systematic sampling plan.

II.2.2. Deep Waters

Deep water sampling is important because some analytes may be stratified at different depths. Stratification is also a problem with estuary and ocean sampling and either surface or deep-estuarine/seawater is comonly sampled.

Depth samples (sub-surface samples from rivers, lakes, estuaries and seas) are collected by using depth samplers, such as Mayer's submergible bottles [10]. This device consits of a ballasted glass bottle and a short cable attached to its neck with a loop to the stopper. The bottle is lowered to the desired depth with the stopper loosly inserted and then, the cable is jerked to release the stopper and the bottle fills. These samplers offer as a main disadvantage the difficulty of opening the bottle in deep water because the excess pressure of the stopper. Flushing samplers, such as the earliest Ruttener's devices or Theiler-Frieinger's devices avoid this problem. These samplers (cointaining around 1 to 2 L) consist of a glass or polyethylene cylinder with a pair of plate valves at each end, which are connected by a metal rod. The sampler is lowered with the end valves open, allowing flushing. At the required depth, the valves are closed by a mechanical messenger (a metal piece which hits the actuaring mechanism allowing top and bottom valves to be locked). Modern Nyskin-type bottles (made of polyethylene and 1.5 L capacity) offer a similar mechanical operation.

Small pumps can be also used to collect deep-water samples, especially in large water reservoirs such as lakes and seas when pumping does not cause disturbance.

In addition to the sampling device, water sampling requires usually auxiliary equipments related to metereological and atmospheric conditions. Measurements of water temperature, pH, conductivity and fluorescence (parameter related to algal bloom) are taken in the field at the time of sampling. For rivers and creeks, additional measurements of flow volumes are also taken at each sampling location. Different specially designed devices for flow volume measurements are commercially available [10].

II.2.3. Groundwater

Groundwater sampling deserves special attention. Groundwater sources include well or piped waters and spring waters, and knownledge of the physical and chemical characteristics of the aquifer (the rock formation containing the groundwater) would be needed before a sampling strategy. The time of the year (sampling before or after rainy season) affects considerably the groundwater composition, and other considerations such as sampling after periods of high agricultural chemical usage must be taken into account. In addition, well water sampling must not be affected by contamination from the loosened topsoil and when constructing and monitoring wells, well materials must not influence the chemical composition of groundwater. In this sense, cement used for pipe joints (pipes made of polyvinyl chloride) must be avoided and threaded pipes are highly recommended. In addition, equipment for monitoring wells must be made of stainless steel or, better, of polytetrafluoroethylene. These devices should be designed to avoid excessive aeration so that analyte oxidation can be minimized. As a general rule, purging wells before sampling is a recommended procedure because stagnant water is eliminated. The method and the rate of purging, time between purging and sampling, and sampling itself depend on several well's characteristics such as the diameter, depth and recharge rate of the well. To establish when the well sample is representative, changes in the pH, temperature or conductance are monitored in consecutive samples.

Sampling devices for groundwater are electric submergible pumps, bailers, suction-lift pumps and positive displacement bladder pumps. Detailed description of groundwater sampling can be found in specific monographs [10].

Standardized methods for water sampling can be found in the ISO 5667 guidance, especificaly guidelines ISO 5667 Part 4, ISO Part 5, ISO 5667 Part 6, ISO 5667 Part 9, ISO Part 10 and ISO 5667 Part 11, related to water sampling in lakes, rivers and creeks, drinking water from treatment works and piped distribution systems, estuaries and seas, waste waters, and groundwater samples, respectively [11-16].

II.3. Sampling for Soils, Sediments and Sludge

II.3.1. Shallow Soils

Scoops or shovels and trowels made of high density polyethylene are commonly used for soil sampling at its surface or at shallow depths (less than 15 to 30 cm). These devices are adequate when non volatile analytes, such as metals, must be determined. The use of these non-disposable sampling equipments creates a potential for cross-contamination between samples, and they must be decontaminated as well as equipment blanks must be collected and analyzed. It could be easier the use of separate sampling devices for each sampling point, and the decontamination process can be carried out later in the laboratory. Surface and shallow soil samples can also be taken with soil punch devices or thin-walled tube devices, the later ones can also be used as shipping containers. Detailed description of these systems can be found elsewhere [17].

II.3.2. Deep Soils

Deep soil samples are usually taken with punch and trowel digged laterally into the soil, or using augers or soil probes and split barrel samplers. Soil probes are useful samplers when collecting small diameter soil samples from the surface to depths of about 75cm. These samplers are made of either chrome molybdenum type 4130 steel or of stainless steel with stainless steel handles fitted with plastic end caps. The wide body slot allows for easy removal of the sample as the tip cuts a core slightly smaller than the inside diameter of the probe, to assist in easy removal of the sample. The top of the slot is closed to prevent soils from being trapped in the upper section of the probe. The probe is pushed and/or twisted into the soil to collect a sample by means of a detachable cross handle or a slide hammer and then the sample is recovered by twisting and pulling up on the cross handle or reverse hammering with the hammer attachment.

Split barrel samplers consist of a steel or stainless steel tube with the tubular section longitudinally splitted into two equal semi-cylindrical halves. The tube is attached to a connector head at the top and a drive tip mounted on a cutting shoe at the bottom. The system is generally used with liners in which the sample is collected for easy removal. The liners can be made of stainless steel or plastic [17]. At the desired depth, the drive tip is pulled and the split barrel sampler (split spoon) is lowered. The drop hammer is then placed on the top and the number of blows to advance the sampler is counted. Then, the drive tip is placed back in the cutting shoe and the device can advance to the next testing depth. This later device avoid cross contamination between soil layers.

Sampling for sediment (lake, river and marine sediment) or sludge are usually carried out with grab samplers, which dredge bottom grab samples (single surficial sediment taken at a specific time or over as short a period as feasible); or with corers, which drill core samples (cylindrical section of a naturally occurring medium consistent enough to hold a layered structure). Grab samples usually consist of the top ten centimeters of the sediment and stratigraphic integrity of the sample is not maintained, so analysis is usually restricted to bulk characteristics. However, the vertical integrity of the different layers underlying the sediment (from surface to more than 50 cm depth) is maintained for core samples.

II.3.3. Surface Sediments

There are several types of grab samplers, such as the Ekman grab sampler and the Van Veen grab sampler. The Ekman type consists of a stainless steel cabinet to which two sprung jaws are attached. A mechanism closes the jaws by remote command, which traps the sediment within the cabinet. Lids on top of the cabinet prevent wash out on retrieval. Other devices require that a messenger is sent down the line tripping when the sampler reaches the bottom. In constrat, Van Veen sediment grab sampler has a clam-shell type scoop. This design allows water to flow through the sampler to prevent premature closure of the jaws when the opened grab descenting to the bottom. When contact is made with the sea/river/lake floor, the stabilizing plates prevent over-penetration, so when the tow-cable is made taught causing the jaws to close, thereby capturing a sample. This system allows that all samples are equally taken (the same top centimeters of sediment) and minimizes disturbance.

II.3.4. Bottom Sediments

The simplest design of bottom sediment sampler is the hand corer. In shallow waters, the sampler (cylinder) can be pushed into the sediment using the handles on a head assembly. In deeper waters, the sampler can be dropped by attaching a line to the clevis, located on the head assembly between the handles. A simple flap valve allows water to flow through the sampler during descent and close tightly when removing the sampler, minimizing sample loss. The corer can be equipped with a liner-type core tube that accepts removable plastic or stainless steel liners, which results convenient when determining trace metals.

Other corer devices are the tube corer and gravity corer types. Tube corer types consist of a stainless steel tube with an internal piston. The device is pushed into the sediment manually by a pole, at the same time the internal piston is drawn up inside the tube. The vacuum created retains the sediment within the corer tube. The device can be removed from the sediment and the sample retrieved. This system can be also operated by attaching with a cable.

Gravity corers are weighted tube mounted within a frame that descends by gravity from the research vessel to the sea floor, where it penetrates the sediment to a given depth, filling the tube with sediment in the process. The hydraulically damped gravity corer has a slow rate of penetration that is controlled by a water-filled piston and disturbance of the water-sediment interface is minimal. A core-catcher on the bottom of the tube moves into place when retrieval begins, trapping the sediment sample in the corer. Gravity corer have replaceable tapered nosepieces on the bottom and balls or other types of check valves on the top, so water passes through them on descent but does not wash out during recovery. They can be also adapted to hold inert liners.

Standardized methods belonging to ISO 5667 guidance, especificaly guidelines ISO 5667 Part 12 and ISO 5667 Part 19, are related to sampling for sediments and marine sediment, respectively [18,19]. Similarly, the ISO 10381 Parts 1 and 2 lists the guidelines for soil sampling [20,21].

II.4. Sampling for Biological Matrices (Biota)

Several aspects such as size difference between specimens, variations within a study population, and mainly tissue differentiation, are some of the factors to be considered when sampling biota. It must be point out that there are seasonal, feeding, spawning and other periodic activities that may influence both the concentration and localization of the target analytes within a specimen. In some cases, each sample must be analyzed individually, but other times, composite samples, after homogenization, can be considered [22]. This is usual when analyzing small animals such as molluscs and composite samples are formed from several specimens located at the same sampling site. Similarly the whole organism can be also homogenized in order to obtain a mean concentration of the analyte. However, tissue differention is commonly carried out, and several samples from the target organs (kidney, liver, flesh, etc., from animals; or leaves, roots, etc., from plants) are separately taken as different subsamples from an organism. In order to obtain a representative sample, ramdom sampling process is the recommended strategy. Ramdomness ensures that nonbiased samples are obtained. In this process, every member of a population to be sampled has an equal probability of being included as a part of the sample [23].

For collecting fish, different devices such as hoop nets, gill nets, or trot lines can be used. However, some disadvantages associated with these samplers are related to the low number of sample obtained and to the fact that some specimen can be killed before retrieval. Other devices which avoids these problems are the electric shockers and slat boxes. Similarly, traps, dart guns and nets capture animals for sampling without any drawback. For collecting, different tools such as saw shears and pruning shears can be used [8].

Special attention must be offered for sampling plankton (phytoplankton and zooplankton). Two main categories of samplers are considered; devices designed for collecting the organism with the water in which they live, and devices able to sieve the plankton in situ [10]. These latter devices provide concentrated samples. Plankton tubes and some pumps allow collecting plankton with water. When sampling plankton in shallow ecosystems, samples can be taken in a transparent tube closed with plugs and both ends of the tube by hang. However, when collecting plankton from deep water, the tube must be attached to a rope or cable. In this case, plankton tubes consist of plastic or organic glass tubes (internal diameter from 40 to 60 cm and variable lengh), with the bottom end closed with a rubber plug but controlled by a cable passing down through the tube, and a handle at the top end of the tube. The open device is vertically immersed and the bottom plug is closed by jercking the cable.

Samplers which sieve the plankton are of different designs but all use a conical net for sieving. The net is made of miller's slik or polyamide fibre, with different mesh sizes (lower than 0.039 mm), and the trap at the end is fitted with an outlet valve. Some devices are the Apstein's sampler, the Clarke-Bumpus's sampler, the Strom's sampler and the Schindler's sampler. Detailed description of these devices can be found elsewhere [10].

Benthic organisims (microbenthos and macrobenthos) can be collected with corers driven by hand (shallow waters) or by their own weight or hydraulically by pushing out of it with a piston. After collecting, the sample must be washed through a sieve so that any organisims are neither forced through the sieve nor damaged. Other devices for sampling are those called as benthic dredges. A typical dredge consists of a heavy triangular metal frame (20 cm of one side) in which a sampling net (a bag around 0.5 m long) is attached. The device is kept on the bottom by attaching an extra weight and the system dredges the surface sediment by the movement of the ship or boat or by thrown out from the shore. Common grabs used for sediment sampling can be used although these devices offer several drawbacks. Other simple devices such as scrapers for collecting growths, sieves on exterdable rods, brushes or notched stoppers with microscope slides can also be adequate for collecting macrobenthos [10].

II.5. Sampling for Air Matrices

Concerning trace metals, negligible concentrations in the vapor phase are expected, and these pollutants are normally associated with aerosol particles. Different aspects such as the efficiency of the sampler and mainly the location and timing of the collection may determine the representativeness of the collected air sample [8]. In addition, different situations have to be taken into account when collecting atmospheric particulate matter from stationary emission sources, such as stacks; when obtaining representative samples from dry deposition; or when collecting indoor samples. In the latter cases, dust can be sampled using a filter bag and a vacuum cleaner. After collection, the material is removed from the filter bag and sieved (200-mesh sieve), then thoroughly mixed and bottled.

Two main collection techniques based on impact (impingers and impactors) and filtration (filters) are used for sampling atmospheric particulate matter [24]. Filtration methods are most used because impactors only allow the collection of size discrete samples. Filters are usually made of Teflon or quartz fiber, although filters made of other materials, such as polytetrafluoroethylene, polycarbonate, polyvinyl chloride cellulose or nylon, are commercially available. These filters are mounted in devices (samplers) equipped with a pump which deliveres the air sample through the filter in which the particles are retined. When working with high sampling flows and large particle sizes, the main particle retention mechanisim is based on the impact of the particles onto the filter. However, other processes such as particles diffusion or sedimentation are important when sampling short particles diameters or when working at low sampling flows.

It is important to know that the characteristics of a filter can vary during the collection time, especially when sampling for long times. This is attributed to the decrease of pressure moreover the filter is clogged. Special attention must be taken when using cellulose filters because their high hygroscopic nature. In general, all filter types can vary their weigth because the humidity. A pre-conditioning stage at a 50% relative humidity is usually recommended [25].

Non continuous samplers based on filtration or impact mechanisims can be classified in function of the pumped volume as high, medium and low volume samplers. High volumen samplers and medium volume samplers work with flows ranging from 20 to 90 m^3h^{-1} and they are highly recommended for sampling non polluted air or when sampling during short times. In constract, low volume samplers work with flows lower than 20 m^3h^{-1} and are used for contaminated air and for sampling during long times.

Although total suspended particles (TSP), offering different sizes, can be sampled, most of the current analytical determinations are based on the analysis of the inhalable fraction, which offers the most useful information when evaluating the interaction between air pollutants and public health [26,27]. The inhalable fraction is commonly refered as PM10 and PM2.5 which are defined as mean particulate matter which passes through a size-selective inlet with a 50% efficiency cut-off at 10 μm aerodynamic diameters for PM10 fraction or 2.5 μm aerodynamic diameters for PM2.5 fraction [28]. Therefore, most of the air samplers are equipped with air entrance systems which allow the aerodynamical retention of PM10 or PM2.5 fractions.

Standardized procedures for sampling PM10 fraction are englobed in the ISO 12341 guidance [29]. For PM2.5 there is a reference method by the European Community [30].

II.6. Sampling for Atmospheric Precipitation

General guidelines for sampling atmospheric precipitation (rain, snow, fog) are difficult because several variations sources can affect the composition of the samples during the sampling event, and also because the variable metereological conditions and an absence of an ideal surface/place for fixing the sampling equipments. This last aspect is important because the volume of precipitation from individual events can differ a great deal, even within relative short distances. By this way, information on metereological conditions when sampling must be given, as well as the sampling time in relation to the total precipitation event [8]. The amount of vertical precipitation is easily gauged by means of the metereological sations network, which locations must be selected as representative ones of the widest possible surrounding areas. The most suitable locations for rain gauges are open plains or slight slopes, and if elevated features are present, the distance from them must be at least four times their vertical height [10].

Sampling devices for precipitation must be made of inert materials for avoiding contamination. Similarly, they must be designed in order to avoid contamination (dust, falling leaves, insects, etc) from the surrounding environment during the sampling operation. Other alterations can occur as consequence of evaporation when finishing the precipitation event and the following transport of the sample. Finally, the aerodynamic properties of the sampling equipments must ensure adequate efficiencies of sampling, which can not be higher than the typical efficiency of the surrounding surfaces.

Simple plastic (polyethylene) buckets, equipped with a tight fitting lid for minimizing evaporation, can be used as manually operated devices for sampling precipitation. Other manually operated systems consist of plastic funnels (30 cm diameter) with the tubular end shortened which are inserted through a bored stopper of a sample container (typically a 2 L polyethylene bottle). This device avoids evaporation of the collected sample.

Automatic sampling equipments are preferable. These systems are equipped with a lid or cover for avoiding dry precipitation. The lid is controlled by a sensing element wich reacts to the first drop of rain (flake of snow), allowing the opening of the collector device. The movement of the lid must be horizontally rather than vertically for avoiding interference from wind. These samplers have a funnel or a cylinder within a hermetically sealed container, which can be thermostatically controlled (just above zero in winter and below 4°C in summer). There are some sampling equipments with two containers for sequentally collecting dry deposition and precipitation. In this case, the lid is moved alternately to positions on one or the other of two containers, depending on whether it is raining or not [10].

The ISO 5667 Part 8 and ASTM D 5012-89 (1994) guidances offers the standardized guidelines for sampling wet precipitation [31,32].

Chapter III

PRESERVATION AND STORAGE. PRELIMINARY OPERATIONS

Due to the low levels of cadmium in environmental materials, all glassware and plastic ware used for sampling and for keeping the collected samples must be previously decontamined (soaked in a 10% nitric acid solution for at least 48 hours and rinsed several times with ultrapure water). Preservation conditions depend on the type of sample.

III.1. CONSIDERATIONS FOR PRESERVING WATER SAMPLES

When collecting waters, metals tend to be adsorbed onto the walls of the glass or plastic containers. This fact can lead to lower cadmium concentrations than those contained in the sample. To avoid this adsorption, nitric acid, at a concentration of 1%, is usually added to water samples, and then, they are commonly stored at 4°C before analysis. In order to know contamination by the addition of acids, reagent blanks and trip blanks must be prepared. It must be taken into account that for speciation studies preservation by acidification must be avoided because the oxidation state of the elements and/or the different metal-complexes ratio can be changed. In these cases, freezing at -18°C is highly recommended for avoiding looses by adsorption. A preliminary operation for waters consists of filtration by using Nucleopore filters of 45 µm to separate suspended particulate matter. This operation is highly recommended when determining metals because they are commonly associated with suspended particles. If possible, this operation must be carried out in the field after water sampling and just before addition of nitric acid for preservation. However, it is not always possible, and the filtration operation is carried out in the laboratory and after acidification. In these cases, the acid can digest the particulate matter, especially organic particulated matter, and changes in chemical composition can be occurred for some target analytes.

The ISO 5667 Part 3 and the APHA.1060 guidances offer the standardized guidelines for handle and preservation of water samples [33,34].

III.2. Considerations for Preserving Soil, Sediment and Sludge Samples

In constrat to preservation techniques for water, there are not prescribed preservation recomendatios for soils, sediments or sludge. It is highly recommended to seal the containers, minimizing the headspace, refrigerate the samples during storage and transportation, and analyze the samples as soon as possible. Dry soils can be kept in decontamined plastic containers at room temperature. However, when sampling wet soils, the samples must be dried for conservation. Water removal minimizes possible biodegradation of the sample as well as oxidation-reduction phenomena. Some times wet soil samples can be air-dried. However, contamination by dust can be occurred during this operation and this drying system is not recommended when trace elements have to be determined. Lyophilisation (freeze-drying) procedure is a more recommended procedure. In this case, wet samples must be previously freezed and then lyophilized at temperatures within the –80 to –40°C and at pressures around a few millibars. After freeze-drying soil samples can be kept at room temperature before analysis. River, lake and marine sediments, and sludge must also be air-dried or freeze-dried to ensure integrity of the sample.

When measuring metals in soils and sediments, there are big differences in the trace elements concentrations in function of the paricle size. Most of the soils and sediments analysis for trace elements are carried out using the $< 63\mu m$ fraction, and a sieving procedure is the first operation before any other sample pre-treatment. Sieving is performed with Nylon sieves (preferred to stainlees steel sieves to avoid metal contamination) manually or by mean of automatic shakers (dried samples). Sieving can be carried out after or before drying. When sieving after drying (dry sieving) some problems derived from compactness of the sample can be occurred. This fact leads to the formation of large aggregates from small particles, which are not efficiently sieved. Wet sieving is carried out by placing the sieve with the adequate mesh in a stable position resting on a bucket. The wet sample is added to the sieve and a minimum amount of ultrapure water should be used to wash the sediment through the bottom sieve and all washing water should be retained in the collection bucket until the sample is allowed to settle. Once the suspended material has settled, excess water on the surface of the sediment should be carefully decanted. Care should be taken to remove only water and not sediment at this stage. After this operation, the isolated fraction can be kept at -18°C or it can be subjected to a freeze-drying process.

The ISO 5667 Part 15 gives general guidelines on the precautions to be taken to preserve and handling sludge and sediment samples [35]. Simmilarly, ISO 23909, ISO 11464, ISO 16720 and APHA 3030 give recommendations for the preliminary treatments of soils and sediment samples [36-39].

III.3. Considerations for Preserving Biota Samples

Biota samples are much more succeptible to decomposition and even inorganic compounds such as metals can be altered, mainly by oxidation-reduction reactions. After sampling, biomaterials must be kept at low temperature (4°C) in inert containers and they must be transported to the laboratory as soon as possible. If biota samples are not be inmediatly treated, samples must be freezed at -18°C. If the analysis requires tissue differentiation, the different parts of the organisim must be obtained and homogenized. Similarly, when dealing with small organisims such as molusks, composite samples must be prepared by mixing and homogenizing the soft tissues from the different specimens. Cutting blade-type homogenizers are commonly used for blending, mixing and homogenizing soft tissues from biota samples. These systems allow homogenize sample volumes from 0.25 to 10 L by using processing speeds from 500 to 20000 rpm. Cutting blades are commonly made of stainless steel, although some equipments work with cutting blades made of other materials to avoid metal contamination. Homogenization can be also obtained by blending (Stomacher blender-homogenizer). In these cases, the biomaterial is placed into an inert bag and the device applies shear forces on the wet tissue without direct contact between the wet sample and the steal paddles.

Once the sample is homogenized analysis can be directly performed with the wet sample. The homogenized sample must be kept at -18°C to avoid decomposition and analyte changes. Freeze-drying is also recommended for biota samples. After this process, biota samples can stored at room temperature. In addition, because water is lost during the freeze-drying process, an analyte preconcentration is achieved after lyophilization. However, moisture of the sample must be previously assessed because some analytical results are usually given in wet mass.

Finally, a further particle size reduction of dried biological samples can be performed. This preliminary operation can be performed by using different grinders made of different materials (zirconia oxide is preferred when metals are determined). Vibrating ball mills or planetary ball mills are commonly used for small amount of dried samples. New equipments allow cryogenic grinding, which avoids analyte looses by increments of temperature during grinding. This operation also allows a further homogenization of the sample.

III.4. Considerations for Preserving Atmospheric Particulate Matter

There are not official guidelines for preserving atmospheric particulate matter samples adsorbed onto filters. Commonly, quartz fiber filters (containing the sampled particulate matter) are kept at -18°C before analysis in accordance to recommendations by the European Community on provisional reference method for the sampling and measurement of PM2.5 [30].

Chapter IV

ANALYTICAL TECHNIQUES TO ASSESS CADMIUM

There are well-established analytical techniques for the determination of metals in all kind of samples. Some of them offer enough sensitivity to carry out directly the analysis; however, other techniques require previous stages for analyte pre-concentration and/or matrix concomitants removal. Because cadmium can be present at very different concentrations in environmental materials, a careful selection of the adequate analytical technique is needed. Several spectrometric techniques, mainly atomic spectrometric detectors and certain electroanalytical techniques, allow a sensitive and reliable determination of trace metals.

In addition to sensitivity, high background signals and non-spectral interferences can be important troubles in some analytical techniques. This is the case of conventional X-ray based spectrometric techniques, which offer high background and scattering radiation when directly analyzing environmental materials. Other analytical techniques, such as neutron activation analysis, are not widely applied because the high toxicity of wastes obtained and the special requirements of installation. However, both X-ray based spectrometric techniques as well as neutron activation analysis are advantageous methodologies because they allow the direct analysis of samples without any sample pre-treatment. This is especially appealing when coping with solid environmental specimens.

Spectrometric methods based on atomic absorption or emission phenomena, as well as those dealing with atomic masses, are widely used because of the inherent high sensitivity, low cost, and fast analysis, which is mainly appealing for routine work.

A description of the theoretical aspects, instrumentation, experimental parameters and application of the most advantageous techniques for trace element determination will be developed along this Chapter.

IV.1. UV-VIS ABSORPTION SPECTROMETRY

The absorption of radiation is the process consisting of the selective attenuation of certain frequencies contained in electromagnetic radiation by chemical species in solution. Because each species has a unique set of energy states, the absorbed frequencies are typical of each chemical species. Therefore, after exposing a molecule in the ground state to a radiation of a certain frequency, it is likely to be absorbed if its energy matches any of the energy jumps quantified for this molecule. Under these conditions, the energy of the photons (radiation) is transferred to the species, whose valence electrons are promoted to a higher energy state, called an excited state. Molecules in the excited state can relax to the ground state by transferring the excess of energy to adjacent molecules, by photochemical decomposition of the excited species and the formation of new molecules, or by the re-emission of energy (fluorescence and phosphorescence phenomena).

Because of a large number of vibrational and rotational energy associated with the electronic states (ground and excited states), several energies, which differ little among themselves, are absorbed by a chemical specie, and the number of transitions is very high, resulting in a wide band. These absorption bands make use of the transitions located in the ultraviolet (10-400 nm), visible (400-1000 nm) and infrared (0.78-1000 μm) regions of the electromagnetic spectrum. Detailed information on UV-VIS spectrometry can be found in the monographs by Heinz-Helmut [40] and Harris and Bashford [41].

As a consequence of the matter-radiation interaction, the beam of radiation can be considered to be attenuated from I_0 to I. Two important terms are then used in absorption spectrometry. One is the transmittance, T, defined as the ratio (%) of the intensity of the radiation beam emerging from the solution (I) to that of the incident beam (I_0)

$$T = I/I_0.$$

The other term is the absorbance, A, defined as the negative logarithm of the transmittance.

$$A = -\log T = \log(I_0/I)$$

Unlike the transmittance, the absorbance of a solution increases as the attenuation of the beam becomes greater. The absorbance of a sample is proportional to the total amount of material that absorbs the incident radiation. The functional relationship between the analytical signal measurement (absorbance) and the analytical parameter of interest (concentration) is known as the Bouguer-Lambert-Beer law and is expressed as

$$A = \log (I_0/I) = abc,$$

where *A* is the absorbance of the solution; *a* is a proportionality constant called absorptivity, which is a property of the material itself as well as the wavelength of the measurement; *b* is the length of the path within a sample; and *c* is the concentration of the material that absorbs the radiation. When the concentration is expressed in mol L^{-1}, absorptivity is represented by *e* and is called the molar absorptivity. This linear relationship between absorbance and concentration is a generalization, and there are several deviations, mainly associated with higher concentrations (up 0.01 M). Other deviations are consequences of the experimental measurements, e.g., the use of polychromatic and dispersive radiation (instrumental deviations), or they may derive from analyte association or dissociation with the solvent or chemical changes to give products with different absorption properties (chemical deviations).

The absorption of UV-VIS radiation normally produces excitation of bonding electrons, and a certain correlation of the absorbed wavelengths with the types of bonds in the absorbing species can be expected. However, the most important application of UV-VIS spectrometry is the quantitative determination of compounds containing absorbing groups. Absorbing species (organic molecules, ions and inorganic anions) can contain π, σ and n electrons, and the energy level of a nonbonding electron normally lies between the bonding and the antibonding π and σ orbitals. Low energy transitions ($n \rightarrow \sigma^*$) are typical in saturated compounds containing atoms with unshared electron pairs, and they lies in the region between 150 and 250nm. Transitions involving more energy exchange ($n \rightarrow \pi^*$ or $\pi \rightarrow \pi^*$) occur in a more convenient spectral region, with the 200 to 700 nm range, and they are the most useful for organic unsaturated compounds. The unsaturated groups in absorbing species which exhibit transitions to a π orbital (unsaturated groups) are usually called as chomophores.

IV.1.1. Instrumentation

Instruments used for transmittance or absorbance measurements consist of five basic elements: 1) a light source of radiant energy; 2) a wavelength selector to isolate certain wavelengths; 3) a sample cell or container; 4) a detector; and 5) a data processing unit.

The most common source used in UV-VIS spectrometry is the tungsten filament lamp, which provides radiation within the 320 – 2500 nm range. The tungsten/halogen lamps, containing a small amount of iodine together with the tungsten filament within the quartz envelope, give higher intensities, and wavelengths up to 240 nm are possible. Other lamps includes the hydrogen lamps, the mercury or deuterium lamps, and the xenon discharge lamps, all providing intense, stable and constant radiations.

Isolation of the wavelength band is commonly carried out by means of interference and absorption filters, or by monochromators (prisms and diffraction gratings). Filters result sufficient to isolate a specific band of wavelengths, but a monochromator is necessary to obtain absorption spectra.

The sample container, cell or cuvette is normally parallelepiped in shape with a standard length of 1 cm and made of glass for the VIS region or quartz (or fused silica) for UV region. The cell has an opening for inserting the sample and a stopper to prevent evaporation.

The detectors used in UV-VIS spectrometry are photon detectors and they can be phototubes, photomultiplier tubes, photovoltaic cells or photodiodes. All these detectors use the photoelectric effect to convert the radiant energy into a measurable (electric) signal.

Finally, a data processing unit consists of an electronic device that amplifies, filters and performs mathematical processes on the electric signal.

These five basic components can be assembled in different ways to produce several instrument designs. Here we consider two general types of spectroscopic instruments: 1) the photometer, which is a simple instrument that uses absorption or interference filters for wavelength isolation and a photoelectric device to measure the radiant power; and 2) the spectrophotometer, consisting of a specialized device capable of recording the various components of complex radiation.

These instruments exist in three different configurations:

1) Single-beam instruments; which consist of a radiation source; a monochromator; two cells for the reference and the sample solutions (alternately inserted in the light path); a detector; an amplifier; a reading device. These instruments require a stable voltage source to prevent errors arising from variations in the beam intensity. Also,

differences between cells, mainly irregularities in the walls, are not easily compensated.

2) Dual-beam instruments, in which the incident light beam is splitted by a rotating mirror-chopper into two separate beams, one of which passes through the sample and the other passes through the reference. The alternating beams reach the detector thus permits a simple mathematical treatment of signals (signal modulation). This design is routinely used and leads to good results, since it minimizes drift of the radiation source and the amplifier.
3) Multi-channel instruments, which are equipped with a photodiode array detection system. The radiation from a tungsten or deuterium lamp is focused on the sample or solvent cell and then it passes to a diffracting grating. The scattered radiation arrives at the diode array, which simultaneously detects and analyses various wavelengths.

IV.1.2. Cadmium Determination by UV-VIS Spectrometry

A typical UV-VIS spectrometric method for cadmium is based on the use of dithizone (HDz) and chloroform or carbon tetrachloride as a chromogenic/extractive reagent. The dithizone (diphenylthiocarbazone) offers a high capacity to form metallic complexes and for cadmium the complex has the formula $Cd(HDz)_2$ [42]. When using either chloroform or carbon tetrachloride as extracting solvents, the complex exhibits an absorption maxima at 517-520nm, and the molar absortivity is around 8.6 10^4 [42]. Improved limits of detection of the dithizone spectrometric method for cadmium determination in river water can be reached by using surfactant induced sensitization with Triton X-100 [43]. The formation of a complex between cadmium and iodide in acid medium is another recurrent approach. After cadmium iodide formation, cadmium is converted to a strongly colored complex by subsequent treatment with a 2,2'-diquinolyl ketone 2-quinolylhydrazone (DQQH) solution in benzene (maximum at 552nm, molar absorptivity of 9.15 10^4) [44], or with Malachite Green (maximum at 685nm, molar absorptivity of 6.1 10^4) [45,46]. The method was successfully adapted for flow-injection analysis [45]. Other reagents proposed for the colorimetric determination of cadmium (Table 1) were N-phenylcinnamohydroxamic acid [47], 2-(5-bromo-2-pyridylazo)-5-diethylaminophenol (5-Br-PADAP) [48,49], triphenylphosphine oxide [50],2-[2-benzothiazolylazo)-5-dimethylaminophenol (BTADAP) [48], N1-hydroxy-N1,N2-diphenylbenzamidine (HDPBA) and 4-(2-pyridylazo)naphthol [51], 1,10-phenanthroline and thymol blue [52], 2-[2-(4-Methylquinolyl)azo]-5-diethylaminophenol (QADP) [53], 4-(2-pyridylazo)-resorcinol [54], phenanthraquinone monophenylthio-semicarbazone (PPT) [55] and p,p'-dinitro-sym-diphenylcarbazid [56]. Reactions between cadmium and other chromogenic reagents in micellar media, such as sodium dodecyl sulfate and cetylpyridinium chloride [57] or Triton X-100 [43,58-60], have also proved to give sensitive spectrometric methods to assess cadmium in environmental samples (Table 1).

Table 1. UV-VIS spectrometric determination of cadmium in environmental samples

Sample type	Chromogenic ragent / solvent	Maximum spectra / molar absortivity	Ref.
Aqueous solutions	DQQH / benzene	552 nm / 9.1 10^4	[44]
Waters	Malachite Green / ----	685 nm / 6.1 10^4	[45]
Tap water, vegetables	Malachite Green[a]	690 nm / ----	[46]
Aqueous solutions	N-phenylcinnamohydroxamic acid / chloroform	380 nm / ----	[47]
Waters	5-Br-PADAP	587 nm / ---	[48]
Aqueous solutions	1-(2-pyridylazo)-2-naphthol / triphenylphosphine oxide	555 nm / ----	[50]
Water	Br-PADAP / o-xylene	590 nm / 3.8 10^4	[48]
Water	BTADAP / o-xylene	600 nm / 4.5 10^4	[48]
Industrial effluents, fly ash	HDPBA	400 nm / 2.0 10^4	[51]
Seawater	1,10-phenanthroline-thymol blue / chloroform	410 nm / ---	[52]
Aqueous mixtures	QADP in aqueous ethanol[b]	569 nm / 1.5 10^5	[53]
Aqueous mixtures	4-(2-pyridylazo)-resorcinol /	510 nm / 2.5 10^5	[54]
Water	PPT / oleic acid[c]	520 nm / 2.4 10^5	[55]
Tap water, seawater	p,p'-dinitro-sym-diphenylcarbazid	630-640 nm / 2.1 10^4	[56]
Aqueous solutions	CCE - sodium dodecyl sulfate and cetylpyridinium chloride	414 nm / -----	[57]
Wastewater	HDAA- Triton X-100	520 nm / 1.5 10^5	[58]
Water, waste water	Cadion Py – Triton X-100	530 nm / 1.9 10^5	[59]
Water, plants	Thiazolylazo chromogenic reagents - Triton X-100	-----	[60]

[a] Cd(II), as a $CdCl_4^{2-}$, is previously retained on an AG1-X8 resin

[b] Cd-QADP on line (FIA-silica gel column) separated from other ions

[c] Cd-PPT complex is floated quantitatively with oleic acid (HOL) surfactant

DQQH, 2,2'-diquinolyl ketone 2-quinolylhydrazone; 5-Br-PADAP, 2-(5-bromo-2-pyridylazo)-5-diethylaminophenol; Br-PADAP, 2-[2-(5-Bromopyridyl)azo]-4,5-dimethylphenol; BTADAP, 2-[2-benzothiazolylazo)-5-dimethylaminophenol; HDPBA, N1-hydroxy-N1,N2-diphenylbenzamidine; QADP, 2-[2-(4-Methylquinolyl)azo]-5-diethylaminophenol; PPT, phenanthraquinone monophenyl-thiosemicarbazone; CCE, N,N'-bis(2-hydroxy-5-nitrobenzyl)-4,13-diazadibenzo-18-crown-6; HDAA, o-hydroxybenzenediazoaminoazobenzene; Cadion Py, 2-pyridinediazoaminoazobenzene

IV.2. Atomic Absorption Spectrometry (AAS)

The physicochemical principles which control measurements by atomic absorption spectrometry (AAS) are similar to those of UV-VIS for molecules; but in AAS, free atoms or ions are involved in the absorption of radiation. Therefore, at a certain resonance line for an element, attributed to an optical transition between atoms in the ground state and the atoms in an excited level, the absorption of radiation can directly be related to the concentration of that element.

An atomic absorption spectrometer consists of a source of radiation providing the radiation to be absorbed, an atomizer or source of free atoms, an optical dispersive element, and a detector and a data acquisition system. The atomizer is perhaps the former characteristic of AAS piece of equipment; and the different types of atomizers (flame, graphite furnace, heated quartz tube) lead to the various available atomic spectrometry techniques in use. The atomization mechanisms in most of these systems involve a thermal atomization by heating at very high temperatures, but some times, a chemical mechanism such as atomization through a chemical reaction which coverts the aqueous analyte into an atomic vapor, is also used.

IV.2.1. Sources of Radiation

The radiation sources for AAS instrumentation are commonly line sources (LSs), which generates a characteristic narrow-line emission of a selected element. However, new designs of continuous sources (CSs), derived mainly from advances in solid-state array detectors and charge-coupled devices (CCD), have led to successful results in new High Resolution Atomic Absorption Spectrometry (HR-AAS).

IV.2.1.1. Line Sources

There are three principal LSs for AAS: the hollow cathode lamp (HCL), the high intensity hollow cathode lamp (HI-HCL), and the electrodeless discharge lamp (EDL) [61,62].

An HCL consists of a tungsten rod (anode) and a hollow cylindrical cathode made of the target element (mono-elemental HCLs) or target elements (multi-elemental HCLs), whose characteristic radiation is required. The cathode and the anode are contained within a glass cylinder with a silica window and filled with Ar or Ne at low pressures (1 to 5 Torr). The characteristic radiation is obtained when a potential of 300-500 V (currents about 5 to 20 mA) is applied across the electrodes, which causes the inert gas contained in the lamp to ionize. Under this high voltage, cations acquire enough kinetic energy to impact on the inner surface of the cathode and to cause metal atoms to sputter out of the cathode cup, producing an atomic vapor. Further collisions of the sputtered metal atoms with the inert gas ions excite these metal atoms, and they emit their intense and characteristic radiation when returning to the ground state. The cylindrical configuration of the cathode enhances the probability of metal redeposition at it rather than on the glass walls, and the cathode is continuously regenerated.

The intensity of the radiation emitted by HCLs is low for cathodes made of volatile elements, such as cadmium, and HI-HCLs or EDLs are preferred. HI-HCLs, also called boosted discharged hollow cathode lamps, BDCHLs, are similar to HCLs, but they exhibit an additional or secondary cathode (boosted cathode). After applying a high voltage, the sputtered atoms from the primary cathode (atomization) are excited by applying an additional voltage (boost discharge) across the secondary cathode (excitation). When both atomization and excitation process occur at two different electrodes, the self-absorption phenomena is minimized and the resulting emission is around 5 to 15 times higher than those provided by standard HCLs.

An EDL consists of a hermetically sealed quartz envelope containing an inert gas (Ar) at very low pressure and the element or salt of the target element (commonly iodides). The lamp does not have electrodes and ionization of the inert gas is obtained after applying microwave radiation (c. 100 MHz) or, as is usually the case, radio-frequency radiation (from 100 kHz to 100 MHz). Commercially available radio-frequency EDLs have a built-in starter, run at 27

MHz, which provides a high voltage spark to ionize the filler gas to initiate the discharge. Similarly to HCLs or HI-HCLs, the inert gas ions acquire high kinetic energies and after collisions with the metal salt, the sputtered and excited metal atoms emit the desired characteristic spectrum.

IV.2.1.2. Continuous Sources

A modern CS is based on a conventional xenon short-arc lamp which has been optimized to run in the so-called `hot-spot mode′ [63]. This discharge mode consists of the appearance of a small plasma spot close to the cathode surface. This plasma is obtained after operating the lamp at a high xenon pressure, using short electrode distances, and new materials and geometries for both anode and cathode rods.

IV.2.2. Background Correction

When using flames and graphite furnace as atomizers it is necessary to discriminate between the absorption by the target element and the background absorption. This latter type of non-specific absorption is attributed to light scattering on particulates formed by recombination of the sample matrix at cold spots, and also by broad molecular absorption signals caused by radicals or molecules vaporized in the atomizer. For this, two main background correction systems are used; the continuous sources based background correction systems and the Zeeman background correction devices [61].

The first system uses a non-specific continuous radiation emitted by additional sources; e.g., a deuterium arc lamp or a hydrogen arc lamp which provide a continuous radiation throughout the UV region (correction below 400 nm), or a tungsten halogen lamp for the VIS region [61]. The system works by passing alternatively the radiation provided by the continuous sources (deuterium or tungsten arc lamps) and the radiation of the HCL or EDL through the atomizer. When radiation from the HCL or EDL passes, the detector records the atomic absorption and the background absorption signals, but when radiation provided by the continuous lamp passes through the atomizer only the background absorption signal is measured. This is because the level of atomic absorption due to the continuous lamp is negligible since the slit width is kept sufficiently wide. Therefore, the signal recorded with the line sources can be subtracted from that recorded with the continuous source, and the background absorption can be removed. This method for background correction is inexpensive, but background signals larger than 0.5 units can not be compensated. In addition, there are other drawbacks associated with degradation of the signal-to-noise ratio and over-corrections in some systems, which occurs when emission lines from other elements in the sample lie close to the characteristic emission line of the target element.

The second background correction system uses the Zeeman effect on atomic energy levels, e.g., the splitting of atomic energy levels when exposing the atomic vapor to a magnetic field, which lead to the formation of several absorption lines for each electronic transition. The normal Zeeman splitting pattern leads to a central or π line, which shows the same wavelength of the electronic transition in absence of the magnetic field; and two σ lines, equally spaced around the π line and very close to the original electronic transition wavelength (π line). Analyte and absorbing species (background) absorb at the π line, while the σ lines are only absorbed by the background absorbing species. The net analyte absorbance signal is obtained by modulating the magnetic field and by using a polarizer for background subtraction. The advantages of this method over continuous lamps are that high background signals (up to 2.0 units) and structured backgrounds can easily be corrected. However, it presents as a main drawback the slight decrease in sensitivity.

IV.3. Flame Atomic Absorption Spectrometry (FAAS)

FAAS is one of the most commonly used atomic techniques in the analytical laboratory mainly because of its simplicity, low capital cost and inexpensive maintenance. A more detailed description of FAAS instrumentation can be found elsewhere [61].

IV.3.1. Atomizer: Flame

The typical atomizer for FAAS is a pre-mixed laminar flame, in which the fuel and the oxidant gases are mixed in an expansion chamber prior to entering the burner. The energy supplied by this atomizer is high enough to atomize the sample and to maintain the atomic vapor within the light path of the spectrometer. The most common flames used in AAS depend on the nature of the oxidant gas: the air-acetylene flame and the nitrous oxide-acetylene flame. With the air-acetylene flame (slot length of 100 mm) is possible to reach temperatures from 2000 to 2500°C, and these temperatures are high enough to atomize elements which oxides or carbides do not offer a refractory behavior. The nitrous oxide-acetylene flame (slot length of 50 mm), which attains temperatures of c. 3000°C, is preferred when atomizing refractory elements or those forming thermally stable oxides or carbides.

IV.3.2. Nebulizer

In flame atomization the liquid sample is introduced into the flame as a fine aerosol. This spray is generated by a nebulizer/expansion chamber arrangement, which converts the sample solution into an aerosol showing droplets of different sizes.

Pneumatic nebulizers are the most widely used in FAAS instruments, especially the pneumatic concentric nebulizer which consists of a concentric stainless-steel tube through which a Pt/Ir capillary tube is located. In this nebulizer the sample solution passes through the capillary by the action of a high-velocity oxidant gas stream parallel to the capillary axis and at the end of the capillary (Venturi effect). When the liquid sample arise the end of the capillary, it escapes through the exit orifice between the outside of the capillary tube and the inside of the stainless-steal concentric tube. In this point, the high-velocity gas breaks the liquid up into a coarse aerosol. Once the aerosol is generated, it passes through the expansion (spray) chamber, where large droplets collect on its walls and drain away. A number of baffles are placed in the spray chamber to ensure that only the smallest droplets reach the flame. Although the combination of a pneumatic concentric nebulizer and the expansion chamber is a cheap and robust system which offer good long-term stability, it offers some drawbacks when working with liquids containing high salts content or particulate matter mainly because the small internal diameter of the capillary. In addition, the efficiency in the aerosol generation by the combination of a pneumatic nebulizer and the expansion chamber is around 10-15%, which offers poor sensitivity. In order to obtain a more concentrated aerosol, an impact bead (made of glass or ceramics) may be placed in the path of the initial aerosol inside the spray chamber [62]. The primary aerosol generated by the nebulizer impacts on the bead and secondary fragmentation takes place; this increases the number of fine drops that can reach the flame and improves the efficiency of nebulization.

Enhancement on analyte transport to the flame can be also obtained by using highly efficient nebulizers, such as high pressure pneumatic nebulizers (HPN) and hydraulic high pressure pneumatic nebulizers (HHPN) [64-66], ultrasonic nebulizers and thermospray nebulizers, which lead to a more concentrated aerosol and a more sensitive FAAS determination. A detailed description of the different nebulizers is given in the following sections devoted to inductively coupled plasma – optical emission spectrometry / mass spectrometry.

IV.3.3. Cadmium Determination by FAAS

Cadmium is a volatile and non-refractory element and therefore an oxidizing air–acetylene flame provides enough energy for cadmium atomization without any interference [61]. However, the main drawback when determining cadmium by FAAS is its low characteristic concentration, around 0.02 mg L^{-1} at the 228.8 nm resonance line for conventional pneumatic nebulization and impact bead. The characteristic concentration at the second main resonance line for cadmium, 326.1 nm, is very high, near 6.1 mg L^{-1}, which results unusuable. This low sensitivity is inherent to the flame atomization process itself, and as commented, it is associated to a poor efficiency of the conventional nebulizer systems and to the short residence time of the analyte atoms in the optical path (typically 10^{-3} s). Since cadmium occurs at very low concentrations, sensitivity given by conventional FAAS is not low enough to directly assess cadmium in most of the biological and environmental samples. To overcome this problem, there have been developed different pre–concentration methods, mainly based on solvent extraction and solid phase extraction. Basic principles of these procedures, as well as current trends, are reviewed in Chapter V.

IV.3.4. Atom Trapping Systems - FAAS

Instead of pre-concentrating the analyte and/or improving sample nebulization, sensitivity of cadmium determinations by FAAS can also be obtained after certain modifications in the atomizer that allow analyte atoms to remain longer in the optical path. This is the case of the use of atom trapping systems and in situ pre-concentration techniques for FAAS [67]. The long-path absorption tube (LPAT) and the Delves' cup microsampling techniques are the first atom trapping systems reported. LPATs are T-shaped or open-ended tubes, firstly made of silica, in which the free atoms from a total consumption burner, and after conventional nebulization, are confined. The tubes or cells are placed in the optical path and just above a flame burning at the T joint (T-shaped cells) or at a side of the open-ended tube. The Delves' cup microsampling technique also uses the idea of collecting the free atoms in an open-ended trapping tube above the flame but here, the sample is directly deposited in a sampling cup or boat instead of being aspirated into the flame. The technique was simplified and improved by incoporating a holder for the Delves' microsampling cup [68]. This improved system uses a nickel or stainless-steel microcrucible (cup) mounted onto a device that enables it to be pushed close to the flame to allow the charring of the sample, and then into the flame to allow atomization. A nickel open-ended tube is mounted in the flame and atoms enter the tube through an opening half-way along its length. Although the Delves′microsampling cup technique was commercially utilized, this approach, as well as LPATs, is not used nowadays and they are not routine laboratory procedures. However, their basic principles and their first applications, reviewed by Matusiewicz [67], can be considered the basis of the current atom trapping systems.

More usable atom trapping systems are the so-called slotted–tube atom traps (STAT) and water cooled atom traps (WCAT). The simplest STAT systems, firstly used by Walting [69,70], are double–slotted tubes which are placed above the burner allowing the entrance of flame gases and analyte atoms through the base slot [67]. Different designs showing entrance and exit slots with 120° or 180° angle between them, as well as different length of the upper and base slots, the internal diameter and the length of the tube, were proposed [71,72]. In addition, modification and optimization of STATs with a row of six exit holes centrally drilled above the entrance slot were also applied [67,72]. STATs are usually made of quartz and can be easily manufactured in laboratory. However, the main drawback of quartz STATs is the divitrification process, attributed to sodium migration into the silica and to metal oxides deposits. This process continuously changes the quartz tube surface and bad repeatability of measurements as well as analytical signal depression is occurred [67,73]. Divitrification process can be prevented by in situ coating of the inside surface with an aluminum sulphate, lanthanum chloride or vanadium (from ammonium vanadate) solutions [67,74]. Alternatively, the use of tubes made of stainless-steel, silicon nitride and graphite has been reported [67]. However, most of the recent applications, including interference studies [73] and new injection systems [75] have been carried out with conventional quartz double–slotted tubes. Although quartz STATs can be easily manufactured in the laboratory, there are several commercial devices based on this atom trapping technique and the atom concentrator tube (ACT-80) by Varian is the most popular.

Table 2 summarizes the applications of STAT to assess cadmium in environmental samples by FAAS. The developed methods were applied to water [76-80], sewage sludge [81], soils [82,83], plants [83,84] and biota [72].

WCAT devices are U-shaped quartz tubes, positioned just below the optical path of the spectrometer, which are normally cooled by a rapid flow of cold water through the quartz tube [67]. This last device can be considered as an in situ pre–concentration system because analyte atoms condense onto the outer surface of the cooled tube when cold water is flowing through the quartz tube. On switching off the flow of coolant, the quartz tube undergoes fast heating and the condensed atoms are released as free atoms. The tube is positioned above the burner and the beam from the radiation source is allowed to pass just above the tube. The tube is commonly made of quartz because it offers a high melting point moreover a low thermal expansion coefficient. Other materials, such as copper, nickel, stainless-steel or titanium, were investigated to trap volatile elements such as cadmium.

Table 2. Atom trap techniques in FAAS determination of cadmium in environmental samples

Atom trap	Sample type	Sample treatment	Ref.
STAT	Drinking water	Dilution with water	[76]
STAT	River and drinking water	Dilution with 0.1% HCl	[77]
STAT	Drinking, dam lake, stream and lake water	Cd-cupferron complex formation and adsortion on activated C	[78]
STAT	Drinking, dam lake, stream and lake water	Solid phase extraction with thiouresulfonamide resin	[79]
STAT	Drinking water	None	[80]
STAT	Sewage sludge	Hot plate acid digestion and sequential extraction	[81]
STAT	Soils	Acid digestion and sequential extraction	[82]
STAT	Soils	Hot plate acid digestion	[83]
STAT	Plants	Dry ashing	[83]
STAT	Plants	Dry ashing and microwave assisted acid digestion	[84]
STAT	Mussels	Microwave assisted acid digestion	[72]
WCAT	Drinking water	None	[80]
WCAT	Water	0.01M HCl acidification	[90]
WCAT	River water	None	[91]
WCAT	Potable water	None	[94]
WCAT	Water and seawater	Acidification	[95]
WCAT	Soils	Acid digestion	[95]
WCAT	Soils	Acid digestion and acetic acid extraction	[96]
WCAT	Soils	0.05M $CaCl_2$ extraction	[97]

STAT, slotted tube atom trap; WCAT, water cooled atom trap.

However, problems derived from the formation of intermetallic compounds and lack of repeatability led to consider these cooled traps as non-usable [85,86]. Using quartz tubes, cadmium as well as other elements, were efficiently trapped and re-vaporized. Recently Korkmaz et al. [87] have investigated the nature of the re-volatilization process when using organic solvents aspiration. It was proved that cadmium is trapped onto the quartz surface as free metal and as oxide, and atomization of the analyte adsorbed onto the quartz surface occurs by evaporation directly from the solid metal or oxide or by melting [88]. In addition, the repeatability of cadmium determinations is improved and the lifetime of the tubes is increased when coating the quartz tubes with vanadium oxide salts [89]. In addition to the firstly described single-tube configurations, double- and triple-tube WCATs were proposed, and the double-tube configuration gave the best analytical performances [90 92]. Recent studies have led to improved dual-tube WCATs that consist of using 'bent' tubes instead of earlier 'flat' tubes, which enables more of the tube to be exposed to the radiation [93]. Table 2 also lists the different applications of WCATs for cadmium determination in environmental materials. Main applications have been developed for assessing cadmium in water [80,90,91,94,95] and soils [95-97].

The combination of STAT and WCAT systems has resulted in a new atom trap, the slotted–tube water–cooled atom trap (STWCAT) which consists of a STAT with a single WCAT passing through it, and mounted on a holder over the burner [98]. This system works in a similar way than WCAT devices but with the difference of released atoms are confined inside the STAT. Therefore, the STWCAT shows higher sensitivity and precision than STATs or WCATs used separately. As suggested by Matusiewicz and Kopras [99], these devices can be considered as integrated atom traps (IAT), and they can work as conventional atom traps or using the termed 'flame alteration technique'. Recent applications deal with the determination of arsenic, selenium and antimony by IAT-FAAS after hydride generation [100,101].

IV.3.5. Beam Injection Flame Furnace Atomic Absorption Spectrometry and Thermospray Flame Furnace Atomic Absorption Spectrometry

As commented, sensitivity in FAAS determinations can be improved by increasing the efficiency in aerosol generation and the residence time of free atoms inside the atomizer. However, sensitivity can be also enhanced if the sample is totally introduced in the atomizer, such as for the use of the Delves's cup system. Recent approaches, firstly proposed by Gáspár and Berndt [75,102], use this idea of total sample consumption, and two different methodologies, beam injection flame furnace atomic absorption spectrometry (BIFF-AAS) and thermospray flame furnace atomic absorption spectrometry (TSFF-AAS), were developed.

IV.3.5.1. Flame Furnace

Both BIFF-AAS and TSFF-AAS use a cell (furnace) heated directly by the flame (flame furnace), but discrete sample volumes are introduced in the flame furnace in contrast to conventional atom trap systems (STATs or WCATs) where the sample is continuously nubulized. In this sense, a better sensitivity when using BIFF-AAS and TSFF-AAS is expected because the whole sample is atomized moreover the increase in the residence time of the free atoms confined in the furnace.

The furnaces or cells are commonly made of pure nickel, although other materials, such as non-porous Al_2O_3 ceramic and quartz, were firstly tested [102]. However, because of memory effects and unsatisfactory analytical results obtained for ceramic and quartz furnaces, nickel, nickel alloys or titanium furnaces are preferred [103]. The furnace is positioned on a standard burner head by means of a stainless steel holder to keep the tube around 5 mm above the burner head. The nickel furnaces are commonly 10 cm long open-ended tubes with 10 mm i.d. and with an opening in the middle of the tube for introducing a 2 mm ceramic capillary, which is used for sample introduction. No flame gases enter the tube through this opening because the plane of this hole is parallel of the flow of the flame gases. This fact originates a lower temperature of the tube, around 200-300°C lower, than the flame temperature. Therefore, a further modification of the furnace implied four or more holes (2 mm diameter) at the bottom side of the furnace and at 90° to the ceramic capillary that allow a partial entering of the flame gases. Higher temperatures in these modified tubes than in the tubes without additional holes were obtained, and the analytical performances for refractory elements were improved.

IV.3.5.2. Sample Introduction System: High Speed Liquid Jet (Beam Injection)

In BIFF-AAS, the liquid sample is transported (introduced) into the flame furnace as a high speed liquid jet. A jet impact nebulization (JIN), droplets around 50 μm in size, occurs when the liquid jet enters the furnace through the ceramic capillary and impacts at the opposite inner wall of the flame furnace. In contract to JIN for sample introduction in ICP-OES [104], the generated droplets in BIFF-AAS are vaporized spontaneously onto the heated wall when impacting. This combined mechanism, jet impact nebulization and spontaneous vaporization, is termed jet impact vaporization (JIV).

The high speed liquid jets are generated by mean of a capillary nozzle (3 cm length) which is fixed in from of the flame furnace. The solvent or the liquid sample are transported through PEEK tubing (0.17 mm i.d.) using a high performance pump, usually a pump for high performance liquid chromatography (HPLC). Additionally, solvent/sample transport can be also carried out at low pressure by use of a peristaltic pump combined with a micro channel smooth jet nozzle (low sample jet generation system) [105]. The sample introduction (discrete volumes) is achieved by using a five- or six-port sample injection valve. Further developments have involved new designs for nozzles and also liquid sub-critical CO_2 as a carrier liquid and as gas-pressure pump for liquid jet generation [106].

IV.3.5.3. Sample Introduction System: Thermospray

TSFF-AAS uses a thermospray as a system for total sample consumption. Thermospray nebulization occurs when passing a liquid through a heated narrow tube, which originates the liquid stream to spontaneously boil. Typical TS required an external electrical heating system to fix the temperature just above the boiling point of the solvent. However, in TSFF-AAS, the TS capillary is directly heated by the flame and additionally by heat transfer based on direct contact of the TS capillary and the furnace [102,107]. The TS capillary (around 10 cm in length) can be made of stainless steal or ceramic and it is partially introduced (ca 2 mm) allowing the sample introduction into the nickel tube. Usually, a peristaltic pump (low flow) [102,108] is used to transport the solvent or the liquid sample, although a high pressure pump can be also used [109]. Similarly to BIFF-AAS, a six-port injection valve is used for sample introduction (discrete volume).

IV.3.5.4. Cadmium Determination by BIFF-AAS and TSFF-AAS

Compared with standard FAAS, a 7- to 17-fold improvement in the power of detection was obtained for several elements, cadmium included. In this sense, recent studies have shown adequate correlations between spray formation (droplets size) and cadmium analytical sensitivity by TSFF-AAS [110].

Scarce applications of BIFF-AAS to assess cadmium in environmental samples have found in literature, e.g. cadmium determination in mussels after ultrasound assisted acid leaching [111]. However, TSFF-AAS has extensively been used. Most of the samples analyzed are waters and seawater [112-116] in some cases after a pre-concentration method [112,113,115,116]; but also the technique has been applied to assess cadmium in biological samples [117-119], plants [120,121], sewage sludge [122], and marine sediments [114].

IV.4. Electrothermal Atomic Absorption Spectrometry (ETAAS)

Electrothermal atomization avoids most of the limitations found in flame atomization, such as the low analyte transfer ratio to the atomizer because the low efficiency of the nebulization process, the short residence time of the free atoms in the flame, the use of an inner environment in the system which avoids interferences, and the possibility of control the different stages before sample atomization.

IV.4.1. Electrothermal Atomizers

The atomic absorption spectrometer for ETAAS is the same as that used for FAAS except that an electrothermal atomizer replaces the flame/burner arrangement in the light pass of the spectrometer. The atomizer uses a refractory material which is electrically heated by low voltage (10 V) and high current (up to 500 A) [61]. Therefore, the material used for electrothermal atomizers must exhibit high electrical conductivity, and also high thermal resistance and high durability. Different materials such as graphite, tungsten or tantalum offer these properties and have been used in electrothermal atomization. Because tungsten- or tantalum- based atomizers present as a main drawback the extreme brightness at high temperatures, graphite atomizers are nowadays the most common used. However, graphite offers as disadvantages a high porosity and a high tendency to form stable carbides with some elements. These problems are partially overcome by coating the graphite tubes with pyrolytic graphite (pyrolytic graphite coated graphite tubes) [61], which minimizes analyte looses by diffusion into the porous material, and prevents analyte carbide formation.

A conventional graphite atomizer for longitudinal heating (graphite tube or graphite furnace) is typically a cylindrical tube (25 to 50 mm long and 5 to 10 mm diameter) with graphite electrodes clamped at either end and held axially in line with the light source. The tube has a central and small opening for sample introduction (discrete volumes between 5 and 100 μl) often with the aid of an autosampler. Graphite tubes for transverse heating (THGAs, transversal heating graphite atomizers) are shorter than those used for longitudinal heating (typically 27 mm long and 5 mm diameter) and the graphite electrodes are clamped transversally, i.e., from the sides, and the electrical current is applied at the central part of the tube. Because these tubes are shorter, sensitivity is decreased and end caps are commonly used.

The electrodes are water-cooled and the whole atomizer is purged with two inert gas (Ar or N_2) streams. The first gas stream is external (sheath gas) and prevents the entrance of air, while the second one flows internally into both ends of the tube and sweeps vapors generated during the mineralization of the sample. When using longitudinal heating, the electrical current is applied at the ends of the tube, and there is a temperature gradient along the graphite tube: the central portion is several hundred degrees hotter than the ends. These non-isothermal conditions can lead to condensation of the analyte or recombination with other species at the cooler ends of the tube. These problems are minimized when using transverse heating, which significantly reduces or eliminates condensation of the sample matrix components and memory effects and improves the atomization efficiency for refractory elements.

For both longitudinal and transverse heating, graphite tubes are commonly modified with a small graphite platform, formally called L′vov platform, only loosely connected to the tube walls and located beneath the sample entrance port [123]. The sample is deposited on this platform and although the graphite tube is directly heated by the electric current, the platform is heated by radiation and convection from the tube walls. There is therefore a time lag between the heating of the tube and that of the platform, and atomization occurs only when the surrounding gas is relatively hot and the whole operation is taking place isothermally. By using the L′vov platform technology (platform atomization), interferences derived from analyte condensation and/or recombination with other species are less significant than those occurred when using wall atomization. In addition, the use of L′vov platform is highly recommended when determining volatile elements, such as cadmium, because platform atomization enhances of sample dissociation and more reproducible absorbance peak schemes are obtained. Standard graphite tubes for both longitudinal and transverse heating are commercially available as graphite tubes with integrated platform.

A typical ETAAS program consists of a controlled four stages program showing the successive drying, ashing, atomization and cleaning steps. Evaporation of the solvent is reached at the drying stage, commonly at a temperature slightly higher than 100°C. Then, the solid residue remaining after drying is heated at temperatures between 300 to 1500°C (in function of the analyte and the sample matrix) during the second step (ashing, pyolysis or charring). Organic matter and also most of the inorganic concomitants of the sample are charred and removed. The programmed increase of temperature between the drying and the ashing steps is established with a temperature ramp of 20 to 30 s to avoid sample sputtering. After ashing, the remaining residue consists of the analyte in molecular form and a small amount of inorganic concomitants. This residue is subjected to the third stage (atomization) at temperatures within the 1200 – 2900°C range. The dissociation of the analyte molecular species occurs at this step and formation of free analyte atoms is obtained. Maximum power, i.e., rapid increase in temperature from the ashing to the atomization step is needed to avoid looses of free atoms. In addition, purge gas flow is stopped during atomization to increase the residence time of the free atoms into the graphite tube. Finally, during the cleaning step, a highest temperature is programmed for removing residues inside the atomizer.

IV.4.2. Matrix Modification

As commented above, a successful ETAAS measurement implies the analyte atomization in absence of matrix compounds. These concomitants must efficiently be removed at the charring of the graphite furnace temperature program. However, the volatility of the target elements and the refractory nature of the matrix concomitants can lead to an insufficient matrix removal to avoid analyte losses, which involves the existence of chemical interferences because of the presence of these concomitants during the atomization step. Other times, the chemical interferences arise from losses of the analyte during the charring step because the formation of volatile compounds between the target element and some component of the sample. Therefore, increasing the difference in volatilization between the analyte and the matrix to obtain a less volatile analyte or a more volatile matrix is desirable to perform a free interference ETAAS measurement. This objective is usually achieved by using the matrix modification approach. Matrix modification, also called chemical modification, was introduced by Ediger [124] and it can be defined as a process aiming to separate the analyte from the matrix, therefore facilitating interference-free determinations. This process consists of the addition of a reagent or a combination of reagents (chemical modifier), which react with the analyte or with the matrix, thus permitting selective volatilization and, consequently, the separation of analyte from the matrix at some point of the graphite furnace temperature program.

A matrix or chemical modifier acts mainly in two ways. Firstly, it removes concomitants (matrix) by incresing matrix volatility. In this case, the chemical modifier reacts with the matrix and the product of this reaction is more volatile so it can be lost at low temperatures, i.e. during the charring step. A classical example of this behavior is exhibited by chemical modifiers such as ammonium nitrate, which transforms a sodium chloride matrix (boiling point 1413°C) into volatile ammonium chloride, which sublimes at 335°C, and sodium nitrate, which decomposes at 380°C. Gas matrix modifiers, such as oxygen or synthetic air [125], or the use of aqueous hydrogen peroxide [126], can be included in this category. These reagents combust a sample with a high organic matrix content at low temperatures, allowing moderate low charring temperatures and avoiding volatile analyte losses.

The second way of action of a modifier is the direct reaction with the analyte to convert it into a phase of higher thermostability, i.e., to decrease analyte volatility. By this way, a more efficient removal of the matrix at higher temperatures at the charring stage, but without loss of the analyte, can occurr. Transition metal ions, e.g. palladium and nickel, usually added as inorganic salts ($Pd(NO_3)_2$ or $Ni(NO_3)_2$) in aqueous solution, are some examples. The mechanism of stabilization of these modifiers implies the formation of thermally stable intermetallic compounds with the analytes. Magnesium, added as magnesium nitrate, is other example; but in this case, magnesium nitrate is thermally decomposed to magnesium oxide at some point of the charring step, and in this process traps analyte atoms within its crystalline matrix, exhibiting a thermal stability until 1100°C. Finally, other conventional chemical modifier is ammonium phosphate, which is mainly used for stabilizing volatile elements such as cadmium and lead. For cadmium, the use of ammonium phosphate leads to the formation of cadmium-oxyphosphorus compounds that shifted the atomization of cadmium to higher temperatures [127]. However, the most reported mixture for matrix modification is $Pd(NO_3)_2$ and $Mg(NO_3)_2$, proposed by Schlemmer and Welz as an universal chemical modifier [128].

As firstly proposed by Hoenig et al. [129] some chemical modifiers behave in other way – they increase the analyte volatility – so that concomitants (matrix) are volatilized during the cleaning step. Organic acids, such as ascorbic acid or citric acid, are some examples. These reagents react with volatile elements, such as cadmium or mercury [129,130], thereby diminishing their volatilities from matrices with a high salts content as seawater. Similarly, other way of chemical modification consists of the decrease of the concomitants volatility. An example of the second type of behavior is the use of ammonium molybdate, which reacts with phosphate ions to form the highly refractory ammonium molybdophosphate. This product is volatilized at the cleaning stage and analyte atomization is free of interferences from phosphate ions [131].

Certain organic reagents exhibiting reducing properties, such as ascorbic acid, citric acid or hydroxylamine chloride, can be used in combination with palladium as a powerful chemical modifier for volatile elements [132]. The formation of thermally stable intermetallic compounds with palladium requires that palladium, usually added as aqueous palladium nitrate (Pd^{2+} ions), to be reduced (Pd^0). The reduction of Pd^{2+} ions to Pd^0 is temperature-dependent and it occurs in absence of a reducing medium at temperatures closed to 1000°C. However, volatile elements such as cadmium, can be lost before palladium ions can be temperature-reduced, and it is results unusable as a chemical modifier for these volatile elements. Although some applications on palladium reduction by in situ thermal decomposition of the palladium nitrate just before sample injection (dry mode) have been described in literature [133], the chemical reduction of palladium nitrate is preferred. The combination of palladium and a reducing agent guarantees the reduction of palladium at an early stage of the temperature program, and palladium can act as a chemical modifier (formation of intermetallic analyte-palladium species) for volatile elements.

Although chemical modifiers are commonly used as aqueous solutions, coating the graphite tubes with the chemical modifier, mainly transition metals, is other possibility for chemical modification [134]. Moreover, the use of metal-coated graphite tubes offers the advantage that chemical modifiers has not to be added to each sample, and also the fact that metal-coated graphite tubes avoids chemical interferences arise from the reaction of some elements with the surface of the graphite tube to form stable carbides. This chemical modification involves coating the graphite tube with a noble metal, such as lanthanum, zirconium and tungsten; and more recently, iridium, platinum and thorium. These modifiers, formally coated permanent chemical modifiers, behave in much the same way as aqueous palladium in that thermally stable intermetallic compounds are formed on the hot inner surfaces of the coated graphite tube.

IV.4.3. The Stabilized Temperature Platform Furnace Concept

The stabilized temperature platform furnace (STPF) concept, introduced by Slavin et al. [135], consists of a series of operating conditions that lead atomization under an isothermal furnace. This conditions implies (1) the use of an isothermal operation mode, i.e., platform atomization and/or transversal heating; (2) the use of the Zeeman effect background correction system; (3) the use of a chemical modifier; (4) the use of measurements based on integrated absorbance signal rather than peak height; (5) the use of a rapid rate heating during atomization (maximum power heating) so that the platform effect can be enhanced; (6) the use of gas stop during atomization; and (7) the use of fast electronics to follow the transient signal without distortion.

Graphite furnace operation under the STPF concept leads to interference-free determinations because an efficient volatilization of concomitants prior atomization, as well as a better control of condensed phase and gas phase chemical interferences is reached. Therefore, these conditions are the basis of the modern ETAAS measurements.

IV.4.4. Cadmium Determination by ETAAS

Similarly to FAAS, the primary resonance line for cadmium in ETAAS is 228.8 nm, while the only alternative line (326.1 nm) results unusable because the low sensitivity [61]. At the most sensitive line, spectral interferences have been reported for the use of ammonium phosphate as a chemical modifier, especially when determining very low concentrations of cadmium in matrices with high iron and arsenic contents [61], and especially with high chloride concentrations [136]. These interferences could not be eliminated by using the Zeeman-effect background correction system, but they can be avoided using palladium-magnesium as a chemical modifier [137]. In general, the characteristic mass for cadmium when using longitudinally heated atomizers is 0.4 pg, while in a transversely heated atomizer it is 1.3 pg [61].

The atomization mechanism for cadmium has not been well-established [61]. A first proposed atomization mechanism implies a carbothermal reduction of the cadmium oxide and the followed direct vaporization of the reduced metal [138]. However, other studies lead to a thermal dissociation of gaseous cadmium oxide [139].

Because cadmium is a volatile element, chemical modification has been shown to be important for a successful determination of cadmium in several matrices. From the early application of ammonium phosphate as a chemical modifier [124,140], and the use of palladium-magnesium as an universal chemical modifier [128,137], different agents have been proposed as chemical modifiers for cadmium determination by ETAAS, such as lithium tetraborate [141], nickel [142,143], silver [144], and even polytetrafluoroethylene [145]. A mixture of scandium-palladium-ammonium nitrate [146] was also used. Determination of cadmium by ETAAS in absence of chemical modifier but using pyrolytic and non-pyrolytic boron nitride platforms and conventional graphite tubes has been proposed [147]. However, most of the reported ETAAS determinations of cadmium require palladium [143,148], reduced palladium [130,133], palladium-magnesium [149-155], magnesium [156,157], or ammonium phosphate [143,157-165] for chemical modification.

Permanent chemical modifiers, obtained after coating the graphite tubes with concentrated noble metal solutions, are also quite popular, especially iridium [153,160-170], rhodium [153,169-173], ruthenium [153,167-170,174-177], tungsten [161,168-173,175,176], zirconium [176] and molybdenum [167]. Electrodeposition was also proposed for coating graphite tubes with palladium and rhodium [178].

IV.5. Inductively Coupled Plasma Based Techniques

IV.5.1. Inductively Coupled Plasma – Optical Emission Spectrometry (ICP-OES)

In atomic emission spectrometry, atoms or ions from a representative amount of an atomized sample are excited by the absorption of thermal energy, and on returning to the ground state, these species emit radiation of the element-characteristic wavelength. The intensity of the emitted radiation is proportional to the concentration of atoms or ions. Atomization cells which provide very high temperatures are required for AES. This is because in addition to the free atom production (atomization) as it does in AAS, these atoms must be promoted to high energetic levels. The high temperatures of atomization cells for AES can eventually produce ionization, enabling ion lines to be used. From all available atomizers, flames (flame atomic emission spectrometry, FAES) and, mainly, plasmas (inductively coupled plasma – optical emission spectrometry, ICP-OES), are nowadays the most important cells for AES. The instrumentation for FAES is similar to that for FAAS, except for the external radiation source, which is not required. The instrumentation for ICP will be discussed in the following section.

ICP-OES is one of the most useful analytical techniques for elemental analysis and one of the most used in routine laboratories. This technique offers as main strengths the possibility of simultaneous or sequential determination of up to 70 elements in theory; several orders of magnitude for the useful working range; and easy of automation, enhancing samples throughput. In addition, the better spectral resolution achieved with the modern instruments has increased the number of potential analytical lines, and some wavelengths which were considered as unusable a long time ago are now used analytically [179]. Complete information on all aspects relating to ICP-OES can be found in some monographs [179-181].

In general, an ICP-OES determination consists of the sample (mainly in solution) introduction into a plasma, temperature of approximately 8000°C, which thermally excited all elements (atoms or ions), allowing the emission of radiation at their characteristic wavelengths. The emitted radiation is passed through a diffraction grating to resolve it into a spectrum of its constituent wavelengths. Each diffracted wavelength is then collected and amplified to yield an intensity measurement that is proportional to the elemental concentration.

As it shown above, the basis of ICP is plasma, sometimes referred to as the fourth state of aggregation. Plasma is a very hot ionized gas, in which co-exist positive ions, electrons and neutral species of an inert gas. Argon is usually chosen as the plasma gas because of its inertness, optical transparency to the UV-VIS spectrum, moderately low thermal conductivity, and high first ionization energy [62]. In addition, the most common plasma sources used in atomic spectrometry are radio-frequency inductively coupled plasma (ICP).

IV.5.1.1. Plasma Torch

The inductively coupled plasma is formed and sustained within a quartz torch placed in a radio-frequency oscillating magnetic field. The quartz torch consists of three concentric tubes. The outer and the intermediate tubes have tangentially arranged entry points, and the outer channel is for the coolant gas (outer, coolant or plasma gas) while the intermediate one is the channel for the auxiliary gas. The plasma gas flows tangentially into the torch along the outer tube until shortly before it reaches the plasma. The gas cools the torch but also maintains the plasma; flow rates between 10 and 20 L min^{-1} are typical. The intermediate tube serves to force the coolant gas to flow tangentially along the outer tube; it also enables another gas (the auxiliary gas) to be introduced. Typical auxiliary gas flow rates are between 0 and 2 L min^{-1}. Finally, the inner tube is a capillary tube through which the aerosol from the nebulization system enters. Therefore, the inner channel is for the sample carrier gas (nebulizer gas). The sample is introduced as an aerosol through the inner tube, also called injector. This tube is usually made of quartz or aluminum oxide. Nebulizer flows are normally low (from 0.6 to 1.0 L min^{-1}). Modern ICP-OES instruments allow both radial and axial viewing (the dual view concept); in them, the plasma lies horizontally and the optics is set up for axial viewing.

IV.5.1.2. Radio-Frequency Generators

Located around the outer tube of the plasma torch is a coil of copper tubing through which water is re-circulated. Power input to the ICP is achieved through this coil, typically within the 0.5 – 1.5 kW range and at a frequency of 27 or 40 MHz. In order to ignite the plasma a high frequency electrical field is applied. Two types of designs for radio-frequency generators can be used: crystal-controlled and free-running generators. In the former, an oscillator circuit incorporating a crystal oscillating at a fixed frequency is responsible for inducing the electrical field. In the latter, changes in power loading are compensated by slight shifts in the frequency of the oscillation circuit in order to bring the whole circuit back into resonance.

IV.5.1.3. Nebulizers

Similarly to FAAS, liquid samples are introduced into the plasma as a fine aerosol. Therefore, a system for converting the liquid sample into suspended fine droplets (nebulizer system) is required. Unlike FAAS/FAES, where solution uptake is by free aspiration, the solution to be nebulized in ICP is usually moved by a peristaltic pump.

There are different types of nebulizers: pneumatic nebulizers (PNs), glass frit (fritted disk) nebulizers, ultrasonic nebulizers (UNs) and thermospray nebulizers (TNs) [62,180-182].

IV.5.1.3.1. Pneumatic Nebulizers

The Meinhard pneumatic concentric nebulizer, which works in a similar way that nebulizers for FAAS/FAES, is the most common pneumatic nebulizers in use today for ICP-OES. Other PNs such as the cross-flow nebulizer and Babington nebulizers are also of interest. The cross-flow nebulizer consists of two capillary needles positioned at 90° to each other. The sample is pumped through one capillary and the argon carrier gas is flowed through the other one. At the exit point both carrier gas and liquid sample are in contact and the force of the escaping carrier gas is sufficient to break the liquid up into a coarse aerosol. V-groove or Babington nebulizers are similar to the cross-flow nebulizer but the modified design allows the nebulization of solutions with high solids contents [181].

Other described PNs are the low-flow nebulizers and the micronebulizers, which deliver a higher mass of analyte to the plasma but have low limits of detection [183,184]; and especially, the dual nebulizer or multimode sample introduction system [185], which permits direct pneumatic nebulization and/or introduction of vapors (hydrides). Recently, the direct injection nebulizers (Vulkan direct nebulizers and direct injection high efficiency nebulizers, DIHN) are becoming popular because they transport nearly 100% of the sample to the plasma, thereby increasing sensitivity [62]. However, DIHNs require very low flow rates of sample uptake and the absence of suspended solids.

IV.5.1.3.2. Glass Frit Nebulizers

Glass frit nebulizers consist of a glass frit or metal grid (grid-type nebulizers) through the liquid sample runs over and through the nebulizer gas passes at a high velocity. This produces that the liquid sample shears into fine droplets. It has been established that these nebulizers have greater transport efficiency than PNs because the very fine droplets generated [62]. However, these nebulizers type suffer from salt deposition when nebulizing solutions containing high dissolved solids.

IV.5.1.3.3. Ultrasonic Nebulizers

An ultrasonic nebulizer (UN) utilizes a piezoelectric ceramic membrane (oscillator) which can vibrate at different radiofrequencies, although 1 MHz is typically the oscillation frequency used for [62]. The piezoelectric material is connected to and AC/DC converter which supplies the electrical energy to be converted into vibration. When the aspirated liquid sample passes near the oscillator, the high frequency vibration makes the liquid to break into very fine droplets. Modern ultrasonic nebulizers also comprise an electrothermal desolvation device instead of a heater-condenser (recirculating chiller) system, which is required to desolvate the aerosol in order to prevent large amount of solvent vapor. UNs convert more of the liquid sample into a usable aerosol.

IV.5.1.3.4. Thermospray Nebulizers

Thermospray nebulizers (TNs) consist of a narrow bore stainless steel tube, which are electrically heated to just above the boiling point of the solvent. The heating causes the liquid stream to spontaneously boil and eject a hot and fine aerosol of fine droplets from the capillary tube mixed with solvent vapor [62]. Instead of electrical heating, microwave irradiation has been also proposed to heat the solvent [186]. Similarly to UNs, desolvation of the aerosol is needed to avoid deterioration in the analytical characteristics of the determinations. Thermostated spray chambers (TSCs) are commonly used to achieve this task [187]. The hot aerosol derived from the TN is introduced into the TSC and part of the solvent vapor is removed from the aerosol stream. Other devices, formally called two-unit desolvation system (TUDS) [187], use a first desolvation system to evaporate most of the solvent, and then a second unit where the solvent vapor is removed from the aerosol.

IV.5.1.4. Spray Chambers

The direct introduction of a fine aerosol generated by a nebulizer into a plasma could extinguish or induce cooling of the plasma. Therefore, the aerosol is directed through a spray chamber (nebulizer chamber) within the large droplets are discharged to waste. In addition to the reduction of the amount of aerosol reaching the plasma, the spray chamber reduces the aerosol particle size and decrease the turbulence associated with the nebulization process [181]. Spray chambers are usually made of glass, quartz or inert polymers (Ryton or several fluorine-based polymers). There are different available designs such as the single-pass type, the double-pass or Scott-type and the cyclonic spray chambers.

The single-pass, also referred as direct or cylindrical spray chamber, consists of a cylindrical tube with an impact bead just in front the nebulizer. The fine aerosol from the nebulizer impact onto the bead and a further fragmentation of the droplets is obtained. The fine aerosol generated after the impact arise the plasma torch and the remaining liquid sample is delivered to waste. This type of spray chamber is usually connected to a desolvation system to reduce sample load into the plasma. The simplest system consists of a thermostated water jacket where the spray chamber is fitted.

The Scott chamber is the most common type. It is formed by two concentric tubes. The nebulizer is fitted in the inner tube and the interaction of the coarse aerosol from the nebulizer with the internal surfaces of the inner and the outer tubes (double-pass) leads to a production of a finer aerosol. This fine aerosol reaches the plasma, while the larger droplets emerge from the tube and, by gravity, exit the spray chamber via a drain tube. The liquid in the drain tube is kept at positive pressure, which forces the small droplets back between the outer wall and the central tube, where they emerge from the spray chamber into the sample injector of the plasma torch. A drawback of this design is the presence of dead volumes which implies large rinsing times between samples to avoid cross contamination [181]. Both single-pass and double-pass types have been superseded by the cyclonic spray chamber mainly because the sensitivity reached (approximately 50% better sensitivity). The cyclonic spray chamber operates by centrifugal force. Droplets are discriminated according to their size by means of a vortex produced by the tangential flow of the sample aerosol and argon gas inside the chamber. Smaller droplets are carried with the gas stream into the plasma torch, while the larger droplets interact with the walls and fall out through the drain. A more concentrated aerosol is obtained with this spray chamber, which implies a higher sampling efficiency and sensitivity. However, the use of the cyclonic chamber has been reported to give slightly inferior precision for certain types of samples [188].

IV.5.1.5. Optics and Detectors

Optical components can be arranged in two principal types of instruments: those that measure all wavelengths at the same time (simultaneous spectrometers), and those in which one wavelength is measured after another (sequential spectrometers). The classical optical mounts based on the Paschen-Runge or Czerny-Turner mounts, which used to be the basis of many simultaneous and some sequential spectrometers, have now been superseded by the echelle mount. This latter configuration gives very good resolution because the mechanically ruled grating, which typically 50-100 grooves per mm. Both a prism and a grating are used as cross-dispersing media, the former to sort the orders of the VIS range, and the latter to perform this task for the UV range. In this configuration, modern ICP-OES instruments use solid state detectors, mainly charge injection devices (CIDs) and charge-coupled devices (CCDs). Complete information of different configurations as well as detectors can be found elsewhere [62,179,181].

IV.5.2. Inductively Coupled Plasma – Mass Spectrometry (ICP-MS)

ICP-MS uses the same sample introduction systems (peristaltic pump, nebulizer and spray chamber) as ICP-OES, as well as a horizontally configured (axial configuration) argon plasma torch. The specific parts of an ICP-MS are the detector (mass analyzer) and an interface between the plasma torch and the mass analyzer (ion detector).

IV.5.2.1. Interface

Because ICP source works at atmospheric pressure and the mass spectrometer is a high-vacuum (c.a. 10^{-9} atm), a suitable interface allowing the coupling of the plasma torch with the ion detector is needed. This interface (ion sampling interface) consists of a water-cooled outer sampling cone which is axially aligned with the plasma torch. The sampling cone is made of nickel (high thermal conductivity and resistance to corrosion) and shows a narrow orifice (1 mm) which reduces the pressure to approximately 10^{-3} atm. The region behind the sampling cone is maintained at a moderate pressure by a vacuum pump. The pressure differential created by the sampling cone allows that ions from the plasma and the plasma gas itself pass through the narrow orifice. Because large flows of plasma gases pass through the sampling cone, a second cone also made of nickel and with a 0.75 mm orifice (skimmer cone) is placed behind the sampling cone. The pressure in the intermediate vacuum chamber after this second cone is maintained at around 10^{-7} atm. At this instant, the ions form a beam that can be electrically focused on to the entrance of the mass analyzer by means of a series of ion lenses.

IV.5.2.2. Collision/Reaction Cells

If the ICP-MS system contains a collision/reaction cell, then this is located after the interface and before the mass analyzer. This system is used in modern instruments to reduce or avoid molecular interferences [189,190]. A reagent (gas) is introduced at this point, which react with the potentially interferences and minimize or avoid their interfering effect on a certain mass-to-charge ratio (m/z ratio). Based on the different ion-molecule reactions (charge exchange, proton transfer, hydrogen-atom transfer, hydride-atom transfer, adduct formation or condensation reactions) [181,191], different gases can be used: He (collision gas), H_2 or NH_3 (charge-exchange gas), O_2 or N_2O (oxidizing gas), H_2 (reducing gas) or CH_4 (other reaction gas).

IV.5.2.3. Mass Analyzers

Mass analyzers act as a filter, transmitting ions with a pre-selected m/z ratio, and allowing ions separation. There are different types of mass analyzers which can offer different resolution and sensitivity. The most common analyzers are the quadrupole mass spectrometer, the sector-field or double-focusing magnetic sector – mass spectrometer, and the time of flight mass spectrometer.

Quadrupole mass analyzers consist of four cylindrical metal rods arranged in parallel and equidistant from a central axis, and operating as electrodes. By applying a combination of radio-frequency (RF) (the RF voltage of each pair are 180° out of phase) and direct current (DC) voltages (the DC voltage can be positive for one pair and negative for the other), only ions in a certain m/z range are allowed to pass. The system can be operated allowing the transmission of only one m/z ratio (single-ion monitoring) or by continuous varying the RF/DC ratio allowing ions of consecutively higher m/z ratio to pass to the detector (multi-element analysis). Mass resolution of quadrupole mass spectrometers is unit-mass resolution.

The sector-field mass spectrometer (double focusing mass spectrometer or high resolution mass spectrometer) consists of an electrostatic analyzer combined with a magnetic analyzer to focus the ion beam. The system also uses narrow entry and exit slits which allow the number of ions to pass through the detector. Ions from the source are accelerated through the entrance slit of the electric sector, which acts as an energy filter. After a narrow band of ions with certain kinetic energies has been selected, the ion beam is focused on to the magnetic sector, where the ions are deflected in accordance with the m/z ratio (a high degree of deflection for ions with high m/z ratios). In the same way as for quadrupoles, a mass spectrum is obtained by scanning the magnetic field and allowing ions of consecutively higher m/z ratio to pass the exit slit of the magnetic sector in the direction of the detector. A better spectral resolution (lower than unit-mass) is obtained with this system, removing most of the interferences from polyatomic species.

Time of flight (TOF) mass analyzers use an electric voltage (voltages from 10^3 to 10^4 V) to accelerate the ions trough a 1 m long analyzer rod, and the velocities of the different ions, which are dependent of the (m/z) ratio, is the basis of its separation: all the ions have the same kinetic energy, their velocities along the analyzer rod must be inversely proportional to the m/z ratio. In this way, those ions with lower m/z ratios reach the detector first. The times to reach the detector (the time of flight) are commonly within 1 to 30 μs. To obtain pulse of positive ions, a device to generate pulse of electrons or photons is needed. The frequencies commonly used are between 10 and 50 kHz.

IV.5.2.4. Ion Detectors

The channel electron multiplier is the usual ion detector for mass spectrometry. The operating principles of this system are similar to those of the photomultiplier tube but in this case, it must be operated at a very low pressure. This device consists of a curved glass tube (1 mm i.d.), with the inner surface of coated with a resistive material (lead oxide); and which flared end of the tube. When the instrument is operating in the pulse counting mode (the most sensitive mode), ions are attracted into the funnel opening by a high applied voltage (up to -3500 V). When these ions collide with the inner coating, a significant number of secondary electrons are ejected from the resistive surface and accelerated down the tube. They then collide with the inner walls of the tube and cause further electrons to be ejected from that surface. As a result, an exponential cascade of electrons is produced, which eventually reaches saturation point and results in a large electron pulse (a gain of 10^7 to 10^8 over the original collision). The second operation mode is the analog mode; this works in a similar way to the pulse counting mode, but lower voltages (between -500 and -1500 V) are applied and the multiplier does not become saturated. The pulses therefore vary in size, and there is a gain of only $10^3 – 10^4$.

IV.5.3. Isotope Dilution Analysis

As any mass spectrometer can measure isotope ratios [62,181], besides to external calibration and the standard addition technique, ICP-MS can perform isotope dilution analysis (IDA) as unique method of calibration [181,192]. IDA is based on the fact that most of the elements have more than one stable isotope and the normal isotopic distribution of those elements is fixed. The analysis consists of measuring the sample (normal isotopic distribution of an element) and the same sample after spiking with a certain isotope (artificially isotopic distribution). After applying a mathematical equation, the concentration of the element in the sample can be obtained. IDA can be applied to those elements with more than one isotope, and it can be used in multi-element determinations. However, this approach offers as main disadvantage the high cost of the enriched stable isotopes and the large time of analysis.

IV.5.4. Vapor Introduction in ICP Techniques

As commented, sensitivity in ICP based techniques, mainly for ICP-OES, is limited by the low efficiency in conventional sample introduction devices (nebulization for liquids). However, the efficiency when introducing vapors is close to 100%, with the high increase on sensitivity. Chemical vapor generation techniques, mainly cold vapor and covalent hydrides generation, will be discussed in section IV.7; however, once the vapor is generated, it can efficiently be introduced in a plasma torch. This section is devoted to vapor introduction in ICP based techniques, and the different modified nebulizer-spray chamber arrangements will be briefly described.

Besides the direct connection of the gas-liquid separator with the plasma torch via a PTFE tubing [193,194], one of the simplest systems, proposed by Feng et al. [195,196], uses a T piece to mix the sample and the reducing agent and a Teflon tubing to transport the vapor to a modified Burgener-type nebulizer (the nebulizer has two tubes; one for a conventional generated aerosol; and other for gaseous species introduction) which is connected to a modified Scott spray chamber (three channels, one additional channel for the vapor introduction). Gas-liquid separation takes place in the spray chamber and the system can be used simultaneously for conventional liquid nebulization and vapor introduction.

Another simple variation consists of using part of the spray chamber as a reduction (vapor generation) cell. Based on this idea, a concentric nebulizer coupled to a commercial Jobin Yvon cyclonic chamber has been used for vapor introduction in ICP techniques. The cyclonic chamber exhibits three entries for sample, hydrochloric acid and reducing solutions, and an exit for liquid waste, all located in the bottom of the chamber. The vapor generation occurs inside the cyclonic chamber and he vapor is transported by the Ar nebulizer to the plasma torch [197-200]. A further device is the called multimode sample introduction system (MSIS), patented by McLaughlin et al. [185], and which consists of a cyclonic spray chamber with two integrally affixed cylindrical ports (one for introducing a nebulizer into the chamber, and the other for facilitating volatilization of analytes within the chamber. Gas-liquid separation also occurs inside the spray chamber and the system allows a conventional use for liquid and sample introduction for vapor [201,202].

IV.5.5. Discrete Sample Introduction in ICP Techniques

IV.5.5.1. Electrothermal Vaporization

Similar to the Delves' cup microsampling approach in FAAS, discrete sample introduction in ICP techniques can be advantageous when analyzing samples that are limited in volume, that contain high dissolved solids or large amounts of organic solvents, or even high solids contents. The whole sample (a discrete sample volume) is introduced into the plasma and all of the analyte is available for analysis, leading to improved sensitivity.

The current technique for discrete sample introduction into plasmas is the electrothermal vaporization (ETV) which is based on pipetting the aqueous (slurried) sample or placing the solid sample onto a graphite furnace. The graphite furnace is subjected to a controlled temperature program (similar to ETAAS) to remove the sample matrix (ashing) and finally to vaporize the analyte. By this way, interferences arising from concomitants of the sample matrix are minimized. From the first use of conventional heated graphite atomizer [203], important improvements on designs and optimization of ETV devices have been carried out [204]. In addition, for analytes which form very refractory oxides the use of matrix modifiers usually assist the vaporization. Fluoride-based modifiers [205], such as PTFE as a suspension or Freon gases, are widely used because they vaporize these analytes as their fluorides (lower vaporization temperatures than those required to volatilize the analyte oxides) [62].

As has been pointed out by Resano et al. [206], ETV results highly convenient for solid sample introduction in ICP-OES/MS (solid and slurry sampling) [207,208]. In addition, ETV also results adequate for liquid microsample introduction in plasma-based techniques [209] and even for speciation studies [210]. Detailed information on the vaporization mechanisims of analytes in ETV as well as chemical modification can be found elsewhere [211].

IV.5.5.2. Laser Ablation

Another method of solids introduction in ICP-based techniques is laser ablation. The approach uses a laser, commonly an Nd:YAG laser operated in the near infrared at 1064 nm, which is focused on to the surface of the sample, allowing a part of the sample to be mobilized (ablated) and transported direct to the plasma torch via the argon carrier. Occasionally, the laser ablation can be occurred either just above or below the surface of the sample. Although Nd:YAG laser operates at 1064 nm, it can work at different wavelengths such as 532, 355, 266 or 213 nm with optical frequency doubling, tripling, quadrupling and quintupling, respectively [181]. Laser ablation has several advantages over other introduction techniques: (1) very limited amount of sample are usually enough to perform the analysis; (2) the technique is applicable to any solid sample; (3) sample pre-treatment is omitted; (4) there are not reagents consumption and any waste is generated; (5) spatial characterization information can be obtained; and (6) by repetitive ablation of the same spot, depth profiling information of a sample can be assessed. However, laser ablation has some limitations, mainly associated to an accurate calibration (it is difficult to find matrix-matched standards with known composition which are very closely matched with the samples). Heterogeneous samples can lead to inaccurate results because the small sampling area (around 10 - 100 μm diameter spot) can not be representative. This fact is also an advantage, as shown above, because a spatial composition can be reached after several ablation spots along the whole sample. Other drawbacks arises from the fact that the amount of sample ablated is dependent upon laser and sample properties and certain fractionation for samples containing low-melting point elements can be occurred [181].

A laser ablation device coupled to a plasma source consists of an ablation chamber, where the sample is placed and fitted with a fused silica window; a lens to allow focusing; an adjustable platform for positioning in the three x-, y- and z- directions; a camera for remote viewing of the sample surface; and a Nd:YAG laser. The detection is usually carried out with a CCD. Once the sample has been ablated, an argon carrier gas transports the ablated materials to the torch.

There are three different calibration strategies when using laser ablation coupled to ICP: matrix-matched direct solid ablation, dual introduction (sample-standard) and direct liquid ablation [181]. Matrix-matched direct solid ablation is the most common approach and consists of using matrix-matched standards (certified reference materials with close composition to the samples) for calibration. Sometimes, it is possible to produce matrix-matched standards, such as after addition of standard solution to a powdered material or by co-precipitation of the analyte over a powdered matrix. Calibration by the dual introduction (sample-standard) system is based on the sequential introduction into the plasma of a conventionally nebulized standard and the laser-ablated material from the sample. A comparison between the analytical response from the ablated sample and the standard is then made. This approach has as a disadvantage the different response when the analyte is from the aerosol (wet) or from the ablated material (dry). Finally, the direct liquid ablation uses aqueous solutions as standards. A reagent, called chromophore, is usually added to the standards to allow the ablation process to occur within the surface layers of the liquid and to avoid aerosol generation.

IV.5.6. Continuous Sample Introduction in ICP Techniques

Flow injection (FI), chromatography and capillary electrophoresis are methods of continuous introduction in ICP-based techniques. FI leads to an automation of the sample introduction and allow stages such as matrix removal or pre-concentration before sample nebulization. FI techniques as well as continuous flow (CF) will be discussed in following section devoted automated sample pre-treatments.

Because the importance of the different toxicological effects of an element based on its chemical form, speciation studies are an active area in modern analytical chemistry. These studies imply a chromatographic / electrophoretic separation, and thus, liquid chromatography (LC), gas chromatography (GC) and capillary electrophoresis (CE) are nowadays common continuous sample introduction systems for ICP-based techniques. High performance liquid chromatography (HPLC), mainly using anion or cation exchange columns, offers few problems with respect to physical coupling with ICP techniques. This is because the HPLC flow rates are compatible with the sample uptake rate of an ICP instrument. Therefore, the coupling consists of a simple connection of the output from the HPLC to the input of the nebulizer through PTFE or PEEK tubing. The only problem that can arise is the mobile phase composition, mainly if high proportion of organic solvent is present, which have the effect of quenching plasma or even plasma extinction. In such cases, a desolvation device can be useful.

Similarly, the interfacing of GC with ICP is not also difficult, although is less straightforward than for LC. This is because the volatile analytes must be transferred from the GC to the plasma at a high temperature to avoid analyte condensation. This is commonly achieved via a heated transfer line at temperatures ranging from 200 to 250°C which allows the volatile analytes and the carries gas to be directly transported to the plasma torch. In this case, nebulization is not necessary and high analyte transport efficiency is obtained (higher sensitivity than that obtained when coupling HPLC).

IV.5.7. Cadmium Determination by ICP-OES/MS

Cadmium can be sensitively determined by ICP-OES (estimated limit of detection of 3 $\mu g\ L^{-1}$) for the emission wavelength 228.802 nm. Other alternative lines are 214.438 and 226.502 nm. All these lines are below 300 nm and spectral interferences from argon emission lines (300 to 600 nm) are not important. In addition, useful emission lines for cadmium are not within the molecular band spectra originate from OH radicals (280 – 330 nm), oscillation of NO (230 nm), N_2 molecular bands (370 nm) or even oscillation of carbon compounds (450 to 650 nm) when high organic matter or organic solvents are introduced in plasmas. Therefore, cadmium determination by ICP-OES can be expected to be free of molecular spectral interferences. In addition, spectral interferences derived from other atomic species with a direct wavelength coincidence or a partial overlapping of the cadmium emission line are also not important because the improved spectral resolution in modern ICP-OES instruments. The classical example of As (at 228.812 nm) partial overlapping of Cd (at 228.802 nm) is minimized with the improved resolution of modern instrumentation. Nevertheless, alternative emission lines (214.438 and 226.502 nm) can be used in ICP-OES with bad resolution to avoid atomic spectral interference by arsenic.

Similarly, cadmium determination by ICP-MS is very sensitive, with a limit of detection of 0.06 µg L^{-1} when using the most abundant isotope for cadmium, ^{114}Cd (28.73%). Other naturally occurring stable isotopes are ^{112}Cd (24.13%), ^{111}Cd (12.80%), ^{110}Cd (12.49%), ^{116}Cd (7.49%), ^{106}Cd (1.25%) and ^{108}Cd (0.89%). There are potential interferences on cadmium determination by ICP-MS when using the isotope ^{114}Cd. Interferences include an isobaric overlap from ^{114}Sn and polyatomic overlaps from molybdenum oxides ($^{98}Mo^{16}O^+$, $^{96}Mo^{18}O^+$), ruthenium oxides ($^{98}Ru^{16}O^+$) and molybdenum hydroxide ($^{97}Mo^{16}OH^+$) [212]. These last interferences are successfully compensated by the reaction cell technology (use of O_2 as a reagent gas) [213], while isobaric interference of ^{114}Sn is usually minimized by adequate correction equations provided by the software. Potential isobaric interferences for ^{112}Cd are also ^{112}Sn, while polyatomic interferences can be observed from zirconium, molybdenum, ruthenium, and calcium and argon ($^{96}Mo^{16}O^+$, $^{96}Zr^{18}O^+$, $^{96}Ru^{18}O^+$, $^{95}Mo^{16}OH^+$, $^{40}Ca^{40}Ca$ and $^{41}K^{40}Ca$) [212].

Because of the multi-element capacity of ICP-OES/MS instrumentation, there are many applications for cadmium determination in different environmental samples. However because cadmium concentrations are very low in many sample types, specially water and seawater, separation/pre-concentration methods are usually required when using ICP-OES. Special attention must be given to liquid samples with high dissolved solid contents, such as seawater, because the high salts content can extinguish the plasma when a direct introduction is performed. Most of the methods for determining cadmium in both liquid and solid environmental samples require a certain sample pre-treatment. These procedures will be discussed in Chapter V devoted to sample pre-treatments. Cadmium determination by methods involving chemical vapor generation will be discussed in section IV.7. However, ICP-OES/MS methods based on special sample introduction approach, such as ETV (for liquid and solid samples) or laser ablation (mainly solid samples without sample pre-treatment), are going to be reviewed as follows.

IV.5.7.1. Cadmium Determination by ETV-ICP-OES/MS

Because of the volatility of cadmium there is risk of analyte losses during conventional sample pre-treatments for solid samples, mainly acid digestion procedures. As it will be commented in Chapter V devoted to sample pre-treatments for solid environmental samples, this has led to the development of methods in AS based on the direct sample introduction into the atomizer. Both approaches, slurry sampling or solid sampling, have been hugely developed, and most of the applications of ETV are related to direct solid sample introduction techniques.

Table 3 lists several applications of ETV for environmental samples. Seawater [214-216] and water [217,218] have been analyzed for cadmium by direct introduction [214,216] or after certain sample pre-concentration treatments [215,217-219]. Similarly, solid environmental samples have been also analyzed after wet decomposition methods [219] or mainly by using the slurry sampling [220-228] or the solid sampling [224,229-233] approaches.

IV.5.7.2. Cadmium determination by LA-ICP-OES/MS

Laser ablation (LA) has widely been used as sample introduction technique mainly for ICP-MS because avoids sample pre-treatment as well as it allows to know the spatial distribution of a target element in a solid sample. Inorganic materials which are difficult to dissolve using conventional pre-treatment methods have been successfully analyzed by LA-ICP-OES/MS.

Cadmium, considered as a chalcophile element has been determined in atmospheric particulate matter, after retention/adsorption onto quartz filters or PTFE-membrane filters, together with other chalcophile elements, and also simultaneously with other siderophile or lithophile metals [234-238]. Other environmental materials such as steelmaking flue dust [239], soils [240,241], sediments [240], mollusc shells [242-244], sewage sludge [245], ice cores [246], or compost [247], have been also analyzed for cadmium by LA-ICP-MS. An interesting environmental application consists of the determination/spatial distribution of trace elements, such as cadmium, in tree rings to find correlation with the pollution history of a certain area [248-250]. LA-ICP-MS has been also used as a powerful detection system to determine the levels of cadmium-binding proteins in biota after electrophoretic separation methods [251,252]. Finally, LA-ID-ICP-MS has been also applied to liquid samples (drinking water) to assess cadmium and other trace elements [253]. In this case, the discrete sample introduction LA approach consisted of quantitatively ablated the liquid sample previously mixed with sodium acetate from a polystyrene substrate.

Table 3. ETV-ICP-OES/MS determination of cadmium in environmental samples

Sample type	Technique	Sample treatment	Ref.
Seawater	ETV-ICP-MS	None	[214]
Seawater	ETV-ICP-MS	on-line separaion	[215]
Seawater	ETV-IDA-ICP-MS	None	[216]
Water	ETV-ICP-MS	Hollow-fiber liquid phase microextraction	[217]
Water	ETV-ICP-MS	Single drop microextraction	[218]
Tap water/ lake water	ETV-ICP-MS	Zr-coated graphite adsorption bar microextraction	[219]
Plants	ETV-ICP-MS	Acid digestion / Zr-coated graphite adsorption bar microextraction	[219]
Fish	ETV-ICP-MS	Slurry sampling	[220]
Fish	ETV-IDA-ICP-MS	Slurry sampling	[222]
Coal fly ash	ETV-ICP-MS	Slurry sampling	[221]
Soil	ETV-ICP-MS	Slurry sampling	[223]
Soil / sediment	ETV-ICP-OES	Slurry sampling / solid sampling	[224]
Coal fly ash	ETV-IDA-ICP-MS	Slurry sampling	[225]
Sediment	ETV-IDA-ICP-MS	Slurry sampling	[226]
Sediment	ETV-IDA-ICP-MS	Slurry sampling (chemical vapor generation and trapping onto Ir tubes)	[227]
Soil	ETV-ICP-MS	Slurry sampling	[228]
Soil / sediment / plants	ETV-ICP-MS	Solid sampling	[229]
Airbone particles	ETV-ICP-MS	Solid sampling	[230]
Particulate matter	ETV-ICP-MS	Solid sampling	[231]
Plants	ETV-ICP-MS	Solid sampling	[232]
Plants	ETV-ICP-MS	Solid sampling	[233]

ETV-ICP-MS, electrothermal vaporization – inductively couple plasma – mass spectrometry; ETV-IDA-ICP-MS, electrothermal vaporization – isotope dilution analysis – inductively couple plasma – mass spectrometry

IV.6. Atomic Fluorescence Spectrometry (AFS)

AFS makes use of the same basic instrumental components as atomic absorption spectrometry; however, it measures the intensity of the light emitted by atoms that have been excited to higher energy levels by absorption of radiation. In contrast to molecular fluorescence spectrometry, the most useful type of fluorescence for atoms is the resonance fluorescence, which involves a fluorescence emission radiation of the same wavelength as that used for excitation. This emitted radiation can be measured with or without being spectrally resolved (dispersive atomic fluorescence spectrometry, DAFS; or non-dispersive atomic fluorescence spectrometry, NDAFS). AFS is very sensitive for many elements, especially for covalent-hydride-forming elements by combining this technique with the vapor generation methods.

IV.6.1. Source of Radiation

Continuous sources, such as tungsten or deuterium lamps, can be used with the advantage of multi-element determinations. However, these sources offer low radiant densities and as the sensitivity in AFS is directly proportional to the source intensity, the limits of detections reached by using these lamps are not low enough for most of the current analytical determinations. Similarly, line sources such as HCLs are also insufficient to guarantee a high intensity of excitation radiation. Previously described EDLs and HI-HCL (BDCHLs) however, can do so; and BDCHls are commonly used in modern atomic fluorescence spectrometers.

Since atomic fluorescence is mainly resonance fluorescence, ICPs into which a concentrated solution is introduced can provide an alternative quasi-monochromatic radiation source for multi-element determinations.

Finally, lasers in the continuous and in the pulse mode provide high intensity monochromatic radiation to saturate atomic transitions. These lasers have several dyes as the active medium to cover the whole wavelength range. For laser-excited AFS (LEAFS) or laser-induced atomic fluorescence spectrometry (LIF) the laser must be capable of generating wavelengths throughout the VIS and UV regions in order to excite as many elements as possible.

IV.6.2. Atomizers

Conventional air-acetylene or nitrous oxide-acetylene flames have usually been used as atomizers for AFS. However, some problems have been found when determining elements with thermally stable oxides; and also, high background signals from the flame are commonly observed. Attempted solutions of this problem have included the separation of the flame with a quartz tube (torch), or with a shear inert gas (argon), which prevents the entrance of air. Despite the use of these approaches, however, all these flames have a high background emission. Nevertheless, flames in which an inert gas (argon or nitrogen) is burned in hydrogen have low background emissions, although the temperature is also low. These flames, called as diffusion flames, are unsuitable for work involving refractory elements, but the temperature in these flames is quite sufficient to atomize hydrides; and as it will be commented in next section, they are normally used in AFS instruments with hydride generation systems for sample introduction.

ICP can be also used as an excellent atomizer which avoids problems for elements with thermally stable oxides. Similarly, graphite furnaces offer high atomization efficiencies and extremely low detection limits [191].

IV.6.3. Cadmium Determination by AFS

Because the high excitation energy for cadmium and the absence of stable oxides for this element, AFS offers very low limits of detection for this element, and the technique results appealing for analyzing environmental and biological samples. Most of the current applications of AFS to assess cadmium use non dispersive instruments (NDAFS) and the chemical vapor generation approach (covalent hydrides) as sample introduction technique. A complete review of these methods is discussed in section IV.7 devoted to chemical vapor generation, while literature related to the remaining methods is summarized in this section.

LIFS has been used for the direct determination of cadmium in Antarctic and Greenland ancient ice and recent snow at ultratrace levels. The atomization unit is an electrothermal atomizer allowing cadmium determinations down to the 0.1 pg g^{-1} levels [254-257]. The technique has been also applied to assess cadmium in snow [258] and river sediments [259]. A microwave plasma torch (MPT) has been also applied as an atomization cell for AFS [260]. A HCL working in the pulsed mode was used as an excitation source, while aqueous samples were introduced by a pneumatic nebulizer. The detection limits reported for cadmium in waters was 0.25 μg L^{-1}. More recently, a single ring electrode radiofrequency capacitively coupled plasma torch (SRTr.f.CCP) operated at 275 W and 27.12 MHz, has been also proposed as atomization cell in AFS [261]. In this case, an EDL was used as primary radiation source, while detection was carried out by a CCD. The limit of detection for cadmium determination in waters and soils by the proposed EDL-SRTr.f.CCP-AFS method was 4.3 μg L^{-1}, which leads to promising cadmium determinations in environmental samples.

IV.7. Chemical Vapor Generation – Atomic Spectrometry (CVG-AS)

In addition to the use of thermal energy, mainly derived from flames, plasmas or electrothermal heating, there are chemical methods capable to produce atomic vapor (atoms) or molecular vapor. The first case is usually termed cold vapor generation, and mercury cold vapor, after a chemical reduction, is the most typical example. Similarly, most of the molecular vapors are also obtained after a chemical reduction, and the reduced products are commonly volatile hydrides. The technique is then referred as hydride generation, and covalent hydride forming elements, such as arsenic, are usually atomized from its vapor state (molecular vapor). However, the generation of volatile species from elements can be reached by using other chemical reactions instead of a chemical reduction. Literature shows several examples, mainly alkylation, halide generation, metal carbonyls generation and chelates or oxide generation [262]. In addition, new developments involving UV assisted CVG approaches or trapping strategies such as solid phase microextraction (SPME) or single droplet microextraction (SDME) have updated the classical cold vapor / hydride generation methodologies. This fact has led to a resurgence of interest in CVG because the number of elements that can be determined after chemical vapor generation has increased in recent years [263-265]. Therefore, moreover the well-known chemical vapor generation for mercury, arsenic, selenium, antimony, lead, bismuth, tin, germanium and tellurium, new elements have been added to the chemical vapor forming elements group, such as cadmium, chromium, manganese, nickel, copper, silver or gold [266,267].

After chemical vapor generation, atomic absorption, atomic or fluorescence emission or mass counting can be monitored, leading the different analytical techniques based on CVG (chemical vapor generation – atomic absorption spectrometry, CVG-AAS; chemical vapor generation – inductively coupled plasma optical emission spectrometry or mass spectrometry, CVG-ICP-OES/MS; and chemical vapor generation – atomic fluorescence spectrometry, CVG-AFS). A monograph by Dědina and Tsalev devoted to hydride generation and atomic absorption spectrometry can be used for understanding principles of CVG methodologies [262].

IV.7.1. Generation and Transport of Vapor

First chemical vapor generation developments were based on the use of tin(II) chloride as a reducing agent for mercury and the metal/acid reducing system (most often zinc / hydrochloric acid), known as Marsch reaction, for covalent hydrides [268]. However, the low reducing power of these reagents and the slow kinetics of the Marsch reaction, have been superseded by the use of sodium tetrahydroborate as a reducing agent [269]. The decomposition of this reagent in acidic solutions is complete within a few microseconds and the resulting nascent hydrogen is capable to reduce analytes fast and quantitatively. The high reactivity of sodium tetrahydroborate is even important at the pH of ultrapure water, and to avoid sodium tetrahydroborate decomposition, this reagent is usually prepared in alkaline solution, mainly in sodium or potassium hydroxide (0.05 to 2.0 %(m/v)).

The design of the chemical vapor generation and also the influence of the oxidation state of the analyte can also have influence on the vapor releasing from the acidified solutions. In batch systems the pH of the resulting solution varies continuously during the reaction because the reducing agent is added for a time period over the acidified sample. This can lead to a hydride re-absortion in the solution and a decrease on the hydride releasing efficiency. In addition, variations in the amount of nascent hydrogen lead to tailing of the atomization signals. In flow systems (continuous flow, CF, and flow injection, FI) the vapor releasing is influenced by the flow rate and the lengths of the reaction coils [61]. However, one of the most critical parameters affecting the chemical reduction is the oxidation state of the analyte. Although chemical reduction of mercury, tin or germanium is not influenced by the oxidation state when using sodium tetrahydroborate as reducing agent, the remaining covalent hydride forming elements are markedly affected. Thus, hydrides from arsenic and antimony give poorer responses when these elements are at highest oxidation state (+5), while the generation efficiency is close to 100 % for the +3 state. Similarly, selenium and tellurium hydride generation is more efficient when these elements are as +4 state (no measurable signals from analytes in the hexavalent oxidation state). In such cases, these elements must be pre-reduced, commonly using potassium iodide, potassium bromide or L-cysteine, or increasing the hydrochloric acid concentration or heating. In contrast, the hydride generation efficiency for lead is poor when this element is as +2 state (the most common oxidation state for lead). In this case, lead is oxidized to a +4 state by using mainly hydrogen peroxide prior reduction [61,62]. This drawback can sometimes be advantageous for speciation studies of inorganic species of arsenic, antimony or selenium based on the direct determination of the inorganic specie at the most favorable oxidation stage, and the total inorganic species after a pre-reduction step. These speciation studies, also referred as first order speciation [270], can be also achieved by combining two different reducing agents, such as tin(II) chloride and sodium tetrahydroborate for inorganic mercury (Hg(II)) after tin(II) chloride reduction and inorganic mercury and methyl-mercury after reduction with sodium tetrahydroborate. In addition, the reaction medium (pH and/or type of acid) has also been found to affect the reduction of inorganic arsenic and even certain organic arsenicals [262].

Reduction can also be carried out electrochemically (electrochemical hydride generation, EcHG). In such cases, an electric current instead of a reducing agent (sodium tetrahydroborate) is used for reduction. Both batch and CF or FI approaches have been used and reduction is occurred in special electrolytic cells. Lower blanks, lower limit of detection and less extension of interferences caused by concomitants (especially transition metals) are the main advantages derived from EcHG methods [262,271].

After an efficient vapor generation (releasing), the vapor must efficiently be transported to the atomizer. In addition to the hydrogen released during the reaction, which can be used as a transport gas, a further inert gas (argon or nitrogen) is used. Besides its role as carrier gas, the inert gas has a second function which consists of enhancing the vapor releasing from the acidic solutions (stripping). Therefore, the purge gas is usually mixed with the reaction solution (acidified sample and sodium tetrahydroborate) in both CF and FI systems. Although first applications used a vapor collection stage (absorption in solutions, adsorption onto solids, collection under pressure or in cold traps), most of the current developments uses a direct transfer to the atomizer or a collection in the atomizer (in situ pre-concentration). The direct transfer of the vapor avoids problems derived from unstable hydrides, which can be decomposed during transport, or problems derived from re-solubilization or adsorption onto tubing surfaces. In general, a moderate gas flow rate must be used to reduce interactions of the vapor with the surfaces and avoid analyte losses. In addition, efficient drying systems for removing moisture (GoreTex membranes or hygroscopic ion exchange membranes) avoid analyte losses by re-solubilization.

Most of the current methods for CVG (CV/HG) are based on continuous operation modes (CF or FI) [262]. In addition, auxiliary lines can be used in flow modes to automatically perform pre reduction or oxidation stages before chemical reduction. CF modes employ a separate delivery of sample and acid to be mixed in a mixing coil, while FI systems use an injection valve to inject a discrete volume of sample to the flow of the carrier (acid). The acidified sample (CF) or the carrier with the injected sample (FI) merge the reducing solution flow upstream a reaction coil where the hydride formation takes place. In the batch mode, the acidified sample is transported to a stirred glass cell containing the reducing solution or to a plastic vessel without stirring but with an entrance line through the reducing solution is pumped. After the chemical reduction, the liberated vapor is flushed with an inert gas into the atomizer.

For CF and FI modes a gas-liquid separator is used to separate the chemical vapor from the liquid reagents before introducing the vapor into the atomizer. Gas-liquid separators can be classified into three basic types: hydrostatic separators, separators with forced outlet and membrane separators [262]. Hydrostatic separators are U-shaped siphons with the liquid column to balance small pressure variations. These separators offer a relatively high dead volume and work satisfactorily at low pressure. Gas-liquid separators with forced outlet consist of small cells (total inner volume around 2.7 mL) with the purge gas entrance upstream the separator vessel and with an outlet connected to a peristaltic pump to draw the reacted liquid to waste. These separators can handle overpressure better than hydrostatic types. Finally, membrane gas-liquid separators use a micro-porous membrane made of polytetrafluoroethylene (PTFE) or silicon rubber, which are permeable only for gases. The separation is then based on the diffusion of the vapor through the membrane while the liquid flow through the separator to waste. The main advantage offered by these separators is the removal of moisture formed in other types of gas-liquid separators [262].

IV.7.2. Atomization of Vapor. Atomizers

Except for cold vapor (monoatomic vapor), once the vapor has been generated, it must be atomized before monitoring the atomic absorption or emission signals, or atomic mass counting. Although conventional flames and non heated flame-in-tube (FIT) atomizers were firstly used for atomization [61,262], current atomizers consist of inert hydrogen diffusion flames, narrow-bore quartz tubes (mainly electrically heated), graphite furnaces and plasmas. As commented, the atomizer can act as a vapor pre-concentration medium just before atomizing. This is the case of the graphite furnace atomizers (in situ trapping) or the silver or gold wires for mercury direct amalgamation.

Diffusion flames are often hydrogen flames, which produce low temperatures (within the 350 – 1000°C range) and which are the preferable atomizers when using atomic fluorescence detectors. Modern diffusion flames have been designed specifically for gaseous sampling and they are quite different from conventional air/nitrous oxide-acetylene flames used for liquid sampling. The miniaturized design consists of a small argon-hydrogen flame burning on a borosilicate glass tube (burner section around 5 mm). This design improves the signal-to-noise ratio as well as minimizes vapor dilution into the flame gases and the background absorption of the flame.

Heated quartz tube atomizers (QTAs) are normally T-tubes with the bar-tube aligned in the optical path of the spectrometer. The central arm of the T-tube serves for delivery of vapor from the gas-liquid separator. These tubes (bar-tube typically about 15 cm long and diameter around 10 mm) are heated by an electrical resistance device (controlled temperature up 1100°C) or by a conventional air-acetylene flame. When using electrical heating, the atomizer is closed at the ends by two removable quartz windows. However, the bar-tube ends are opened when the quartz tube is heated by a flame. In such cases, quartz windows are replaced by attached graphite rings at the ends of the bar-tube to prevent ignition of hydrogen.

Plasmas have also been used as excellent atomizers for vapors. The sensitivity of plasma-based instrumentation for the determination of vapor forming elements increases markedly when compared with conventional nebulization because the improved transport efficiency (close to 100%). The vapor carried by a flow of gas from the gas-liquid separator can be directly introduced into the plasma, although some approaches use combined vapor generator-gas-liquid separator devices for vapor introduction into the plasma. These developments were briefly commented in the section devoted to plasma-based methods section IV.5.4).

Graphite furnaces have been used as efficient atomizers for vapors, especially hydrides, because their inherent high atomization temperatures. First attempts used a direct transfer of hydride from the generator to the furnace, which was previously heated at the optimum atomization temperature (on-line atomization). However, hydride losses by adsorption on the cooler metal or graphite parts of the atomizer occurred, which implied a poor sensitivity. In addition, the life time of the hydride transfer tube was short because it was subjected at a high temperature (atomization temperature). Therefore, an alternative approach based on an in situ trapping of the hydrides in the graphite tube at low temperatures (or even at room temperature) was developed. In this case, the graphite furnace (graphite tubes) is used as a trapping medium and also as an atomizer. This approach offers several advantages over on-line atomization, such as a low temperature for trapping (from room temperature to 600°C), which avoids vapor transfer tube damage and minimizes vapor losses by capturing on cooler parts of the atomizer. In addition, the graphite tube acts as a pre-concentration medium and a high amount of vapor from large volumes of samples can be efficiently trapped before atomization. Therefore, graphite furnace has offered appealing perspectives for chemical vapor atomization [264].

Current vapor transfer tubes in modern CVG-ETAAS are fine capillaries made of quartz which are coupled to the arm of conventional autosampler devices. After a fixed temperature of the graphite tube is reached (trapping temperature), the transfer tube is automatically inserted through the sampling port of the furnace until finishing vapor evolution from the generator. Then, the transfer tube is automatically removed and the following stages of the graphite furnace temperature program (atomization) are run [272].

Although pyrolytic coated graphite tubes with L'vov platforms were firstly used, better trapping efficiencies were then obtained when using coated graphite tubes, mainly zirconium coated graphite tubes (permanent coating) or graphite surfaces treated with palladium (non permanent coating). Other noble metals such as iridium, platinum, ruthenium or rhodium were also investigated as efficient trapping medium for hydrides [61,262]. The mechanism of hydride trapping in untreated graphite tubes differs when the trapping occur in heated and nonheated graphite furnaces. Sturgeon et al. [273] proposed an analyte deposition in the preheated graphite tube via thermal decomposition. The deposited metal is then reoxidize during the trapping step, and then it is atomized identically to that occurring for conventional ETAAS. The mechanism for unheated graphite tubes (or trapping at low temperatures) implies chemisorption of hydride to the surface rather than a thermal decomposition [274]. Trapping mechanisms when using palladium coated graphite tubes are explained taking into account the ability of palladium to participate in hydrogen abstraction reactions [262].

The atomization mechanisms are strongly dependent on the atomizer used, but for most of available atomizers (diffusion flames and heated quartz tubes) atomization occur via free hydrogen radicals generation. These hydrogen radicals are produced by the reaction of hydrogen (excess of nascent hydrogen) with oxygen (from the air dissolved in the working solutions). This mechanism explains the decrease on the atomization efficiency in heated quartz tubes, even at high temperatures, in absence of hydrogen. In addition, hydrogen radicals formation also explain the efficient atomization in quartz tubes at low atomization temperatures when using oxygen mixed with the purge gas [262].

As previously commented, atomization mechanism in graphite furnaces occurs identically to that occurring for conventional ETAAS. This fact is well accepted when high atomization temperatures (above 1600°C) are used. However, a mechanism based on hydrogen radicals formation has been also proposed when atomizing a low temperatures (e.g. about 1200°C for selenium) [61,262].

IV.7.3. Cadmium Determination by CVG-AS

Although the first approaches for cadmium vapor generation were reported in 1989 [275,276], chemical vapor generation has not extensively been applied for the determination of this element [277]. The reasons are attributed to the inherent instability of cadmium hydride (some authors state that volatile cadmium species are cadmium hydride, and other ones that it is a cold vapor); and difficulties with generation, which implies very narrow conditions, mainly hydrochloric acid concentration, for an efficient generation. Cadmium vapor generation has been carried out by hydride generation (presumable CdH_2) using sodium tetrahydroborate [194,202,276,278-307], although some reports use an alkylation reaction [275,308-310], mainly ethylation with sodium tetraethylborate ($Cd(C_2H_5)_2$) [275,308,310]. An increase on the efficiency of cadmium vapor generation has been obtained by using different reagents such as thiourea [281,282,287,288,292,293], L-cysteine [301], 8-hydroxiquinoleine [194,296], phenanthroline [296], sodium iodate [297], as well as organized media (mainly vesicles of didodecyldimethylammonium-bromide, DDAB) [278-280,283,284,289,290], and certain metallic catalysts (cobalt [194,281,282,285-287,292,301], gallium [285,286], or nickel [288,292,296]). In addition, EcHG for cadmium has been also tested [311]. Table 4 lists the different methods for cadmium determination based on vapor generation. It can be seen that cadmium is sensitively assessed in environmental samples (water, sediments, soils and biota) by coupling the vapor generation technique with all the available atomic detectors (AAS, AFS, ICP-OES and ICP-MS).

IV.8. X-Ray Fluorescence Spectrometry

X-ray fluorescence spectrometry (XRF) is a non-destructive analytical technique used to identify and determine the concentrations of major and trace (at the $\mu g\ g^{-1}$ level) elements present in solid, powdered and liquid samples. One of its main advantages is that minimal sample-pretreatment is required, being appealing methodologies for routine analsyis.

Table 4. CVG-AS determination of cadmium in environmental samples

AS technique	CVG conditions	Sample type	Ref.
Batch-AFS (Ar-H_2 diffusion flame)	Ethylation ($NaBEt_4$), pH 2.15	Tap water, seawater	[275]
Batch-AAS (flame heated QTA)	$NaBH_4$ in DMF and HCl	Aqueous solutions, DDC extracts	[276]
Batch-ICP-OES/ FAAS	$NaBH_4$ in HCl	Seawater	[278]
Batch/CF-AAS (QTA)	$NaBH_4$ in HCl and vesicles (DDAB)	Aqueous solutions	[279]
Batch/CF-AAS (QTA), CF-ICP-OES	$NaBH_4$ in HCl and vesicles (DDAB)	Aqueous solutions	[280]
CF-AFS (FIT)	KBH_4 in HCl, thiourea, Co(II)	Tap water, sediments	[281]
CF-AAS (QTA)	KBH_4 in HCl, thiourea, Co(II)	Tap water, sediments	[282]
CF-ETAAS (palladium coated graphite tubes)	$NaBH_4$ in HCl and vesicles (DDAB)	Aqueous solutions	[283]
CF-ETAAS (palladium coated graphite tubes)	$NaBH_4$ in HCl and vesicles (DDAB)	River water	[284]
FI-ETAAS (iridium, tungsten and zirconium coated graphite tubes)	$NaBH_4$ in HCl, Co(II) or Ga(III)	Seawater	[285]
FI-ETAAS (iridium coated graphite tubes)	$NaBH_4$ in HCl, Co(II) or Ga(III)	Seawater	[286]
FI-IDA-ICP-MS	$NaBH_4$ in HCl, thiourea, Co(II)	River water, seawater, sediments	[287]
FI-AAS (QTA)	$NaBH_4$ in HCl, Ni(II)	Water, soils, geological materials	[288]
Batch/CF-ETAAS (conventional pyrolytic coated graphite tubes – the pyrocoating was mechanicallyremoved)	KBH_4 in HCl, vesicles (DDAB), thiourea, Co(II)	Seawater, soils, sediments, plants	[289]
FI/CF-AAS (QTA)	$NaBH_4$ in HCl, vesicles (DDAB)	Tap water, seawater	[290]
FI-AAS (QTA)	$NaBH_4$ in HCl	Sewage sludge, waste water	[291]
FI-AAS (Pyrex glass cell)	$NaBH_4$ in HCl, thiourea, Ni(II)	Plants, soils	[292]
FI-AAS (Pyrex glass cell)	$NaBH_4$ in HCl, thiourea, Ni(II)	Plants, soils	[293]
FI-AAS (QTA)	$NaBH_4$ in HCl, KCN	Sewage sludge, Antarctic krill	[294]
FI-NDAFS (Ar-H_2 diffusion flame)	KBH_4 in HCl, thiourea, ascorbic acid	Seawater, soils, total airborne particulate matter	[295]
CF-NDAFS (Ar-H_2 diffusion flame)	$NaBH_4$ in HCl, 8-hydroxiquinoline/ phenanthroline, Ni(II)	Soils, plants	[296]
FI-NDAFS (FIT)	KBH_4 in HCl, NaIO4	Aqueous solutions	[297]
FI-AAS (trapping into heated QTA)	$NaBH_4$ in HNO_3, Trapping at 350°C (1.4V) and revolatilization at 1030°C (4.1V)	Seawater, oyster	[298]

Table 4. Continued

AS technique	CVG conditions	Sample type	Ref.
FI-AAS (QTA, Multimode Sample Introduction System (MSIS))	$NaBH_4$ in HCl, thiourea, Co(II), Ni(II)	Water, mussels	[202]
FI-AAS (QTA)	$NaBH_4$ in HNO_3, thiourea/L-cysteine, Co(II)	Waters, plants, soils	[299]
FI-IDA-ICP-MS, CF-ETV-IDA-ICP-MS (iridium coated graphite tubes)	$NaBH_4$ in HCl	River sediments, marine sediments	[300]
FI-ETAAS (iridium/zirconium coated graphite tubes)	$NaBH_4$ in HNO_3	Plants, whale liver	[301]
CF-AFS (Ar-H_2 diffusion flame)	$NaBH_4$ in HCl	Plants, soils	[302]
FI-AAS (QTA)	$NaBH_4$ in HCl	Water	[303]
CF-ICP-OES	$NaBH_4$ in HNO_3, 8-hydroxiquinoline, Co(II)	Aqueous solutions	[194]
CF-AFS (Ar-H_2 diffusion flame)	$NaBH_4$ in HCl	Seawater	[304]
CF-AAS (QTA) (trapping into a heated tungsten coil)	$NaBH_4$ in HCl	Seawater, oyster	[305]
AFS (QTA)	KBH_4 in HCl, thiourea	Plants	[306]
FI-AFS (QTA)	$NaBH_4$ immobilized onto cellulose, HCl	Waters, seawater	[307]
CF-AFS (Ar-H_2 diffusion flame)	Ethylation ($NaBEt_4$), pH 8.0 – 8.5	Tap water, sludge	[308]
CF-ETV-ICP-OES	In situ alkylation $(C_2H_5)MgBr$	Mussel, plant, rice, vehicle exhaust particulates	[309]
FI-IDA-ICP-MS	Ethylation ($NaBEt_4$) in HNO_3	Water, sediments, lichens	[310]

DMF, dimethylformamide

DDC, diethyldithiocarbamate

DDAB, didodecyldimethylammonium-bromide

AAS, atomic absorption spectrometry; AFS, atomic fluorescence spectrometry; CF-AAS, continuous flow – atomic absorption spectrometry; CF-AFS, continuous flow – atomic fluorescence spectrometry; CF-ETAAS, continuous flow – electrothermal atomic absorption spectrometry; CF-ETV-IDA-ICP-MS, continuous flow – electrothermal vaporization – isotope dilution analysis – inductively couple plasma – mass spectrometry; CF-ETV-ICP-OES, continuous flow – electrothermal vaporization - inductively couple plasma – optical emission spectrometry; CF-ICP-OES, continuous flow – inductively coupled plasma optical emission spectrometry; CF-NDAFS, continuous flow – non-dispersive atomic fluorescence spectrometry; FAAS, flame atomic absorption spectrometry; FI-AAS, flow injection – atomic absorption spectrometry; FI-AFS, flow injection – atomic fluorescence spectrometry; FI-ETAAS, flow injection – electrothermal atomic absorption spectrometry; FI-NDAFS, flow injection – non-dispersive atomic fluorescence spectrometry; FI-IDA-ICP-MS, flow injection –isotopic dilution analysis - inductively couple plasma – mass spectrometry; ICP-OES, inductively couple plasma – optical emission spectrometry; FIT, flame in-tube atomizer; IDA-ICP-MS, isotopic dilution analysis - inductively couple plasma – mass spectrometry; QTA, quartz tube atomizer

IV. 8.1. General Aspects

XRF is based on the emission of a characteristic radiation in form of photons (fluorescence) of an atom. However, the physics of XRF differs from AFS in the fact that the emission of the radiation is attributed to transitions of electrons from higher energy levels to lower energy orbitals after the removal of tightly-held electrons from the inner orbitals of the atom (ionization). Short-wavelength X-rays or gamma rays are energic enough (energy greater than the ionization potential of most of the elements) to expel electrons from the inner orbitals of an atom. The removal of these electrons renders the electronic structure of the atom unstable, and electrons in higher orbitals fall into the lower orbitals. Therefore, the released energy (photons) is equal to the energy difference of the two orbitals involved and it is characteristic of the atom. However, when removing an inner electron from an atom, there are only a limited number of ways in which electrons from outer or higher energy orbitals can drop into its place. Taking in mind the nomenclature of N, M, L and K energy levels (from higher to lower energy), the main electronic transitions are referred as Kα, which implies a L→K transition (an electron from the outer L energy level to the inner K level); Kβ, which is related to an M→K transition; Lα, which implies a hole in the L energy level and a M→L transition; and so on. As consequence, different fluorescent photons, with a characteristic energy equal to the difference in energy of the initial and final orbital, will be obtained, leading to a characteristic XRF spectrum for each element.

IV. 8.2. Instrumentation

XRF instruments comprise an X-ray source for ionization, a wavelength selector for restricting the wavelength range of incident radiation, a sample holder, a detector and a signal processor. Wavelength selectors (filters or monochromators) lead to wavelength-dispersive instruments (WDXRF or WDS); X-ray photometers or X-ray spectrophotometers, respectively. However, instead of separating the wavelengths of the radiation, the fluorescent radiation can be analyzed by sorting the energies of the emitted photons. In this case, the instruments are commonly referred as energy-dispersive systems (EDXRF or EDS). A third type spectrometer, which is considered as a special variant of EDXRF [312], is the total reflection X-ray fluorescence spectrometry (TXRF). TXRF occurs when X-ray beam exhibits total reflection for a particular angle of incidence. In this case, the sample is excited by the primary X-ray beam at a glancing angle less than the critical angle at which total external reflection occurs, and the primary exciting radiation is incident on a plane polished surface, which serves either as a sample support or is itself the object to be examined. A detailed description of XRF and other X-ray based techniques can be found elsewhere [313,314].

IV.8.2.1. X-Ray Sources

Conventional X-ray tubes, also called Coolidge tube, consist of a cathode (tungsten filament) and an anode made of pure metals (silver, chromium, molybdenum, rhodium or tungsten) mounted in a vacuum glass tube, although homogeneous alloy of molybdenum and tungsten has also used [315]. When the cathode is heated by establishing an electrical current from a high voltage power source (typically from 30 to 150 kV), a thermoionic emission of electrons from the cathode is obtained, electrons which are accelerated to the anode. X-ray emission from the anode is a consequence of high energy electrons passing through the tube which are progressively decelerated by the material of the anode, and the continuous X-ray radiation obtained is referred as a 'bremsstrahlung' radiation. Anyway, the X-ray spectrum obtained depends on the anode material and the accelerating voltage. Galodinium anodes have been demonstrated to generate high energy polarizing X-ray beam [316,317], and the technique is commonly referred as high-energy polarized X-ray fluorescence spectrometry

(HE-P-XRF). Modern X-ray generators use pulsed electrical currents, typically between about 1 ms to 1 s, rather than high voltage power sources, which allow a more accurated control of the X-ray unit and reduced X-ray exposures.

Gamma ray sources can also be used without the need for an elaborated power supply. In these sources, different radioactive substances, which give continuous or lines spectrum, can be used. The most common radioisotopes are tritium adsorbed onto non-radioactived titanium ($^{3}_{1}H$-Ti) and $^{147}_{61}Pm$-Al, which give continuous X-ray spectrum. Other isotopes, such as $^{56}_{26}Fe$, $^{57}_{27}Co$, $^{109}_{48}Cd$, $^{125}_{53}I$ or $^{210}_{82}Pb$, give lines X-ray spectrum. The X-ray intensities provided by radioisotopes are several orders of magnitude weaker than conventional X-ray tubes. Although their limited sensitivity, these sources are mainly used in portable instrumentation.

There is a new X-ray source type consisting of a synchrotron which offers an intensive, polarized and collimated X-ray radiation (synchrotron radiation, SR) and has led to synchrotron radiation induced X-ray fluorescence spectrometry (SR-XRF) [318]. Although some applications can be found in literature, the major drawback of SR is the limited access to the facility.

IV.8.2.2. Wavelength Selectors

Filters and monochromators can be used in WDXRF, although filters are limited by the few number of target-filter combinations. Most of the WDXRF use a monochromator consisting of a single crystal mounted on a goniometer which separates the photons by diffraction. Crystals with simple structure tend to give the best diffraction performance. Different materials are used, e.g. germanium (Ge), graphite, lithium fluoride (LiF), sodium chloride (NaCl), indium antimonide (InSb), ammonium dihydrogen phosphate (ADP), tetrakis-(hydroxymethyl)-methane, penta-erythritol (PE), potassium hydrogen phthalate (KAP), rubidium hydrogen phthalate (RbAP) or thallium(I) hydrogen phthalate (TlAP). By varying the angle of incidence and take-off on the crystal, a single X-ray wavelength can be selected according with the Bragg equation. It must be noticed that there are not a dispersing crystal useful for the whole X-ray spectrum, and commercially monochromators are commonly available with at least two interchangeable crystals. Instead of crystals, some modern instruments use layered synthetic microstructures as dispersing systems. These sandwich structured materials comprise successive thick layers of low atomic number matrix and monoatomic layers of a heavy element, and they diffract any desired long wavelength.

Two different configurations (flat crystal with Soller collimators or curve crystal with slits) can be used to precisely define the incident and exit angles. In the first configuration, a Soller collimator is a stack of parallel metal plates, spaced a few tenths of a millimeter apart. The use of beam collimators offers some drawbacks such as intensity reduction and scattering, however, the simplicity of the geometry and low cost is especially useful for variable-geometry monochromators. In the second configuration, the Rowland circle geometry ensures that the slits are both in focus, but in order for the Bragg condition to be met at all points, the crystal must first be bent to a radius of 2R (where R is the radius of the Rowland circle), then ground to a radius of R. This arrangement allows higher intensities with higher resolution and lower background. But the mechanics of keeping Rowland circle geometry in a variable-angle monochromator is extremely difficult.

There are not dispersing elements for EDXRF. As it will be commented in following sections the fluorescent X-rays emitted by the sample are directed into a solid-state detector which produces a continuous distribution of pulses, the voltages of which are proportional to the incoming photon energies. This signal is processed by a multichannel analyzer (MCA) which produces an accumulating digital spectrum that can be processed to obtain analytical data.

IV.8.2.3. Detectors

IV.8.2.3.1. Detectors for WDXRF

There are different types of detectors for WDXRF which offer high pulse processing speeds in order to cope with the very high photon count rates.

Gas-Filled Transducers

These devices consist of a chamber containing a gas (ionization chamber), which it can be sealed (sealed gas detectors) or it can be open, allowing the gas to flow through it continuosly (gas flow proportional counters). For sealed gas detectors, an inert gas (argon, krypton or xenon) at a few atmospheres pressure is used. However, a mixture 90% argon (helium or neon) and 10% methane (P10) is used for gas flow detectors. Methane is commonly used to suppress the formation of fluorescent photons caused by recombination of the argon ions with stray electrons. A wire anode made of tungsten or nichrome (20-60 μm diameter), a cathode and a X-ray transparent window are the other components of these devices. The window needs to be conductive, thin enough to transmit the X-rays effectively, but thick and strong enough to minimize diffusion of the detector gas into the high vacuum of the monochromator chamber. Materials often used are beryllium metal, aluminized PET film and aluminized polypropylene.

For both systems, radiation (X-ray photons) enters a chamber through the transparent window and the gas is ionized (loss of outer electrons) as consequence of X-ray photon-gas atom interactions. This photoelectron has a large kinetic energy and loses this excess of energy by ionizing several hundred additional atoms of the gas. Photoelectrons move toward the central wire anode while cations are attracted toward the cylindrical metal cathode under a fixed applied potential. Depending of the applied potential, the number of photoelectrons that reach the anode is different. The number of electrons attacting the anode can be constant and they represent the total number formed by a single photon. This is the case of the called ionization chamber counters. However, the number of electrons can rapidily be increased with the applied potential due to a secondary ion-pair production caused by collisions between the accelerated electrons and gas molecules. In this case, the system works as a proportional counter. Finally, a third type of system (Geiger detector) can be obtained if the electrical signal is amplified.

Gas flow proportional counters are used mainly for detection of longer wavelengths while sealed gas detectors are used for detecting wavelengths in the 0.15-0.6 nm range.

Scintillation Counters

These detectors consist of a scintillating crystal, typically made of sodium iodide doped with a little amount of thallium iodide (around 0.2%) wihich is attached to a photomultiplier. Organic scintillators such as stilbene, anthracene and terphenyl can also be used. These crystals must be protected with a relatively thick aluminium/beryllium foil window, which limits the use of the detector to wavelengths below 0.25 nm. When the incoming radition traverses the crystal, the X-ray photons are absorbed and after a certain time (about 0.25 μs), this excess of energy is released as photons of fluorescence radiation. Thousands of these fluorescent photons are produced by each primary X-ray photon absorbed. The flashes of light produced in the crystal are transmitted to the photocathode of the photomultiplier tube, which convert the the radiation energy in an electrical pulse which is amplified and counted.

Semiconductor Detectors

Semiconductor detectors (solid-state detectors) can be used for both WDXRF and EDXRF, although historically their use for WDXRF had been restricted by their slow response. The lithium-drifted silicon detector, Si(Li), is one of the most semiconductor detectors used. Si(Li) consists of a 3-5 mm thick three layers silicon junction (p-i-n diode): a p-type semiconducting layer that faces the X-ray source, a central intrinsic zone (the lithium-drifted non-conducting i-layer), and an n-type layer. Under a negative applied voltage, when an X-ray photon enters the detector, its energy is aborbed and it causes a swarm of electron-hole pairs to form (electrons are promoted from the valence to the conduction band, leaving positive hole in the valence band). To obtain sufficiently low conductivity, the detector must be maintained at low temperature, and liquid-nitrogen must be used for the best resolution. With some loss of resolution, the much more convenient Peltier cooling can be employed.

More recently, high-purity wafer silicon with low conductivity have become routinely available. Cooled by the Peltier effect, this provides a cheap and convenient detector, although the liquid nitrogen cooled Si(Li) detector still has the best resolution.

IV.8.2.3.2. Detectors for EDXRF

In energy dispersive analysis, dispersion and detection are a single operation. Proportional counters or various types of p-i-n diodes (Si(Li), Ge(Li) or silicon drift detector, SDD) can be used. They all work as previously commented and its critical factor is the detection speed because the spectrum must be built up by dividing the energy spectrum into discreet bins and counting the number of pulses registered within each energy bin.

IV.8.2.4. Signal Processors

Signal processing is achieved by pulse-heigth selectors which reject pulses of about 0.5 V or less, avoiding the transduction and amplification of noise. There are also electronic circuits which reject not only low-heigth pulse but also those above a maximum level. The pulses generated by the detector are processed by pulse-shaping amplifiers. It takes time for the amplifier to shape the pulse for optimum resolution.

IV.8.3. X-Ray Fluorescence Spectrometers

As commented, X-ray fluorescence spectrometers can be of two types: wavelength dipersive and energy dispersive instruments.

IV.8.3.1. Wavelength Dispersive Spectrometers

WDXRF uses a Coolidge tube as a source of radiation due to losses suffered in the monochromator. There are single channel (sequential) and multichannel (simultaneous) configurations.

Sequential spectrometers have a single variable-geometry monochromator (with interchangeable dispersing crystals) and a single detector assembly (but usually with more than one detector arranged in tandem). The instrument is programmed to move through a sequence of wavelengths, in each case selecting the appropriate X-ray tube power, the appropriate crystal, and the appropriate detector arrangement. The main disadvantage of these systems is the relatively long analysis time, particularly when many elements are being determined, not only because the elements are measured in sequence, but also because a certain amount of time is taken in readjusting the monochromator geometry between measurements.

Simultaneous spectrometers have a number of channels dedicated to analysis of a single element. Each channel consists of a fixed-geometry crystal monochromator, a detector, and processing electronics. This allows a number of elements to be measured simultaneously, and in the case of high-powered instruments, complete high-precision analyses can be obtained in short times (less than 30s). Another advantage of this arrangement is that the fixed-geometry monochromators do not have ontinuously-moving parts, and so, they are very reliable spectrometers. In addition to the relatively high cost of these instruments, simultaneous detection offers other disadvantages such as the number of elements that can be measured is limited to 15-20, because of space limitations on the number of monochromators that can be arranged.

IV.8.3.2. Energy Dispersive Spectrometers

An EDXRF consists of a Coolidge tube or a radioactive material as a source of radiation, a semiconductor detector and different electronic componens for energy discrimination. These systems are very simple because monochrmoators are not included in the design and avoid large distances between the sample and the detector, which results in a 100-fold increase sensitivity. These instruments are commonly multichannel spectrometers, allowing the simultaneous measurement of all the emitted X-ray lines.

IV.8.4. Limitations of XRF

Although XRF offers several advantages such a minimum sample pre-treatment, high speed of analysis, relative simple spectrum with few lines and absence of spectral line interferences, XRF method have some limitations. Sensitivity is not high enough for the determination of concentrations at trace levels and detection of lighter elements (atomic number below 23) is difficult because the simultaneous competitive process (Auger emission) which reduce the fluorescence intensity.

In addition, matrix effects can be also found as consequence of the number of photons leaving he sample can be affected by the sample itself (physical properties of the sample). The most important matrix effect are derived from the fact that the emitted X-ray radiation from a target element can partially be absorbed by other elements in the sample, and this can lead to the same element concentration gives different analytical signals in function of the sample components. These types of 'mass absorption' are well known for many matrices and corrective calculations can be made when knowing the composition of the sample.

Sometimes, matrix effects can arise from X-ray enhancement that occurs when the secondary X-rays emitted by a heavier element are sufficiently energetic to stimulate additional secondary emission from a lighter element. This phenomenon can also be modeled, and corrections can be made.

The third type of matrix effects is attributed to sample macroscopic effects, which consist of effects of inhomogeneities of the sample, and unrepresentative conditions at its surface. These artifacts are inherent to the method of sample preparation and they can not be compensated by theoretical corrections, and must be 'calibrated in'. This means that the calibration materials and the samples must be compositionally and mechanically similar, and a given calibration is applicable only to a limited range of materials.

IV.8.5. Cadmium Determination by XRF

Cadmium can be determined by XFS at the spectral Cd K line of 0.05357 nm (line $K_{\alpha 1}$). This line is the most intense and free of lines overlaps for cadmium determination. Solid environmental materials can be directly analyzed by XRF techniques, but sometimes sample preparations involving grinding the sample into a powder that is either analyzed as a loose powder or pressed into a pellet is required. This procedure increases uniformity and homogeneity of the sample. In some cases, fusion can be needed to pre-treat the solid sample in order to minimize or avoid matrix effects. Liquid samples can be directly analyzed when the liquid is compatible with the sample cup and sealing film, however, a sample pre-concentration stage can be needed for some applications.

Detailed data on XRF methods for different material types can be found in a review by Potts et al.[318]. In this section, brief information of recent applications of XRF for assessing cadmium in environmental materials is summarized.

WDXRF has been used for the analysis of aerosol loaded onto filters [319,320], atmospheric dust and vehicles exhaust particles [321], marine sediments [322], soils [323], plants [324], and water [325-327] and waste waters [328]. EDXRF was also applied for cadmium determination in soils [329,330], rocks [329], marine sediments [331,332], biota [333], liquid wastes [334], and waters [335] and seawater [316,317] after certain pre-concentration stages. TXRF was also used for domestic sludge [336], marine sediment [337], atmospheric particulate matter [338,339], biota [340,341] and waters (river water [342,343] and rainwater [338]).

IV.9. NEUTRON ACTIVATION ANALYSIS

Similar to XRF techniques, neutron activation analysis (NAA) is a non-destructive analytical methodology which gives quantitative and qualitative information of a number of major components and trace elements (at the $\mu g\ g^{-1}$, or even $ng\ g^{-1}$ levels) in different samples. Moreover the non-destructive character as well as its inherent high sensitivity, NAA enables the simultaneous determination of many elements. However, NAA is not a common methodology in many laboratories because the high cost of the equipment, and mainly because the special conditions required for laboratories and highly qualified staff to cope with radioactivity.

IV.9.1. General Aspects

NAA consists of irradiating the sample (which contains the naturally occurring stable isotopes of the target elements) with neutrons. After the element-neutron interaction (neutron capture), the target elements are transformed into radioactive isotopes. The compound nucleus of the activated elements can almost instantaneously be de-excited into a more stable configuration through emission of one or more characteristic prompt gamma rays. However, in many cases, this new configuration can yield a radioactive nucleus which also decays by emission of one or more characteristic delayed gamma rays, but at a much slower rate according to the unique half-life of the radioactive nucleus. Depending upon the particular radioactive species, half-lives can range from fractions of a second to several years. Therefore, with respect to the time of measurement, NAA falls into two categories; prompt gamma-ray neutron activation analysis (PGNAA), where measurements take place during irradiation; or delayed gamma-ray neutron activation analysis (DGNAA), where the measurements follow radioactive decay. This latter operational mode leads to the most common NAA technique.

Instead of neutron capture, other nuclear processes such as those based on (n,gamma) reactions can be used for activating nuclides.

Qualitative information from NAA is based on the energy of the emitted gamma quanta (E) and the half life of the nuclide (T½). Simmilarly, the quantitative data is given by the intensity (I), which is the number of gamma quanta of energy E measured per unit time.

IV.9.2. Instrumentation

NAA instrumentation consists mainly of a source of neutrons and a gamma-ray detector.

IV.9.2.1. Irradiation (Neutron) Sources

Kinetics of neutron activation process is out of the scope of this Chapter and a brief description of the different neutron sources is only given. Detailed description of the neutron activation mechanisms can be found elsewhere [344, 345].

IV.9.2.1.1. Isotopic Neutron Sources

An isotopic neutron source uses an alpha emitting radioactive material which is mixed with beryllium, allowing an (,n) reaction for generating neutrons as follows:

$$^{9}_{4}Be + ^{4}_{2}He \rightarrow ^{12}_{6}C + ^{1}_{0}n + 5.7\ MeV$$

Different radioisotopes such as ^{227}Ac (T½ of 22 years), ^{226}Ra (T½ of 1620 years), ^{239}Pu (T½ of 2.4 10^4 years) or ^{210}Po (T½ of 138 days) can be used as emitters, giving neutron energies around 4 MeV.

Other isotopic neutron sources are based on certain artificially produced radioisotopes, such as ^{252}Cf (T½ of 2.6 years) which undergo spontaneously fission, producing 3.76 neutrons of 1.5 MeV per event (one milligram of ^{252}Cf emits 2.28 10^9 neutrons per second).

These neutron sources generate a stable neutron flux but its intensity is low when comparing with other neutron sources.

IV.9.2.1.2. Neutron Generators

The neutron sources consist of accelerators where a convenient target material is bombarded by accelerated charged particles, commonly deutrons (deuterium ions), and the neutrons are produced in a nuclear reaction. Tritium, an isotope of hydrogen, previously adsorbed onto titanium or zirconium, is the most frequently target material used in these devices and the nuclear reaction carried out is denoted as $^{3}H(d,n)^{4}He$:

$$^{2}_{1}H + ^{3}_{1}H \rightarrow ^{4}_{2}He + ^{1}_{0}n + 14.0\ MeV$$

The monoenergetic neutrons flux produced by this accelerator is of approximately 100 neutrons $cm^{-2}\ s^{-1}$.

IV.9.2.1.3. Nuclear Reactors

Nuclear reactors for uranium fission offer the highest available neutron fluxes, and hence the highest sensitivities for most elements. Different types of reactors and different positions within a reactor can vary considerably with regard to their neutron energy distributions and fluxes due to the materials used to moderate or reduce the energies of the primary fission neutrons. Among the three principal components (thermal, epithermal, and fast) of neutron energy distributions, the thermal neutron component is the most useful as a source for NAA. This component consists of low-energy neutrons (energies below 0.5 eV) in thermal equilibrium with atoms in the reactor's moderator. In most reactor irradiation positions, 90-95% of the neutrons that bombard a sample are thermal neutrons. In general, a one-megawatt reactor has a peak thermal neutron flux of approximately 1 10^{13} neutrons $cm^{-2}\ s^{-1}$.

IV.9.2.2. Gamma Ray Detectors

IV.9.2.2.1. Scintillation Counters

Scintillating crystals based on sodium iodide doped with a little amount of thallium iodide, described in section IV.8.2.3 as X-ray detectors, can also be used to detect gamma ray. Similarly to X-ray detection, the incoming gamma ray is absorbed by the crystal, and after a certain time the absorbed energy is released as photons of fluorescence radiation. After attaching the scintillating crystal to a photomultiplier tube, the released fluorescent photons are converted in an electrical pulse which is amplified and counted. The efficiency of these detectors is quite high, but their energy resolution is poor comparing to other gamma ray detectors.

IV.9.2.2.2. Semiconductor Detectors

Most of the detectors for gamma ray are semiconductor detectors, especially made of germanium, HP-Ge. These detectors consit of large volume cylinder of coaxial geometry in which the outer surface serves as one electrode and a central conducting cylindrical core acts as the other electrode after applying a high voltage between them. When gamma rays interact with semiconductor, free electrons and holes are produced (electrons are promoted from the valence to the conduction band, leaving positive hole in the valence band). Under the applied voltage, the electrons and holes behave as negative and positive charges and are collected at the electrodes and the amount of charge collected is correlated with intensity of the gamma ray. One practical drawback of HP-Ge detectors is that it must be operated at liquid nitrogen temperatures to keep a low thermal noise.

Semiconductor detectors based on silicon can be also used for detecting low-energy gamma-ray. Other semiconductor materials, such as cadmium telluride (CdTe) or mercuric iodide (HgI_2) can be used in some applications. However, these materials offer poorer energy resolution than germanium and silicon, although they can be operated at room temperature.

The energy resolution of HP-Ge detectors is around 1.9-2.2 keV, which is quite high than that obtained for scintillating counters (approximately 90 keV).

IV.9.2.3. Signal Processors

As commented for XRF, pulses derived from both scintillation counters and semiconductor detectors are selected by means of electronic discriminators which reject those electronic pulses that fall below or above a specific voltage level. After amplification, these devices are connected to single-channel analyzers (SCAs) or multichannel analyzers (MCAs), which are computer based systems for an automatic spectrum evaluation.

A SCA uses two electronic discriminators, called upper and lower level discriminators. Pulses from the amplifier are fed to the analyzer, and if the pulse height falls between the lower and upper discriminators the usual logic is to allow such a pulse to be counted. The voltage levels of the two discriminators are adjustable so that the gap between them corresponds to a group of pulse heights within a fixed energy interval. These groups of pulses are related to a particular radionuclide and the voltage gap between both discriminators is frequently referred to as an energy window. By varying the position of the window (i.e., changing the voltage levels of each of the discriminators) it is possible to measure gamma rays of different energies (from different activated elements). This process offers as a disadvantage the large time when measuring many different energy gamma rays from several elements. For these cases, a MCA is a much more efficient system to use.

MCA consists of an analog to-digital converter (ADC) which takes the analog voltage pulse from the detector and converts it to a digital signal proportional in magnitude to the height of the analog pulse. The MCA contains a number of equal width storage bins (called channels, up to a maximum of 8192). The width of the storage bins corresponds to a specific energy interval and can be set by the user, depending on the application.

IV.9.3. Analysis by NAA

IV.9.3.1. Absolute Analysis

Since the theory of NAA is well founded, an 'absolute' standardisation procedure can be applied. In this case, the quantitative measurement can be obtained by determining the neutron flux and counting the absolute gamma rays, according to equations related to the measured intensities of gamma ray of certain energy. However, these absolute methods can offer systematic errors due to uncertainties of nuclear data taken from literature, especially those on decay schemes and activation cross-section.

When using nuclear reactors as neutron sources, there is another approach (k_0-standardization technique of NAA) originally developed by Simonits et al. [346] for multielement analysis, which also avoids the use of standards for multi-element determinations. K_0-standardization involves the simultaneous irradiation of a sample and a neutron flux monitor, such as gold, and the use of a composite nuclear constant called k_0-factor. The k_0-factor is independent of irradiation and measuring conditions [347]. For the calculation of elemental concentrations, the k_0-NAA uses input parameters such as subcadmium to epithermal neutron flux ratio, the epithermal neutron flux shape factor, and the absolute efficiency of the detector. These parameters are dependent on each irradiation site in a reactor and counting facility, and they must be determined for standardization.

IV.9.3.2. Classic Relative Methods

Classic relative methods, based on the simultaneous irradiation of samples and standards under identical positions, and followed by measuring the induced intensities of both the standard and the sample in a well known geometrical position, are also convenient methodologies. Based on this method, there are two types of NAA determinations according to a destructive or non-destructive process of the sample. As commented, for both analysis types, several standards, and also the sample, contained in sealed quartz or polyethylene vials, are simultaneously irradiated for a fixed time (from several minutes to hours). After finishing the irradiation stage, both standards and sample are let for a certain time (cold-down time) in order to allow desactivation of those radionuclides with short T½ which can be a source of interferences. This time can be within few minutes to several hours.

IV.9.3.2.1. Non-Destructive Methods

In non-destructive methods, pulses (counts) from sample and standards are directly counted after the cold-down stage. In this case, the detector must guarantee a good selectivity for distinguishing gamma rays of different energy. The amount of the analyte (analyte weigth) in the sample is obtained by applying the following equation:

$$w_x = w_s\ (R_x/R_s)$$

where w_x and w_s are the analyte weight in the sample and the standard weight, respectively; and R_x and R_s are the counts for the analyte in the sample and for the standard, respectively.

Non-destructive methods are very simple and avoid any sample manipulation. However, the success of the analysis are related to the high energy resultion of detectors and also to the complexity of the sample and absence of certain elements which can lead gamma rays of similar energy than those generated from the target analytes.

IV.9.3.2.2. Destructive Methods

In destructive methods, the anlyte must be isolated from the sample just before gamma ray counting. Analyte isolation commonly occurs after dissolving the irradiated sample and extracting the analyte by precipitation, liquid-liquid extraction, ionic exchange or chromatography. Once the analyte and the standard are isolated, the gamma ray counts are recorded and the amount of target element is calculated by the equation shown above.

IV.9.4. Cadmium Determination by NAA

There are reported several applications of NAA to assess cadmium and other trace elements in environmental materials, especially solid environmental samples. In addition to NAA based on thermal neutron activation, different NAA methodologies based on epithermal activation (ENAA), cyclic activation (CNAA), Prompt gamma activation (PGNAA) with or without Compton suppression, have been implemented in several laboratories, and they have been used in environmental studies. Some recent NAA determinations include the analysis of soils [348-350] and sediments [348,351-356], atmospheric particulate matter [357-366], fly ashes [367-368], and biota (molluscs [369], seaweeds [370], linchens [371-374], moss [372,375-380] and plants [354]).

IV.10. Electrochemical Techniques: Anodic Stripping Voltammetry and Potentiometric Sensors

Voltammetry comprises a large group of instrumental techniques in which the potential of an electrode (indicator or working electrode) immersed in solution is controlled so as to force a redox reaction to occur at the electrode surface. The rate of this reaction is proportional to the current flowing through the electrode, and as conditions are chosen so that the rate of the redox reaction is controlled by the rate of transport of the reacting species to the indicator electrode surface, the current flowing through the cell is proportional to the concentration of the reacting species. For these techniques, the plot of the measured electric current versus the potential of the electrode is commonly named as voltammogram.

Usually, a set of two or three electrodes is dipped into the analyte solution, and a regularly varying potential is applied to the working electrode relative to the second electrode (the other half of the cell). The working electrode is an ideally polarizable electrode (i.e., the electrode shows a large change in potential when an infinitesimally small current is passing through), such as the dropping mercury electrode (DME), with surfaces as small as possible so that there is a high concentration of active species in the surrounding area, and therefore a strong polarization. The second electrode has two functions; firstly it must have a known potential with which to gauge the potential of the working electrode; and secondly, it must balance the electrons added or removed by the working electrode. However, it can be difficult for this second electrode to maintain a constant potential while passing current to counter redox events at the working electrode. To solve this problem, two electrodes are used, the reference electrode which is a half cell with a known reduction potential and which acts as reference in measuring and controlling the working electrodes potential and at no point does it pass any current; and the auxiliary electrode, which passes all the current needed to balance the current observed at the working electrode. To achieve this current, the auxiliary electrode will often swing to extreme potentials at the edges of the solvent window, where it oxidizes or reduces the solvent or supporting electrolyte. A three electrodes set (working, reference, and auxiliary) is the basis of the modern voltammetry instruments.

Potential modulation is usually carried out by a simple staircase ramp, but modulation based on pulses offers advantages in sensitivity and speed. There are three pulse techniques for modulation, which leads to three voltammetry methods, the normal pulse voltammetry (NPV), the differential pulse voltammetry (DPV) and the square-wave voltammetry (SWV). NPV uses a series of potential pulses of increasing amplitude and the current measurement is made near the end of each pulse. DPV also works with potential pulses but these pulses are of fixed amplitude and are superimposed on a slowly changing base potential. In this case, current is measured just before the application of the pulse and at the end of the pulse (two points for each pulse). Finally, SWV uses a symmetrical square-wave pulse of amplitude E_{SW} superimposed on a stairscase waveform of step height ΔE, where the forward pulse of the square wave coincides with the stairscase step.

There is a wide variety of voltammetry techniques such as linear sweep voltammetry (LSV), anodic stripping voltammetry (ASV), cathodic stripping voltammentry (CSV), or adsorptive stripping voltammetry (AdSV). Polarography is also a voltammetric technique for which the working electrode is a DME. This electrode type is useful for its wide cathodic range and renewable surface. Detailed information for voltammetry methods is given in monographs by Bard and Faulkner [381], Wang [382] and Brainina and Neyman [383].

IV.10.1. Anodic Stripping Voltammetry

Anodic stripping voltammetry (ASV) is the electroanalytical technique that has the lowest limits of detection (ppb or lower) for metal determination. The technique is based on the electrochemical deposition or accumulation (reduction) of the analyte at the working electrode (anode or cathode) of constant surface and controlled potential electrolysis; and the subsequent stripping (oxidation) from the electrode by a voltammetric technique or chemical reaction. A certain waiting time between deposition and stripping (typically 30 s) is needed. The accumulation of the analyte onto the surface of the working electrode implies a pre-concentration of the analyte (pre-concentration factor within 100 to 1000). This is the main advantage of ASV over other voltammetric techniques and it is directly related with its inherent high sensitivity. If the experimental conditions during the pre-concentration stage are constant, the measured voltammetric answer (the peak area or peak height of the voltammogram) is related with the concentration of the active species and it can be used for quantification. Moreover, the potential for re-dissolving the deposited metal can be used for cualitative purposes. With more than one metal ion in the sample, the ASV signal may sometimes be complicated by formation of intermetallic species. This may distort the stripping signal for the analyte of interest. These problems are minimized or even avoided by adjusting parameters such as the deposition (oxidation) time and the deposition (reduction) potencial.

ASV is used for the determination of amalgaman-forming metals (metallic ions). Other similar techniques are CSV, which is employed for determining species that form insolubles salts with mercury on the electrode surface. These species are typically inorganic anions, such as halides, or oxyanions or certain organic compounds. In addition, AdSV is quite similar to ASV, but the pre-concentration step of the analyte is accomplished by adsorption on the electrode surface or by specific reactions at chemically modified working electrodes rather than accumulation by reduction. This stripping voltammetric type is commonly used for organic targets.

IV.10.1.1. Instrumentation

Conventional instruments (cells) consist of three electrodes: a working electrode that should be easily polarizable (usually a microelectrode based on mercury MDE); an auxiliary electrode, which usually is a platinum wire or a small mercury deposition; and a reference electrode, which usually is calomel electrode (Hg/Hg_2Cl_2, KCl). The MDE is a micro working electrode, which the mercury drop is formed at the end of a glass capillary (length 10 – 20 cm, inside diameter 0.05 mm). The mercury droplets are highly reproducible diameter and of a life time ranging from 2 to 6 s. MDEs offer several advantages over other working electrodes (also referred as solid electrodes), such as the constant renewal of the electrode surface, the fact that the charge transfer overvoltage of hydrogen ions present in the aqueous solvent is high on Hg, and that many metals are amalgamed by Hg. There are two types of MDEs; the hanging mercury drop electrode (HMDE), most often used; and the static mercury drop electrode (SMDE). The HMDE consists of a very fine capillary tube connected to a mercury-containing reservoir. Mercury is forced out of the capillary by an arrangement controlled by a micrometer screw, which allows drops of very reproducible surface areas. SMDE uses a method of drop formation in which the mercury drop is dispensed rapidily and then allowed to hang stationary at the capillary tip. This second electrode allows more reproducible and stable drops.

Because the high toxicity of mercury, different working electrodes based on more environmentally friendly materials have been developed and successfully applied for the voltammetric determination of several metals. The most popular electrodes are those based on bismuth films (BiFEs, bismuth film electrodes) or glassy carbon electrode (GCE) electrodes. Advances in nanotechnology have led to the development of different electrodes based on nanoparticles, mainly bismuth nanoparticles and carbon nanotubes. Similarly, the screen-printing technology has found an appealing development area in producing disposable or single-use electrodes, commonly refrerred as screen-printed electrodes (S-PEs). These electrodes avoid the memory effect between samples as well as the electrode fouling phenomenon.

To carry out the measurements, an excess of a strong electrolyte (KCl) is added to the sample. Thus, the ionic strength and the conductivity remain constant. The dissolved oxygen must be removed because oxygen is easily reduced and it can be a significant interference. Oxygen removal is performed by bubbling an inert gas (N_2) for a few minutes and then keeping a stream of nitrogen on the surface to prevent oxygen re-dissolution.

Table 5. ASV determination of cadmium in environmental samples

Sample type	Working electrode	Technique	Ref.
Seawater	HMDE	DPASV	[385]
Soil extracts	HMDE	DPASV	[386]
Plants	HMDE	DPASV	[387]
Water	HMDE	DPASV	[388]
Atmospheric particulate matter (PM10)	HMDE	DPASV	[389]
Waste water	HMDE	SI-DPASV	[390]
Seawater, hydrothermal fluids	HMDE /BiFE	Single-run ASV	[391]
Plants	HMDE	LSASV	[392]
Estuarine water	Poly-L-lysine-poly(sodium 4-styrenesulfonate)-coated Hg film electrode	SWASV	[393]
Water	Hg-coated glassy carbon	DPASV	[394]
Spicules of marine sponges	Hg-film electrode plated on an epoxyimpregnated graphite rotating-disk support	SWASV	[395]
Soil extracts	BiFE	LSASV / SWASV	[396]
Tap water	BiFE	SWASV	[397]
Tap water / soil	BiFE	SWASV	[398]
Tap water / plant extracts	BiFE	DPASV	[399]
Tap water / seawater	BiFE	ASV	[400]
Water / plant extracts	BiFE	DPASV	[401]
Rocks	BiFE	SI-SWASV	[402]
Tap water	BiFE	LSASV	[403]
Tap water	BiFE	SWASV	[404]
Iron ores	BiFE	FI/SI-SWASV	[405]
Water	BiFE	SWASV	[406]
Tap water	BiFE	SWASV	[407]
Environmental materials	Fibrinogen-BiFE	SWASV	[408]
Tap water / snow	Poly(sodium 4-styrenesulfonate)-BiFE	SWASV	[409]
River water	Carbon nanotubes-BiFE	SWASV	[410]
Water	Carbon nanotubes-BiFE	SWASV	[411]
Tap water	Bismuth/glassy carbon composite	SWASV	[412]
Waste water	GCE	DPASV	[413]
Soil / indoor atmospheric particulate matter	GCE	SWASV	[414]
Estuarine water	GCE	SWASV	[415]
Tap water	Modified carbon paste electrodes	SWASV	[416]
Syntehic samples	Modified carbon paste electrodes	DPASV	[417]
Water	GECE	DPASV	[418]
Seawater	Bare gold-disk micro-electrodes	SWASV	[419]
Tap water	Tin film electrodes	DPASV	[420]
Standards	Bi-BDD electrode	SWASV	[421]
Soil extracts	S-PE	LSASV	[422]
Water / soil	S-PE	DPASV	[423]
Seawater	S-PE	SWASV	[424]
Fish	S-PE	SWASV	[425]

BiBDD, bismuth boron doped diamond; BiFE, bismuth film electrode; DPASV, differential pulse anodic striping voltammentry; FI-SWASV, flow injection - squared wavelength anodic striping voltammentry; GCE, glassy carbon electrode; GECE, graphite-epoxy composite electrode; HMDE, hanging mercury drop electrode; LSASV, linear sweep anodic stripping voltammetry; SI-SWASV, sequential injection - squared wavelength anodic striping voltammentry; S-PE, screen-printed electrode; SWASV, squared wavelength anodic striping voltammentry

IV.10.1.2. Cadmium Determination by ASV

Table 5 lists some recent applications of ASV methods to assess cadmium in environmental samples. It can be seen that the conventional HMDE [385-392] is used for several determinations, although coated mercury film electrodes [393-395] are also used. BiFEs using different supports [391,396-407] and modified BiFEs with fibrinogen [408], poly(sodium 4-styrenesulfonate) [409], or carbon nanotubes [410,411], have also been used for cadmium determination. A bismuth/glassy carbon composite material by incorporating bismuth powder in the composite electrode has also been described [412]. GCEs [413-415] modified carbon paste electrodes [416,417], graphite-epoxy composite electrodes (GECEs) [418], bare gold-disk micro-electrodes [419], tin film electrodes [420], bismuth nanoparticles modified boron doped diamond (Bi-BDD) electrode [421], and S-PEs [422-425] are also available (Table 5).

IV.10.2. Pontentiometric Sensors: Ion Selective Electrodes (ISEs)

Pontentiometry consists of the measurement of a thermodynamic equilibrium potential in an electrochemical cell, without net current flows. The direct determination of the activity results from the thermodynamic equilibrium relationship between the activity of an ion and the potential of a cell. These techniques provide a method for distinguishing between free or ionized and bound or complexed ions in a sample, an also between the activities of different oxidation states of an ion. Therefore, several dissolved ions or certain dissolved gases can be measured with high selectivity and sensitivity and these electrodes are also called sensors. When ions in solution are monitored, the sensors are known as ionic selective electrodes (ISEs). The use of ISEs in environmental analysis offer several advantages over other methods of analysis, mainly the cost of initial setup to make analysis is relatively low. Detailed information on ISEs can be found in the monographs by Bard and Faulkner [381] and Evans [426].

IV.10.2.1. Instrumentation

The basic potentiometric sensor setup includes a meter (capable of reading millivolts) and two electrodes, the first one is the indicator or membrane electrode (ion specific electrode surrounded by a thin film of an intermediate electrolyte solution and enclosed by an ion permeable membrane) and other one is the reference electrode. Both electrodes are immersed in the sample solution. The indicator electrode consists of an internal reference electrode with an internal filling solution and a membrane located at one end. The membrane separates two liquids, the internal filling solution and the sample solution. The liquid that fills the membrane electrode is a solution containing the ion which will be analysed in the sample, and which concentration remains constant. The sample solution, outside, also contains those ions, but the concentration is unknown. The membrane allows that these ions (and only these ions) migrate through the membrane from the two sides. The speed of migration is dependent on the existing concentrations on such ions at each side of the membrane, which creates an imbalance of charges on both sides of it. This fact is reflected in the emergence of a potential. Since potentials due to the reference electrodes are constant, signal variations are only attributed to the potential dependent on the ion concentration in the sample solution. All these considerations and the implementation of the relevant equations lead to the conclusion that the potential measured by a potentiometer connected to the two electrodes is a constant value E_K plus a value depending on the concentration of ion in the sample solution, according to the following Nernstian expression

$$E_{measured} = E_K + (RT / nF) \ln C,$$

where C is the analyte concentration, the value E_K, encompasses all potential constants and it can be determined after calibration; R and F are constants (8.314 JK^{-1} mol^{-1} and 96485 °C mol^{-1}, respectively) and T is he temperature expressed as K.

Among the different working electrodes, membrane electrodes are quite popular for ISEs, and several types of membranes (glass membranes, solid-state membranes, liquid membranes, gas-permeable membranes or biocatalytic probes) can be used. Glass membranes-based electrodes, including chalcogenide glasses, are one of the most used. In addition to glass membrane electrode for pH measurements, there are several glass membrane based electrodes commercially available for different metallic ions such as cadmium (Cd^{2+}). In this case, the cadmium ISE consists of a membrane glass of $AgS_2(CdS)$ with an internal filling solution of potassium nitrate. Solid-stated membranes consist of crystal or pressed pellets containing the salts of the target ion (e.g., silver halides or metal sulfides) which has a low solubility and is electrically conductive. Solid-stated ISEs are typically used to measure ions like halides and certain cations such as cadmium (Cd^{2+}). This sensor is equipped with an Ag/AgCl reference electrode and an electrolyte filling solution of potassium nitrate. As reviewed by Bakker and Pretsch [427], hydrophobic polymeric membranes doped with active sensing ingredients, typically a lipophilic ion exchanger and a highly selective ionophore, are appealing solid-stated membranes for ISEs because the high sensitivity achieved (ppb or lower). In these membranes, an ion-exchanger is responsible for attracting a fixed concentration of the analyte into the membrane phase, while an ionophore selectively binds this analyte ion.

IV.10.2.2. Cadmium Determination by ISEs

There are described several Cd^{2+}-ISEs in function of the membrane type. In addition to chalcogenide glass ($CdSAgIAs_2S_3$) membranes [428], CdSe based- solid-state membranes are also used [429]. Poly(vinyl chloride) (PVC) or polyestirene containing different membrane carriers (Table 6) have been also synthesized and applied for selective cadmium measurements [430-432]. Similarly, coated surface graphite rod electrodes containing cetylpyridinium-tetraiodocadmate or cetylpyridinium-tetrabromocadmate have been also proposed [433]. Other group of ISEs are those based on composite materials formed by the combination of an inorganic ion exchanger of multivalent metal acid salts and an organic conducting polymer (polyaniline, polypyrrole, polythiophene, etc.), which provides a new class of organic-inorganic hybrid ion exchangers. An example of these systems is the ion exchanger polyaniline Sn(IV) tungstoarsenate [434] or the cation-exchanger polyaniline Ce(IV) molybdate [435]. Other polymeric membranes (PVC) containing certain neutral ionophores, such as Schiff bases, have also been proposed [436-439].

As commented, ISEs based on polymeric membranes containing an ionophore and an ion exchanger offer the highest sensitivities and can be applied for the determination of traces.

For cadmium, polymeric membranes based on the ionophore, N,N,N',N'-tetradodecyl-3,6-dioxaoctanedithioamide (ETH 5435) have been described, showing LOD of 11 μgL^{-1}

[440]. Other polymeric membranes containing different ionophores (Table 6) have also been described [441-446].

Table 6. ISE determination of cadmium in enironmental samples

Membrane type	Some remarks	Ref.
Glass membrane	Membranes based on a chalcogenide glass material ($CdSAgIAs_2S_3$) deposited as thin film onto different transducer structures	[428]
Glass membrane	CdSe instead of CdS. The reference electrode is an Ag/AgCl	[429]
Solid-stated membrane	PVC containing [1,1'-bicyclohexyl]-1,1', 2,2'-tetrol as a membrane carrier	[430]
Solid-stated membrane	PVC containing tert-Bu thiacalix[4]arene or thiacalix[4]arene as membrane carriers	[431]
Solid-stated membrane	Polyestyrene containing 3,4:12,13-dibenzo-2,5,11,14-tetraoxo-1,6,10,15-tetraazacyclooctadecane as a membrana carrier	[432]
Solid-stated membrane	Graphite rod electrode coated with cetylpyridinium-tetraiodocadmate or cetylpyridinium-tetrabromocadmate	[433]
Solid-stated membrane	Polyaniline / Sn(IV) tungstoarsenate as organic / inorganic composite material as a cation exchanger	[434]
Solid-stated membrane	Polyaniline / Ce(IV) molybdate as organic / inorganic composite material as a cation exchanger	[435]
Solid-stated membrane	PVC containing 5-[((4-Me phenyl) azo)-N-(6-amino-2-pyridin) salicyladimine]	[436]
Solid-stated membrane	PVC containing 5-[((4-Me phenyl) azo)-N-(2-diamino-2-cyano-1-Et cyanide) salicylaldehyde]	[436]
Solid-stated membrane	PVC containing N,N'-[bis(pyridin-2-yl) formylidene]butane-1,4-diamine	[437]
Solid-stated membrane	PVC containing N-(2-yridinylmethylene)-1,2-benzenediamine	[437]
Solid-stated membrane	PVC containing N'-[1-(2-furyl)methylidene]-2-furohydrazide	[438]
Solid-stated membrane	PVC containing N,N'-(4-methyl-1,2-phenylene)diquinoline-2-carboxamide	[439]
Solid-stated membrane	PVC containing N,N,N',N'-tetradodecyl-3,6-dioxaoctanedithioamide (ionophore), sodium tetrakis[3,5-bis(trifluoromethyl)] (ion-exhanger)	[440]
Solid-stated membrane	8-hydroxyquinoleine (ionophore), sodium tetra-Ph borate (ion-exchanger)	[441]
Solid-stated membrane	8-hydroxyquinoleine (ionophore), sodium tetra-Ph borate (ion-exchanger)	[442]
Solid-stated membrane	tetrathia-12-crown-4 (ionophore) phenylborate (ion-exchanger)	[443]
Solid-stated membrane	dicyclohexano-18-crown-6 (ionophore), sodium tetra-Ph borate (ion-exchanger)	[444]
Solid-stated membrane	dicyclohexano-24-crown-8 (ionophore), sodium tetra-Ph borate (ion-exchanger)	[445]
Solid-stated membrane	K hydrotris[N-(2,6-xylyl)thioimidazolyl] borate (ionophore), sodium tetra-Ph borate (ion-exchanger)	[446]
Solid-stated membrane	K hydrotris(3-phenyl-5-methyl-1h-pyrazol-1-yl)borate (ionophore), sodium tetra-Ph borate (ion-exchanger)	[446]

IV.11. Automatic Analyzers

The emergence on the market of automated systems for analysis (automatic analyzers) has been one of the major advances in analytical chemistry in recent decades. Automatic analyzers provide analytical data with minimal operator intervention. Automation implies the partial or complete replacement of human involvement in an operation or sequence of operations.

Initially, these systems were designed to address the needs of clinical laboratories, but at present they are used in diverse areas as industrial process control (process analyzers) or routine determinations of various substances in the air, water, and soil (environmental monitoring). A monograph by Valcárcel and Luque de Castro [447] offers detailed descriptions of the different automatic analyzers.

Advantages of automation can be summarized as follows:

1) Rapidity of the process, which allows the treatment and analysis of large number of samples. This is useful in environmental monitoring and process analysis.
2) Reduction of human participation, which avoids errors and increases analytical quality (high repeatability and accuracy for long time periods is obtained).
3) Increase on human and environmental security because automation avoids dangerous handled manually operations, and decreases human accidents and environmental risks.
4) Lower consumption of sample and/or reagents, which is very important for scare of precious samples or expensive reagents (reduction of costs).

Automatic analysis can be classified according to the way in which samples are transported and manipulated. Therefore, there are two modes, batch and continuous modes.

1. Discrete or batch analysis. In this operation mode, each sample preserves its integrity in a vessel transported mechanically to various parts of the analyzer, where the different analytical stages are carried out in a sequential manner. Each sample finally arrives at the detector where the relevant signals are recorded.

Batch analysis can be classified according to the manner in which the sample is transferred to the final operation, that is, with or without final transfer of the sample. In analyzers with final transfer, the reaction mixture is transported to the detection system, where the analytical measurement is carried out in a fixed cuvette. In analyzers without final transfer of the reaction mixture, all the stages of the process take place in the same vessel.

2. Continuous analysis. These systems are characterized by the use of a continuous stream of liquid. The samples are sequentially introduced at regular intervals into a channel carrying a liquid that does or does not merge with other channels carrying reagents, buffers, masking agents, and so on, upon reaching the detector. The final reaction mixture yields an analytical signal that is recorded. Continuous analysis can be adapted for analysis of one or several analytes by using a single-channel (for the determination of one analyte) or multi-channel configuration (for several analytes).

There are two types of continuous systems: segmented flow analysis (SFA) and un-segmented-flow analysis. SFA are characterized by a segmented flow of air bubbles, the aim of which is to preserve the integrity of samples. Un-segmented-flow analysis, mainly flow injection (FI) and continuous flow (CF), provide a flow that is not segmented by air bubbles.

IV.11.1. Un-Segmented-Flow Analysis: Flow Injection and Continuous Flow

FI is a simple, versatile and low cost methodology, and it allows fast analysis (lower than 1 minute) which improves the sampling frequency (300 - 400 h^{-1}). The rapidity of FI is due to neither physical equilibrium (homogenization of a portion of the flow) nor chemical equilibrium (reaction completeness) is attained by the time the signal is detected. Hence, FI can be considered a fixed-time analytical methodology. As consequence of these characteristics, batch analysis and SFA has been replaced by FI. As it will be shown in Chapter V, automatic sample pre-treatment for liquid and solid samples have been extensively adapted to FI methodologies.

In FI the sample and reagents are incompletely mixed and there is a concentration gradient that varies along the system as a function of time. The continuous detectors provide a transient signal, which is recorded. The physical foundation of FI is related to the behaviour of the sample plug inserted in the flow, which is characterized mathematically by means of the so-called dispersion (the shape of the profile yielded by the injected sample portion along the system).

The main elements of these analyzers are:

1. A propelling system, which is a peristaltic pump to propel fluids to the different elementary units along the system. It must provide a pulse-free and perfectly reproducible flow of constant rate. Flexible plastic tubing is squeezed by a series of rollers (8 or 10 rollers in circular configuration), which starts the flow of the enclosed liquids as a result. These systems can work with several streams (between 1 and 16, but 4, in most cases), and the flow rate is determined by the internal diameter of the tube (i.d. 0.25 – 4.0 mm, which allow flows of 0.0005 – 40 ml min^{-1}) and the pump rotation speed (> 30 rpm).
2. An injection system, which is a six-way rotary valve (three inlets and three outlets) inject the sample (sample volume of 5.0 to 200 μL) into the carrier stream in a highly accurate and reproducible manner without disturbances. This operation is fast to achieve a high sampling rate. The rotator valve, controllable by an electric motor, can adopt two positions. In the filling position, the sample fills the loop; in the injecting position, the carrier sweeps the sample toward the reactor.
3. A separation/pre-concentration system, which permits continuous separation techniques, for example, filters or knotted reactor (KR) (on-line co-precipitation), membrane or gravitational liquid-liquid separators (on-line liquid-liquid extraction) and adsorbent columns or exchange resins (on line solid phase extraction), as it described in Chapter V. Simmilarly, microwave or ultrasounds devices can be also coupled for assisting solid-liquid extraction (Chapter V). As it will be commented, in these last cases flows of extractants are passed several times through a column containing the solid sample.
4. A transport and reaction systems, which are helically coiled PTFE tubes of a given diameter (i.d. 0.1 and 2 mm) that are connected to the other elements of the system.

This system can be thermostated if necessary (reaction development must be carried out at a fixed temperature). The length of the tubes and the flow rate fix the reaction time.

5. A detector system, which can be an optical or an electroanalytical instrument.

CF is similar to FI. The main difference between CF and FI is the long analysis time required to stabilize the hydrodynamic conditions. In addition, other differences are the use of large i.d. tubes, faster flow rates, and special requirements for the rinsing stages.

IV.11.2. Segmented-Flow Analysis

In SF, the bubbles prevent contamination between successive samples, reduce sample dispersion, and facilitate mixing of sample and reagents in a way that enables physical and chemical equilibrium to be attained before the sample reaches the detector. The main elements of this analyzer are as follows:

1. A sampling system, sampling is carried out with the aid of a moving aspirating tip, air bubbles are also aspirated between samples
2. A propelling system, a peristaltic pump
3. A separation/pre-concentration unit
4. A reaction system
5. A debubbler, which is a high-precision device that removes the bubbles. This device is not essential if certain software treatments that discriminate signal parasites caused by bubbles are used
6. A detection system similar than FI or CF.

Many of the automatic analyzers described have been adapted to automatic water and air analysis (off-line and on-line water analyzers) to facilitate the analysis and monitoring of constituents and contaminants of water and air.

Chapter V

SAMPLE PRE-TREATMENT PROCEDURES FOR ENVIRONMENTAL SAMPLES

V.1. GENERAL ASPECTS

Sample pre-treatment procedures are necessary to isolate the desired components from complex liquid and solid matrices, and also because most of the analytical instruments can not directly handle the liquid or the solid matrix. The basic concept of sample pre-treatment procedure is to convert a real matrix sample into a sample suitable for analysis. This process almost inevitably changes the interactions of analytes with their matrices. These interactions are determined by the physical and chemical properties of both analytes and matrices, and they affect the applicability of different sample pre-treatment techniques and analytical methods as well as their efficiency and repeatability.

Sample pre-treatments can be achieved by using a wide range of techniques; including the most simple and inexpensive methods, which do not require special piece of equipment, such as distillation and liquid-liquid extraction (LLE) procedures; and the most sophisticated pre-treatments such as pressurized liquid extraction (PLE), which piece of equipment, and also mantainance, is of high cost. Independenly of the degree of sophistication, sample pre-treatment methods have the same goals:

1) The increase on the selectivity of certain analytical techniques by removing potential interferences, e.g., high saline matrices (Cl^- and Na^+ and, to a lesser extent, SO_4^{2-} and Mg^{2+}) form seawater samples; high concentrations of ions such as Ca^{2+} and HCO_3^- from river waters and Ca^{2+}, HCO_3^-, Na^+, Mg^{2+}, SO_4^{2-} and Cl^- from groundwaters; organic matter form biological samples or mineral matrices (silicates and aluminum and iron oxides) from sediments, soils or atmospheric particulate matter. In this sense, the matrix components removal can also avoid the potential deterioration of the analytical instruments.
2) The improvement of the sensitivity of the analytical techniques by increasing the analyte concentration (pre-concentration methods), e.g., trace metals concentration in open seawater and atmospheric particulate matter are in the range of ng L^{-1} and ng m^{-3}, respectively.
3) If necessary, the analyte convertion into a more suitable form, e.g. metals derivatization to organic chelate by using adequate ligands in liquid-liquid extraction,

liquid-liquid micro-extraction or solid phase extraction (SPE) methodologies; organometallic compounds derivatization into apolar volatile species by Gridnard reaction (ethylation, propylation or pentylation) in LLE or solid phase micro-extraction (SPME) methodologies; etc.

4) The transfer of the analyte to a compatible solvent for a certain analytical technique to allow the analytical determination, e.g., back-extraction LLE procedures into aqueous / acid solution for further metal quantification by atomic or emission spectrometric methods, or a change from a polar to a non polar solvent or viceversa for organic compounds determination by chromatographic techniques.
5) If possible, the isolation of certain species (speciation studies), e.g. chromium can be determined by the reaction between Cr(VI) and 1,5-diphenylcarbazide, which forms a complex that could be extracted into an adequate organic solvent or adsorbed into adequate solid support, allowing Cr(VI) to be determined in the presence of Cr(III).
6) The establishment of a robust and reliable method that is not affected by variations in the sample matrix.

Although, many traditional sample pre-treatment methods are still in use (e.g. precipitation, co-precipitation, liquid-liquid extraction, solid phase extraction, acid extraction or acid digestion), there have been trends in recent year towards:

1) The use of small initial sample sizes, e.g. sample volumes around few milliliters are usual for single-drop micro-extraction (SDME) or dispersive liquid-liquid micro-extraction (DLLME) procedures.
2) The ease of automation, which minimizes the sample and extract handling and avoids analyte losses and sample contamination; e.g. microwave or ultrasound assisted extraction techniques and pressurized liquid extraction.
3) The miniaturization, e.g., miniaturized devices available for operations such as liquid-liquid extraction and pre-concentration into solid supports by flow injection analysis (FIA) and new miniatured LLE procedures such as single-drop micro-extraction (SDME), dispersive liquid-liquid micro-extraction (DLLME) or hollow fibre liquid phase micro-extraction (HFLPME).
4) The use of non exhaustive extraction procedures in which only a small fraction of the analyte is extracted / pre-concentrated for further analysis, e.g., in SDME or DLLME pre-treatments the analyte extraction is incomplete; the high repeatablity of measurement of time in the sequence of operations (as a result of automation) ensures that the sample and the patterns are exactly the same process and exactly in the same period of time, achieving a high reproducibility.
5) The high selectivity in the extraction, e.g. the development of new solid phases for SPE such as molecular / ionic imprinted polymers for organic compounds and metal pre-concentration from liquid samples, respectively; carbon nanotubes, nanometer-sized materials, ordered mesoporus materials, etc.
6) The reduction of the pre-treatment time; e.g. microwave assisted extraction (MAE) and pressurized liquid extraction (PLE) dramatically shorten the extraction times, (completed extractions after 1 to 5 minutes).
7) The increase on the extraction efficiency without degradation or losses of the target analytes, and decrease on the number of extraction cycles, e.g. 2 - 3 extraction steps

are typical for LLE methodologies, while one extraction step is enough for SDME or DLLME; the extraction cycles for organic compound isolation by soxhlet (typically around 24 cycles) are dramatically reduced by using PLE (typically 1-3).

8) The high throughput performance, which implies that the number of samples that could be pre-treated per unit of time should be similar to the number of samples per unit of time that modern analytical detectors can analyze. MAE and PLE procedures allow the pre-treatment of high number of sample per unit of time; thus, these procedures could be on-line coupled with analytical instruments.
9) The on-line coupling with analytical instruments, transferring the treated sample to the detection systems; e.g. metals pre-concentration from liquid matrices by using ion-exchange columns in a flow injection analyzers coupled to atomic absorption or emission spectrometry; or metals determination from solid samples by using continuous digestion / extraction devices coupled to spectrometric instruments.
10) The increase on the operator safety and the decrease on the environmental impact of the analytical methodologies. Four top priorities concerning the elimination or reduction of organic solvents and/or highly toxic or ecotoxic reagents, the prevention of waste generation and the reduction of energy consumption, were identified in the analytical laboratory [448], which connects with the principles established by the Analytical Green Chemistry (AGC): *the use of chemistry techniques and methodologies that reduce or eliminate the use or generation of feedstock, products, by-products, solvents, reagents, etc. that are hazardous to human health or the environment* [449].

V.2. Sample Pre-Treatment for Liquid Samples

The extraction of trace cadmium and other metals at low concentrations in complex environmental aqueous matrices, such as saline waters, has often been difficult for analytical chemists. There are two important problems:

1) The saline composition of water matrix (mainly for seawater samples), which produces a high background signal and non-spectral interferences when spectrometric techniques (most usual) are used for quantification. In addition, the occurrence of a high background and scattering radiation is a limiting factor for direct analysis in most natural waters by XRF.
2) The low trace metal concentration in environmental waters, e.g., trace metal concentrations around few ng L^{-1} to 10 µg L^{-1} are typical for waters. Cadmium concentration in surface seawater is between less than 5 ng L^{-1} (for nutrient-depleted surface water) and 15 ng L^{-1}. This value could be increased up to 100 ng L^{-1} in deep seawaters and to 50 ng L^{-1} in seawater from up-walling areas [3,4,6,7]. High average concentrations for cadmium, between 1.0 ng L^{-1} and 10 µg L^{-1}, are found in river waters, while values less than 100 ng L^{-1} are reported in lake waters.

Most of the chemical elements are present in seawater, but most of them are at trace level. Six components (chloride, sodium, sulphate, magnesium, calcium and potassium) account for over 99% of the composition of solutes. Table 7 lists the most abundant components of seawater of 35‰ average salinity [450]. The composition of fresh water (ponds, lakes, rivers and streams, underground water, rain and snow) contains less than 0.5 parts per thousand of dissolved salts, being bicarbonate (48%) the major component of river waters.

Table 7. Water composition according to reference [450].

	Concentration (mM)	
Ion	Seawater (salinity of 35 ‰)	River water
Chloride (Cl^-)	545	0.16
Sodium (Na^+)	468	0.23
Magnesium (Mg^{2+})	53.2	0.15
Sulphate (SO_4^{2-})	28.2	0.069
Calcium (Ca^{2+})	10.2	0.33
Potassium (K^+)	10.2	0.03
Bicarbonate (HCO_3^-)	2.38	0.86

Ions such as Ca^{2+}, Na^+, Mg^{2+}, SO_4^{2-} and Cl^- are present in fresh waters as minor components. Table 7 also shows the average composition in river waters [450].

A comparison of the low concentrations of metals in natural waters (cadmium in the range of less than 0.045 to 90 nM) to the high concentrations of the saline matrix (Table 7), shows very high matrix / analyte ratios.

To overcome these problems, early emerged extraction / separation techniques offer not only the ability to isolate the metals from the matrix solution, thus they reduce, control and even eliminate the interferences, but also the opportunity for metals to be pre-concentrated and determined at very low levels.

An ideal method for pre-concentration of trace metals from natural waters should have the following characteristics:

1) It should simultaneously allow the isolation of the analyte from the matrix and yield an appropriate enrichment factor.
2) It should be a simple process, and the use of few reagents must be minimize to avoid contamination, hence low sample blank and low detection limit can be obtained.
3) It should produce a final solution that is readily matrix matched with solutions of the analytical calibration method.

Sample pre-treatments, such as evaporation or lyophilization, precipitation or co-precipitation of cadmium from waters and cadmium electrodeposition on wires, were the first applied procedures to achieve this task. Liquid-liquid extraction (LLE) and solid phase extraction (SPE) have been extensively used for metals pre-concentration from natural water, and most of the standardized and official methods for metals determination in waters enclosed these techniques.

Although the high applicability of LLE and SPE, new extraction techniques are nowadays available, which allow a huge versatility for selecting the most suitable approach for an analyte. Single-drop micro-extraction (SDME), dispersive liquid-liquid micro-extraction (DLLME) and hollow fibre liquid phase micro-extraction are new miniaturized and not exhaustive sample LLE procedures which main advantages are their speed, negligible volume of solvents used and their ability to detect analytes at very low concentrations.

A description of the theoretical aspects, experimental parameters and advantages and disadvantages of these sample pre-treatment methods will be developed in the following sections. In addition, the application of those techniques to the isolation and pre-concentration of cadmium from environmental liquids samples will be also commented.

V.2.1. Non-Boiling Evaporation

V.2.1.1. General Aspects

Evaporation procedure is the easiest and simplest pre-treatment procedure for liquid samples. In non-boiling evaporation procedures, the sample (in the range of 50 to 1000 mL) is transferred into a Erlenmeyer flask, vessel or bulb, and placed into a thermostat containing several fluids (water, water-ethylene glycol (1:1), triaryldimethane, etc) or placed directly on a heating plate and, evaporated then to a drop (c.a. 50 – 100 µl). Temperature setting is adjusted to avoid any boiling, and the process occurs at temperatures within the 80 – 90°C range. In some cases, the water vapor evolved can be transferred using an inert gas flow rate to an ice bath where water vapor is condensed (distillation). The evaporation is continued to near dryness, and for a typical 70 mL sample volume, this procedure takes about 2 – 15 hours. After evaporation, few milliliters of ultra-pure water or diluted mineral acids are added to residue for dissolution. Metals in the final solution are then analyzed.

Evaporation is a slow process and it is generally carried out in open systems and therefore, it is subjected to possible contamination from the laboratory environment. In addition, there are risks of contamination from container walls (mainly from glass vessels) and possible loss of volatile metals. Various improvements have been proposed for allowing a more efficient control of contamination problems when coping with low metal concentrations. They include:

1) The use of clean room conditions, such as work stations fitted with special high efficiency particulate (HEPA) filters, or carrying out the whole evaporation procedure inside an all-plastic vertical laminar-flow clean bench fitted with HEPA filters.
2) The choice of Teflon vessels for the evaporation containers, which avoids the possible dissolution of several elements from glass vessels. In addition, sample droplets adhesion on the Teflon's walls is negligibly small [451-453].
3) The use of cleaning procedures, the Teflon beaker is first taken out of three heated acid cleaning bath and rinsed with ultra-pure water.
4) The use of ageing and pre-conditioning procedures, by filling the cleaned Teflon beaker with ultra-pure water and heating it for several minutes. Thus, heavy metals contamination released from walls of beaker is discarded.

V.2.1.2. Non-Boiling Evaporation Methods for Cadmium Pre-Concentration

Water removal by non-boiling evaporation techniques has not extensively been applied for the pre-concentration of cadmium and other trace metals from environmental samples due to the saline matrix is also pre-concentrated. This approach is more useful for metals pre-concentration from high purified water. Thus, there are few references on non-boiling evaporation for cadmium pre-concentration: snow samples [451-453], river water [454], and rain water [452]. Table 8 summarizes the main analytical characteristics of these procedures.

V.2.2. Lyophilization

Lyophilization, also called freeze-drying , is a method commonly used for drying biological samples to stabilize and preserve them for storage and transport. Also it has been used for drying natural waters as pre-concentration technique.

V.2.2.1. General Aspects

The freeze-drying process is essentially a low temperature, low pressure distillation of water which sublimes from the surface of the ice and condenses in a cold trap. Ice is directly converted to water vapor, without passing through the intermediary stage of a liquid. Thus, no heating is required for water removal from sample. In the case of solid samples, the result is a sample whose structure is largely preserved, which can be stored at room temperatures and pressures. For natural waters, the result is a residue which is then irradiated with thermal neutrons flux for neutral activation analysis (NAA) or dissolved in few milliliters of water or mineral acid for further analysis by atomic spectrometric techniques. The heat of sublimation removed from the sample, together with the low pressure conditions and the insulation provided by vacuum, is sufficient to keep water with salinities below 9 mg g^{-1} frozen during the period even when the apparatus is at room temperature.

Table 8. Non-boiling evaporation procedures for cadmium pre-concentration from natural waters

Brief description	Sample	Pre-concentration time / h	LOD	Concentration factor	Analytical technique	Ref.
Use of Teflon bulbs in presence of HF and HNO_3 (for solubilization of alumino-silicate particles) and clean room conditions.	Snow	9 – 12	-	60 - 100	ETAAS	[451]
Use Teflon sample vessels and conventional rotary evaporator	Rain water	7	-	100	ICP-OES	[452]
Use of Teflon vessels, sophisticated cleaning, ageing and pre-conditioning process	Snow and ice	7 – 8	0.05 pg g^{-1}	-	FAAS	[453]

In lyophilization procedures, the temperature of water sample is first lowered to the freezing point. Then, the sample is inserted into a vacuum chamber of the lyophilizer (freeze dryer) and the system is subjected to vacuum (10 Pa of residual pressure, approximately). The water vapor obtained by sublimation is automatically removed from the vacuum chamber and collected into a condenser attached to the vacuum chamber. The extremely low pressure causes water molecules to be drawn out of the sample.

V.2.2.2. Lyiophilization Methods for Cadmium Pre-Concentration

Simmilarly to evaporation techniques, there are few references about cadmium (and other metals) pre-concentration form natural waters by lyophilization [455-459]. This is because the matrix sample is also pre-concentrated after this procedure, and lyophilization is completed after a long time (about several days). However, possible contamination from laboratory environment and the risk of volatile metals losses are avoided, except for Hg [456].

V.2.3. Precipitation and Co-Precipitation

V.2.3.1. General Aspects

Precipitation and co-precipitation procedures are based on a selective or specific precipitation of the analyte by the addition, under certain conditions (pH, temperature, stirring, etc.), of different precipitating and/or co-precipitating agents. The precipitate obtained is then separated from aqueous medium by filtration, centrifugation or flotation, and it is re-dissolved in a small volume of mineral acids (nitric acid, hydrochloric acid, hydrobromic acid), hydrogen peroxide or organic solvents (acetone, ethanol, 4-methyl-2-pentanone, IBMK) for further analysis by atomic spectrometric techniques. Solid particles could also be introduced as solid or slurry into atomic spectrometric atomizers or irradiated with thermal neutrons flux for further NAA. Great care must be taken to ensure that precipitate is not lost at either filtration or centrifugation steps.

Precipitaion procedures are mainly used for isolating elements which are present at high concentrations in the sample. Because of the relative high concentration of the target metals, precipitation is achieved by directly adding an organic or inorganic precipitant. However, co-precipitation approaches are preferred when the target metals are at low concentrations. In this case, carrier element and an organic co-precipitant are added to sample. Carrier elements are metals which easily form insoluble hydroxides, oxides or which can be easily reduced. Some examples of carrier elements are copper, zinc, iron, lanthanum, indium and magnesium. As organic co-precipitants, reagents such as dithiocarbamate, diethyldithiophosphate, heterocyclic azo-dyes, quinoline, etc., are used. In this case, co-precipitation involves two steps: (1) first, the trace elements react with an organic / inorganic compound and form a solid phase; and (2) the carrier element reacts with the precipitant to form chelates.

A quantitative precipitation of the analyte from solution must be reached. Colloidal suspensions (particle diameters from 10^{-7} to 10^{-4} cm) due to poorly soluble substances must be avoided. The relative super-saturation affects the particle size and it is expressed as $(Q-S)/S$, where Q is the instantaneous concentration of the added species and S is the equilibrium solubility of the compound that precipitates. Particle size seems to be inversely proportional to relative super-saturation. The electric double layer formed during precipitation keeps the colloidal precipitate particles from coming into contact with each other, thus preventing further coagulation. The formation of a colloidal suspension can be minimized or prevented by several methods, for instance, under constant stirring, or by carrying out the precipitation at a temperature close to the boiling point of water. This is because heating increases overall thermal motion, and it affects the mobility of adsorbed ions and colloidal precipitate particles. The final effect of temperature is that there are collisions among particles that increase the particle size as a result of coagulation.

When the flotation approach is used for isolating the precipitate, the formation of a colloidal suspension does not give any problem. In this technique, precipitate isolation occurs in a flotation cell in which a bubbler produces numerous tiny N_2 bubbles (diameter below 0.5 mm). Several surfactants (such as sodium oleate and sodium dodecylsulfate) are added to the sample while it is gently stirred to obtain an efficient mixing and the flotation of the precipitate with the aid of numerous tiny bubbles. This operation usually takes several minutes. The foam containing the precipitate from the flotation cell is sucked off into a sampling tube, and ethanol is usually added to dissolve the foam.

As commented, temperature is a critical parameter. However, precipitation and co-precipitation procedures are affected by other experimental parameters such as the stirring time and the standing time, the pH, the amount of precipitant and/or co-precipitant and the sample volume. In addition, centrifugation rate and time (for those procedures that use a centrifugation step for isolating the precipitate from aqueous samples) and the amount of surfactants, stirring time and bubbles flow (for flotation procedures) are also important factors.

As disadvantages, these procedures can involve sample contamination because they require the addition of large amounts of reagents, which result in high concentrations when comparing to trace elements concentrations (generally at $\mu g\ L^{-1}$ and $ng\ L^{-1}$ levels). In addition, these procedures require a high volume of sample and they are tedious and time consuming. The use of thermoresponsive polymers can reduce or remove some of these drawbacks (inherent to off-line coprecipitation procedures). In addition, it is desirable the implementation of rapid, convenient and sensitive on-line procedures by flow injection systems, thereby reducing sample and reagents consumption while enhancing throughput.

V.2.3.2. Precipitation and Co-Precipitation Methods for Cadmium Pre-Concentration

Cadmium has been pre-concentrated mainly by co-precipitation (Table 9) using both inorganic and organic precipitates. As inorganic precipitants, $In(OH)_3$ [460-464], $Hf(OH)_2$ [465], $Fe(OH)_3$ [466], $Sn(OH)_4$ [467], $La(OH)_3$ [468,469], $Zr(OH)_4$ [470], $Mg(OH)_2$ [471], $Dy(OH)_3$ [472], $Y(OH)_3$ [473], $Cu(OH)_2$ [474] and MnO_2 [475] have been fully applied. In addition, cadmium retention onto the surface of reduced iron and/or palladium (reductive precipitation by sodium tetrahydroborate) [476,477], or reduced palladium by using H_2 gas flow [478] has been also addressed.

Simmilarly, diferent organic compounds precipitates from chelating agents such as phenanthraquinone monophenythiosemicarbazone (PPT) [55], ammonium pyrrolidin dithiocarbamate (APDC) [479-481], 1-(2-pyridylazo)-2-naphthol (PAN) [482], hexacyano-ferrate [483], dibenzylammonium dibenzyldithiocarbamate (DDDE) [484], thionalide [485], 8-hydroxiquinoline (8-HQ) [486,487], ditizone (H_2DZ) [488], 1-(2-thiazolylazo)-2-naphthol (TAN) [481], N-nitroso-phenylhydroxylamine [481], dibenzyldithiocarbamate (DBDTC) [335] and diethyldithiocarbamate (DDC) [489] have been used. In addition, the combination of a chelating agent and an inorganic metal as a carrier element has been extensively used. Therefore, APDC after Co(III) [490,491], Zn(II) [492] or Bi(III) [493] addition as carriers; hexamethylene-ammonium hexamethylene dithiocarbamate (HMA-HMDTC) after adding Fe(II) [494,495] or Co(II) [496]; oxine after adding Mg(II) [497]; DDC after adding Cu(II)

[498], Co(II) [499,500], or Ni(II) [501,502]; 8-HQ after adding La(III) [503]; heptyldithiocarbamate (HpDTC) after adding Pb(II) or Co(III) [504]; and rubeanic acid after adding Cu(II) [505] carrier; have been used for cadmium pre-concentration from natural waters.

V.2.3.3. Flow Injection On-Line Pre-Concentration Systems by Co-Precipitation

As commented, one of the major advances in analytical chemistry in recent decades has been the development of automated systems for analysis, i.e. batch and continuous analysers (mainly FI and CF), which provide analytical data with minimal operator intervention. Thus, several separation methods based on ion exchange, liquid-liquid extraction, chemical vapour generation and precipitation and co-precipitation have been successfully adapted to FI and CF on line pre-concentration thereby resulting a variety of rapid, convenient, sensitive and reliable spectroscopic methods for trace metal analysis.

V.2.3.3.1. Flow Injection Methods with Filtration Stage

FI co-precipitation systems with filtration (in-line filters) often use a stainless steel HPLC screen or a PTFE membrane on a polypropylene support as filtering devices. After complete co-precipitation and retention of the precipitate, a small volume of an acid is used to dissolve and flush the target element to the detector. FI systems based on the use of filters have been found in literature for assessing cadmium (Table 9). In these cases, indium hydroxide [461,464] and pyrrolidinedithiocarbamate (PDC) chelate of cobalt (carrier element) [491], were used as reagents.

The use of filters in these on-line procedures can be a problem if large amounts of precipitate are formed and clogg the filter. In these cases, large volume filters can be used, although large volume filters impair the attainable enrichment factor through dispersion and retarde the dissolution of the precipitate.

V.2.3.3.2. Flow Injection Methods without Filtration Stage

There are two different approaches of FI methods for on-line co-precipitation which avoid the use of in-line filters: the use of knotted reactors (KR) and the filter-less magnetically assisted collection.

KRs are generally made of PTFE, although other materials can be used for on-line precipitation or co-precipitation. They present a 100-150 cm length and 0.5-1.5 mm inner diameter and are made from tubing by tying interlaced knots. The KR produced from PTFE tubes was able to retain metal complexes under appropriate experimental conditions. Two factors are responsible for the retention of the complex molecules on the wall of the KR. The first factor is related to the fact that the molecular species are launched at the inner wall of the KR´s by centrifugal forces generated by secondary flows in three dimensionally disordered systems. The second factor is attributed to the nature of the material that forms the KR and the nature of the complex.

After adsorption of the precipitates from hydrophobic complexes of metal ions on the walls of the KR, an elution/dissolution step using an acid solution is mainly used to sweep the target metal to the detector.

Table 9. Precipitation and co-precipitation procedures for cadmium pre-concentration from natural waters

Brief description	Sample	LOD	Concentration factor	Analytical technique	Ref.
Flotation by addition of 2.0 mL of PPT (5 10^{-4} M), 2.0 mL of acetate buffer (pH 6.5) and 3.0 mL of oleic acid (4 10^{-5} M) to the sample. After heating (40°C for 5 min) the floated layer is measured spectrophoomterically (520 nm)	Tap water, river water, seawater	--	--	UV-visible spectrometry	[55]
Co-precipitation by addition of 8-HQ (100 mg as 8-HQ) and magnesium ions (20 mg) to 100-400 mL of sample at pH 9.0 (ammonia solution). After ageing (1.0 h at 70°C on a hot plate) and filtration on a glass filter, the filter was dried at 110°C for 1 h. A portion of the precipitate was used for direct solid sampling technique.	Drinking water	--	-	EDXRF	[335]
Co-precipitation by addition of 1.0 ml of In (10 mg mL^{-1}) to 1200 mL of sample at pH 9.5 (sodium hydroxide, 0.3 M). Precipitate was floated with the aid of tiny N_2 bubbles and supported by the stable foam layer of sodium oleate (2 mL, 1.0 mg mL^{-1}) and NaDDS (1.0 mL, 4.0 mg mL^{-1}). After floats collected in a small sampling tube, the precipitate was dissolved with 2.0 ml ethanol (99.5% (v/v)) and 2.0 ml of HNO_3 (14.0 M) and dilute to 5 mL.	Seawater	-	240	ICP-OES	[460]
Co-precipitation by addition of 1.0 mL of In (120 mg) to 20 L of sample at pH 9.5 (30 mL of hydrogenocarbonate / carbonate buffer). Precipitate was floated by the addition of ethanol (10 mL) with the aid N_2 bubbles (at 1.0 mL min^{-1}). After foam-precipitate collected in a 300 mL beaker, the precipitate was dissolved with 60 - 70 mL of HCl (8.5 M) and dilute to 10 mL.	Fresh water	-	2000	FAAS	[461]
Co-precipitation by addition of 1.0 mL of In solution (10 mg mL^{-1}) in 0.1M HNO_3 to 100 mL of sample at pH 9.5 (sodium hydroxide solution 3.0 M). After centrifugation (3500 rpm for 10 min), the precipitate is dissolved in 5.0 mL of HBr (0.5 M).	Tap and river water	-	20	ETAAS	[462]
Co-precipitation by addition of 1.0 mL of In solution (0.1 mg mL^{-1}) to 100 mL of sample at pH 9.5 (sodium hydroxide solution 0.1 M). After filtration on 1.0 µm Nuclepore membrane filter, precipitate was ultrasonically dissolved (5.9 min) in 0.5 mL of HNO_3 (1.0 M). The solution was diluted to 1.0 mL.	Fresh and river water	2.0 ng L^{-1}	100	ETAAS	[463]
Continuous flow on line co-precipitation with In and filtered. The filtered precipitate is continuously dissolved in HNO_3 (3.0 M) and	Water	1.43 µg L^{-1}	10	ICP-OES	[464]

transported directly into the nebulizer. The sample throughput is 20 per hour.					
Co-precipitation by addition of 10.0 mg of Hf (as $HfOCl_2$ solution) to 50-400 mL of sample at pH 9.5 (ammonia solution 50%). After filtration through a 0.45μm cellulose membrane by vacuum filtration, the precipitate is dissolved in 1.0 mL of conc. HNO_3 (for ETAAS) or 2.0 mL of conc. HCl (for DPP) and dilute to 25 mL.	River water	-	2-16	ETAAS / DPP	[465]
Co-precipitation by addition of 5.0 mL of Fe(III) (2000 mg L^{-1}) to 1000 mL of sample at pH 9.0 (ammonia solution). After filtration through a 0.45 μm cellulose membrane by vacuum filtration, the precipitate is dissolved in 2.0 mL of HNO_3 conc. and dilute to 50 mL.	Seawater	-	20	ETAAS	[466]
Co-precipitation by addition of 1.0mL of Sn (20 mg mL^{-1}) to 200 mL of sample at pH 5.0-6.0 (sodium hydroxide solution 5.0 and 0.3 M). After centrifugation (1000 rpm for 10 min), the precipitate is dissolved in 0.5 mL of HNO_3 (14.0 M) with the aid of ultrasonic irradiation. The solution was diluted to 10 mL.	River water	0.5 ng[a]	20	ETAAS	[467]
Co-precipitation by addition of 3.0 mL of La (0.1 M) to 1000 mL of sample at pH 9.5 (sodium hydroxide solution). Precipitate was floated with the aid of tiny N_2 bubbles and supported by the stable foam layer of a 1:8 mixed of sodium oleate and NaDDS (each 0.1 % (w/v)). After floats collected in a small sampling bottle, the precipitate was dissolved with HNO_3 and dilute to 25 mL.	Waste water	-	40	FAAS	[468]
Co-precipitation by addition of 1.0 mL of La (10 mg mL^{-1}) to sample at pH pH 9.5 (ammonia solution). After ageing for 30 min and filtration, the precipitate was dissolved with 1.8 mL of HNO_3 (2.0 M).	Lake water	0.2 pg mL^{-1}	25	ICP-MS	[469]
Co-precipitation by addition of 5.0 mL of Zr (10 g L^{-1}) to 100-1000 mL of sample at pH 9.0 (ammonia solution 7.0 M). The precipitate was aged (1 h at 90°C) and then collected on a membrane filter (8.0 μm). The precipitate is direct analysed by AAS.	Natural water	0.38 ng L^{-1}	-	ETAAS	[470]
Co-precipitation by addition of 20-40 μL of NH_3 into 1.3 mL of sample. Under these conditions Mg present in the sample was precipitated as $Mg(OH)_2$. After centrifugation, the precipitate is dissolved in 100 μl of HNO_3 (5.0 %).	Seawater	5.0 pM	13	ICP-MS	[471]

Table 9. Continued

Brief description	Sample	LOD	Concentration factor	Analytical technique	Ref.
Co-precipitation by addition of 1.0 mg of Dy(III) to 100 mL of sample at pH 11 (ammonia solution). After centrifugation (3500 rpm for 10 min), the precipitate was dissolved with 1.0 mL of conc. HNO_3 and diluted to 2 mL.	Natural water	14.1 $\mu g\ L^{-1}$	50	FAAS	[472]
Co-precipitation by addition of 1.0 mL of Y (1.0 g L^{-1}) after addition of 3.7 mL of sodium hydroxide solution (3.0 M) to 100 mL of sample. After ageing (10 min) and centrifugation (3500 rpm for 5.0 min), the precipitate was dissolved with 5.0 mL of HNO_3 (1.0 M).	Seawater	0.06 $\mu g\ L^{-1}$	20	ICP-OES	[473]
Co-precipitation by addition of 1.0 mL of copper(II) solution (0.1 % (w/v)) to 200 mL of sample at pH 9.0 (sodium hydroxide). After centrifugation (3500 rpm for 10 min), the precipitate was dissolved with 0.5 mL of conc. HNO_3 and diluted to 2 mL.	Tap water	3.0 $\mu g\ L^{-1}$	100	FAAS	[474]
Co-precipitation by addition of 10.0 mL of $KMnO_4$ (1 % (w/v)) and 20 mL of 0.1% D-glucose to 2000 mL of sample at pH 3.5-4.0. After filtration on filter paper, the precipitate (MnO_2) is dissolved in 2.0 mL of HCl (50 % (v/v)) and 20 mL of H_2O_2 (30 % (v/v)). The solution was diluted to 25 mL.	Ground water	1.0 $\mu g\ L^{-1}$	80	ICP-OES / FAAS	[475]
Reductive precipitation by addition of 5.0 mL of sodium tetrahydroborate (6.0%), and Fe^{3+} and Pd^{2+} (each a 2.0 mg L^{-1}) to 900 mL of sample at pH 8.0-9.0 (ammonium solution). After ageing (15 h) and filtering through a 0.45 µm cellulose nitrate membrane, the precipitate is dissolved with HNO_3/HCl conc. (3.0 mL of each) and dilute to 25 mL.	Seawater	1.0 ng L^{-1}	36	ETAAS	[476]
Flow injection on-line co-precipitation with sodium tetrahydroborate (2% at 0.5 ml min^{-1}) and Fe^{3+} and Pd^{2+} (30 $\mu g\ mL^{-1}$) in ammonium hydroxide (0.09 M at 1.7 ml min^{-1}) medium. After magnetic collection, the precipitate was dissolved in HCl (2.0 M)/HNO_3 (2.0 M) at 0.25 mL min^{-1} and transported direct to the graphite furnace.	Seawater	2.8 pg[a]	400	ETAAS	[477]
Co-precipitation by addition of 1.0 mL of Pd (10 $\mu g\ mL^{-1}$), which are subsequently reduced by introduction of H_2 gas (for 5min), to 100 mL of sample at pH 2.0 (nitric acid solution). After filtration, the precipitate is dissolved with HNO_3/HCl (1:1). The solution was diluted to 2.0 mL.	Seawater	-	50 - 500	ICP-OES	[478]

Co-precipitation by addition of 2.0 mL of APDC (2.0 % (w/v)) to 150 mL of sample. The precipitate is filtered off, and dissolved in HNO_3, and the solution heated to near dryness and the residue is dissolved to 2 mL in 1% HNO_3 - 1% $HClO_4$	Seawater	-	75	ICP-OES	[479]
Co-precipitation by addition of 2.0 mL of TAN, APDC and ammonium salt of N-nitroso-phenylhydroxylamine (8.0 % (w/v)) to 500-1000 mL of sample at pH 6.0-7.2 (ammonium solution). After filtration the precipitate was irradiated.	Tap and seawater	-	~10^4	NAA	[481]
Flow injection on-line co-precipitation with PAN. After precipitate collection in a knotted reactor, the precipitate was dissolved/eluted in hydrochloric acid and transported directly into the nebulizer. Sampling frequency of 48 h^{-1}.	Tap water	0.10 $\mu g\ L^{-1}$	18	FAAS	[482]
Co-precipitation by addition of 10 mL of DDDE (1.0 % (w/v)) to sample at pH 5.0 (ammonium hydroxide solution), after reduction step by heating (to boiling) with 25 mL of conc. HCl, 5.0 mL of KI (4 % (w/v)), 5.0 mL of ascorbic acid (5 % (w/v)) and 2.0 mL of sodium thiosulfate (0.7 % (w/v)). Direct analysis, after filtration.	Tap water	-	-	EDXRF	[484]
Co-precipitation by addition of 5.0 mL of 8-HQ (2.0 %) to 100-500 mL of sample at pH 7.0-8.5 (ammonia solution). After heating for 3 h at 70°C and several hours of ageing, the precipitate was collected on a glass filter. A portion (0.1 to 1 mg) is used for direct solid sampling technique.	Seawater	1.4 ng L^{-1}	-	ETAAS	[486]
Precipitation by addition of 2.0 mL of DBDTC (1.0 %(m/v) in methanolic solution) to 200 mL of sample at pH 4.0 (0.1 M potassium hydrogen phthalate buffer). After ageing (15 min at room temperature) and vacuum filtration through a 25 mm diameter 0.45 µm pore size cellulose nitrate filter paper, the filter was air dried and mouned for XRF analysis.	Fresh water	0.6 ng L^{-1}	-	ETAAS	[487]
Flow injection on-line co-precipitation with DDC (0.025 % (m/v) at 0.9 mL min^{-1}). After precipitate collection in a KR, the precipitate was eluted in ethanol (6.0 mL min^{-1}) and transported directly into the nebulizer. Sampling frequency of 30 h^{-1}.	Rain and seawater	3.44 $\mu g\ L^{-1}$	33	FAAS	[489]
Co-precipitation by addition of 2.5 mL of Co(III) (0.02 % (w/v) and 5 mL of APDC (2 % (w/v)) to 1000 mL of sample at pH 4.0 (ammonium acetate 34 % (w/v) solution). After filtration through a fine-porosity funnel under vacuum, the precipitate is dissolved in 5.0 mL of HNO_3 and dilute to 25 mL.	Tap water	1.0 $\mu g\ L^{-1}$	40	FAAS	[490]

Table 9. Continued

Brief description	Sample	LOD	Concentration factor	Analytical technique	Ref.
Flow injection on-line co-precipitation with Co(III) (20 μg mL^{-1} at 3.3 mL min^{-1}) and APDC (2.0 % (m/v) at 0.4 mL min^{-1}). After precipitate collection in a PTFE membrane on a polypropylene support filtering device combined with a 1.5 m reaction coil, the precipitate was dissolved in (1:1) conc. HNO_3 and conc. H_2O_2 and transported directly into the nebulizer. The sample throughput is 20 h^{-1}.	Rain water	0.7 μg L^{-1}	28	ICP-OES	[491]
Co-precipitation by addition of APDC and Zn to sample. After filtration, the precipitate is dissolved in HCl (for ASV) or direct placed into the atomizer (for ETAAS).	Seawater	-	-	ETAAS / ASV	[492]
Co-precipitation by addition of APDC and Bi(III) to the sample at pH 4.0. After filtration, the precipitate was irradiated.	Tap, river and lake water	-	-	NAA	[493]
Co-precipitation by addition of 1.5 mg of Co(II) and 2.0 mL of HMA-HMDTC (0.01 M) to 1000 mL of sample at pH 6.0 (potassium hydroxide solution). Precipitate was floated with the aid of tiny air bubbles and supported by the stable foam layer of NaDDS (1.0 mL, 0.5%). After floats collected, the precipitate was dissolved with 5.0 mL HNO_3 (65%) and diluted to 25 mL with HNO_3 (4.0 M)	Natural water	3.0 ng L^{-1}	40	ETAAS	[496]
Co-precipitation by addition of 2.5 mL of oxine solution (20 mg mL^{-1}) and 0.5 mL of magnesium solution (10 mg mL^{-1}) to 200 mL of sample at pH 9.0-9.5 (ammonia solution, 1.0 and 5.0 M). After filtration on a 1.0 μm Nuclepore membrane filter, the precipitate is dissolved in 5.0 mL of HNO_3 (1.0 M). The solution was diluted to 10 mL.	River and seawater	1.0 ng L^{-1}	20	ETAAS	[497]
Flow injection on-line co-precipitation with Cu(II) (120 mg) and DDC (2.1 % (w/v)) in HCl (0.5 M) medium. After precipitate collection in a KR, the precipitate was dissolved in IBMK and transported directly into the nebulizer.	Tap, river and waste water	0.23 μg L^{-1}	26	FAAS	[498]
Co-precipitation by addition of 0.6 mL of cobalt solution (1.0 mg mL^{-1}) and 4.0 mL of Na-DDC (5.0 % (m/v)) to 200 mL of sample at pH 5.7 (2.0 mL of ammonium acetate buffer). After filtration (0.45 μm cellulose nitrate membrane and vacuum filtration), the membrane with the precipitate was digested with 0.5 mL of conc. HNO_3 and the residue diluted with 2 mL.	Seawater BCR-403	-	100	ETAAS	[499]
Co-precipitation by addition of 600 μg of	Seawater	4.0 μg	200	FAAS	[500]

Co(II) and 200 mg of Na-DDC to 400 mL of sample at pH 6.0 (acetate/acetic acid buffer). After filtration (cellulose nitrate membrane filter and vacuum), the precipitate and the membrane are dissolved in 2.0 mL of conc. HNO_3, the solution was evaporated near to dryness and the residue was dilute to 2 mL with HNO_3 (1.0 M).		L^{-1}			
Flow injection on-line co-precipitation with DDC-Ni(II) in HNO_3 medium. After precipitate collection in a KR, the precipitate was dissolved in IBMK and transported directly into the nebulizer. Sampling frequency of 60 h^{-1}.	River water	0.2 µg L^{-1}	65	FAAS	[501]
Co-precipitation by addition of Ni (3.0 mg) and 2.0 mL of DDC (2.0 % (m/v)) to 500 mL of sample at pH 9.0 (ammonia solution). After filtration on a sintered-glass filter, the precipitate was dissolved with HNO_3/acetone (1:1) and dilute to 10 mL.	River and seawater	1.2 pg mL^{-1}	50	ETAAS	[502]
Precipitation by addition of 8-HQ and La^{3+} to 1000 mL of sample at pH 9.0 (ammonia solution, 2.0 M). Precipitate was floated with the aid of tiny N_2 bubbles and supported by the stable foam layer of sodium lauryl sulfate. After floats collected on a fritted glass filter with a suction, the precipitate was dissolved with 5.0 mL ethanol and 1.5 mL of conc. HNO_3 and dilute to 25 mL.	Water	-	40	FAAS	[503]
Co-precipitation by addition of 1.0 mg of copper (II) (1 ml of 1000 mg L^{-1}) and 2.5 mg of rubeanic acid (0.5 ml 0.1 % (w/v)) to 50 mL of sample at pH 7.0 (phosphate buffer). After centrifugation (2500 rpm for 20 min), the precipitate was dissolved with 0.5 mL of HNO_3 conc. and diluted to 2 mL of Milli-Q water.	Tap, bottle mineral and lake water	0.58 µg L^{-1}	25	FAAS	[505]

[a] Absolute detection limit

APDC, ammonium pyrrolidin dithiocarbamate; ASV, anodic stripping voltammetry; DBDTC; dibenzyldithiocarbamate; DDC, diethyldithiocarbamate; DDDE, dibenzylammonium dibenzyldithiocarbamate; DPP, differential pulse polarography; EDXRF, energy dispersive x-ray fluorescence; ETAAS, electrothermal atomic absorption spectrometry; FAAS, flame atomic absorption spectrometry; HMA-HMDTC, hexamethylene-ammonium hexamethylene dithiocarbamate; IBMK, 4-methyl-2-pentanone; ICP-OES, inductively coupled plasma-optical emission spectrometry; KR, knotted reactor; NAA, neutron activation analysis; NaDDS, sodium dodecylsulfate; PAN, 1-(2-pyridylazo)-2-naphthol; PQPD, 5,8-polyquinolyl polydisulfide; PPT, phenanthraquinone monophenythiosemicarbazone; PTFE, poly tetrafluoroethylene; TAN, 1-(2-thiazolylazo)-2-naphthol; UV, ultraviolet; 8-HQ, 8-hydroxiquinoline.

KRs offer several advantages for collecting the precipitate: (a) relatively high capacity as consequence of the large internal surface area; (b) low back pressure (even for high flow rates) due to the removal of any filter; (c) contamination are avoided due to the inert tube material; (d) the reactor is easy and cheap to make, and, (c) the lifetime is unlimited and the reactor requires no maintenance [61].

After the first application of an on-line co-precipitation system with KR by Fang et al. [506] for lead determination, several methods have been developed for the pre-concentration of trace elements from waters. For cadmium (Table 9) FI co-precipitation systems using KR were based on the use of HMA-HMDTC and Fe (carrier element) [494], DDC and Cu (carrier element) [498], (DDC)-Ni [501], DDC [489], and 1-(2-pyridylazo)-2-naphthol (PAN) [482].

The filter-less magnetically assisted collection system consists of a 2.0 m length of Teflon, microbore tygon or silicone tubing (1.27 mm i.d.), where the coprecipitate analytes (by tetrahydroborate reductive precipitation of added iron and palladium in alkaline medium) are collected. This tube is wrapped around a (5.7 x 1.4 x 1.5 cm) rare earth cobalt ceramic magnet of 8.58 KG field strength block. After collection, the precipitate is dissolved in a 20 mL volume of mixed acid and directly transported to measure instrument. This approach has been applied for the determination of several trace metals, such as cadmium, in seawater by FI on line ICP-OES and ETAAS [477].

As commented above, the heart of the manifold used for on-line co-precipitation is the KR or the filtering device in which the precipitate is collected. Other components of the manifold are the peristaltic pumps and the injection valve (these components have been described in section IV.11).

V.2.3.4. Polymer-Mediated Extraction

Recently, an appealing alternative to classical co-precipitation methods consists of using polymer-mediated extractions of hydrophobic metal chelates for pre-concentration. This approach is based on the thermoresponsive precipitation phenomenon of several water soluble polymers.

V.2.3.4.1. General Aspects

Thermoresponsive polymers are water-soluble at room temperature, but become sparingly soluble above their critical solution temperatures (ca. 32°C) to form gum-like precipitates (polymer phase) which can be separated from the bulk aqueous solution. During the precipitation, any hydrophobic substances (derivatized phosphate as molybdophosphate-melachite green, fatty acids, phospholipids, polycyclic aromatic hydrocarbons, alkylphenols, chlorobenzenes, chlorophenols, pesticides and polyethoxylated derivatives) in the aqueous solution can be solubilized and incorporated into the polymer assembly [507-511]. However, hydrophilic components remained in the bulk aqueous solution. In these methods, sample and polymer solutions were mixed and the analyte-bound polymer was precipitated in a solvent, e. g. acetone. Because of the analyte and the polymer are thoroughly mixed and interacted in the solution phase, the analyte is completely bound to the polymer. As consequence, the polymer phase is highly condensed, and the concentrations of analytes are easily increased up to 100-fold.

This phenomenon can be applied for the pre-concentration of heavy metals only after converting metal ions into hydrophobic metal chelates or ion-pairs of charge chelates.

Thermoresponsive polymers such as poly (N-isopropylacrylamide) polymer [PNIPAAm] and poly(vinyl methyl ether) polymer [PVME] have been extensively used for metals pre-concentration. On the other hand, efforts on the introduction of chelating functionalities into the thermoresponsive polymers have been performed for achieving a more efficient collection of heavy metals ions from water matrices. Poly(N-isopropylacrylamide-co-4-vinylpyridine) polymer, [PNIPAAm-co-methacrylhydroxamic acid] copolymer and [PNIPAAm-Im], [PNIPAAm-COOH] and [PNIPAAm-IDA] (a [PNIPAAm] polymer with imidazole, carboxylic acid and iminodiacetic acid groups, respectively), have been also applied.

An important advantage of the the polymer mediated extraction is the ability to concentrate a variety of hydrophobic analytes (with high pre-concentration factors) by very simple and fast procedures and without interferences because the analytes are determined in a simple matrix (dissolved polymer). The polymer precipitates could be condensed into a small pastelike solid, which can be easily removed from the solution by just heating and shaking the solution for few seconds. Thus, instruments, such as centrifuges, are not necessary for the phase separation. The polymer phase was re-dissolved at room temperature with a small amount of water or organic solvent and all the analyte bound to polymer is completely transfered into the solution phase. Because of the ease of pyrolysis of the polymer, this phase can be sampled in graphite furnaces for ETAAS determinations.

Other advantages of this methodology are related to safety because it requires only very small amount of relatively non-flammable, non-volatile, and non-toxic reagents, and it avoids the use large volumes of organic solvents. The non-toxic properties of PNIPAAm have been clearly observed and have a negligible effect on the user´s health and the environment.

V.2.3.4.2. Polymer-Mediated Extraction for Cadmium Pre-Concentration

As shown in Table 10, cadmium, together with copper and nickel, as a pyrrolidindithiocarbamate chelate, was co-precipitated with [PVME] [512], while using decyltrimethylammonium chloride (DTMAC), cadmium was pre-concentrated with[PNIPAAm-Im], [PNIPAAm–COOH] and [PNIPAAm–IDA] [513]. Other water soluble polymers, but without thermoresponsive properties, have been also proposed for cadmium pre-concentration. This is the case of polyacrylic acid polymer [514] which has been applied for the simultaneous pre-concentration of cadmium, copper and zinc from waste water.

V.2.4. Electrochemical Deposition

As commented, electrochemical pre-concentration is an essential part of various electrochemical methods for trace element determination such as ASV. This modality utilizes the electrolysis laws in which cationic species are deposited on the electrodes surface into an electrolytic cell. The application of a controlled potential allows the selective separation and pre-concentration of trace metals.

V.2.4.1. General Aspects

Electrolytic cells are constructed in stationary [515,524] or flow-through [525,526] arrangements. The electrolysis can be accelerated by rigorous mixing of the solution and by maximizing the ratio of electrode surface to sample volume. This can be achieved in a flow system by passing the sample solution through a porous working electrode made from crushed reticulated vitreous carbon or glassy carbon [525]. However, large volumes of the eluent are required for washing out metals in these cells. Therefore, the cells with porous electrodes are often used for on-line sampling of solutions in atomic absorption spectrometry. In this case, a flow-through cell with a thin space between the electrodes is advantageous when the distance of electrodes determines the volume of the cell. Thus, electrochemical micro-cell [525,526] (3 µL volume), with a thin stream of sample solution flowing between two electrodes, is used.

Table 10. Co-precipitation procedures by polymer-mediated extraction for cadmium pre-concentration from natural waters

Brief description	Sample	Concentration factor	Analytical technique	Ref.
Water sample (50-100 mL) was treated with NH_3 (4.0 M) to pH 2.0-3.0. The sample was treated with 1.0mL of APDC (1.0 mg mL^{-1}) and 5.0 mL of a PVME (6.0 % (v/v)) with stirring at 50°C for 10 min. Aggregates were beaker with 10 mL H_2O and dissolved in 1.0 mL acetone followed by 3.0 mL of HNO_3 (0.1 M). After combining the washings, the solution was diluted to 5.0 mL with HNO_3 (0.1 M).	River and seawater	10-20	ETAAS	[512]
Water sample (20 mL) was treated with 0.1 g of PNIPAAm-COOH or PNIPAAm-IDA, 0.017 g of sodium nitrate and 5.0 mL of DTMAC (0.2 M) with stirring at 50°C. Aggregates were beaker with N,N-dimethylformamide and the solution was diluted to 1.0 mL.	River and seawater	20	ETAAS	[513]
Co-precipitation by addition of 1.0 mL of polyacrylic acid (0.56 g) y 50 mL of acetone. The pH of the solution (50 mL) was adjusted at pH 3.0 by using hydrochloric acid or sodium hydroxide (0.1 M) solution. After decantation, the precipitate was dissolved with 5.0 mL of water.	Waste water	10	FAAS	[514]

APDC, Ammonium pyrrolidin dithiocarbamate; DTMAC, Decyltrimethylammonium chloride; [PNIPAAm-COOH], poly (N-isopropylacrylamide) polymer with carboxylic acid groups; [PNIPAAm-IDA], poly (N-isopropylacrylamide) polymer with iminodiacetic acid groups; PVME, Poly(vinyl methyl ether) polymer

In general, electrolytic cells are composed of three electrodes. The reference electrode, which is usually a saturated calomel electrode (SCE) and saturated Ag/AgCl, the working electrode, and the auxiliary electrode. Analyte deposition after reduction ocuurrs at the working electrode (cathode), which can be made of different materials, such as high melting metals wires (platinum, iridium or tungsten) [515,516,518,522,524], graphite rods, probes, disk electrodes [520,521,523,526] or glassy carbon [519,525]. Conventional HMDE can be also used as a working electrode [517]. Finally, the auxiliary electrode is commonly the platinum electrode.

The analytical procedure consists of three steps:

1) Electrochemical deposition of the analyte at the working electrode by keeping it at a suitable potential (~ -1.0 V) for a fixed time. During this step, the metal ions are separated from the sample matrix and are deposited on the electrode.
2) Dissolution of the analyte from the electrode by setting the potential of the cathode at 0 V. This second step (dissolution of the analyte from the electrode) can be avoided when using atomic spectroscopy techniques due to the fact the working electrode could be inserted into the flame (FAAS determination) or into the electrothermal furnace (ETAAS determination).
3) Analyte measurement by electroanalytical techniques such as ASV or by atomic spectrometry techniques. When using spectrometric techniques, atomization is carried out by an electrical heating of a W-wire in a quartz tube or after inserting of

the electrode with deposited analyte into the graphite atomizer with or without touching the wall of the atomizer. There are commercially available or specially constructed graphite electrodes, e.g graphite disk electrodes, graphite ridge probes, graphite rods or tungsten wires for sampling introduction in ETAAS. Other times, the electrode can also form a part of the atomizer as a graphite platform or it can be the graphite tube itself. In this case, the electrode with the deposite (graphite tube) is fixed between the contacts of the graphite furnace. It is also possible the dissolution of metal deposits and the subsequent injection into the atomizer as a liquid solution. When an electroanalytical technique is used for final determination, the deposition potential of the metal has a significant influence on the signal to background ratio. In addition, the presence of oxygen, surface active compounds and inert salts in the sample can give several interferences.

Effective electrochemical pre-concentration has some important requirements:

Electrodeposition of most of heavy metals competes with the reduction of the hydrogen ions. To avoid extensive hydrogen ion reduction, an electrode with high hydrogen overvoltage is required. Mercury is the best material for this purpose. Other materials with significant high hydrogen overvoltage are glassy-carbon and glassy-carbon covered by a mercury film. Both materials are useful, but in the later case the mercury film deposited on the solid electrode provides a larger hydrogen overvoltage than that on uncovered glassy-carbon. Once the metal is deposited, great care must be taken of the mercury film on the glassy carbon surface and it must be kept all the time at a secure negative potential, with no guarantee that it will keep constant activity.

Reduction of the hydrogen ions, together with metal ions, during the pre-concentration process means that the current efficiency of the pre-concentration is small, but analyte deposition directly on the graphite electrode is possible. This is all that is needed for the subsequent determination by atomic spectroscopy techniques. However the hydrogen overvoltage on pyrolytic graphite or on electrographite is smaller than on glassy-carbon and the performance of these electrodes may significantly change due to surface corrosion during atomization cycles. As the roughness of the graphite surface increases, the electrochemical deposition of more electronegative ions such as Cd^{2+} becomes more difficult.

As commented above, an important factor in electrochemical pre-concentration is a ratio of the working electrode surface area to the volume of the electrolysed sample. To achieve high electrochemical recoveries while depositing from flowing solution, the mass transfer of the analyte species to the electrode surface should be as great as possible. To obtain this condition, an electrode with the maximal area must interact with the minimal volume of sample solution. Bulk pre-concentration usually requires few hours to transfer most of the metals from the solution to the electrodes. A more effective approach is the use of flow-through cells with a thin channel between the electrodes or large surface area working electrodes, where the solution circulates over the electrode.

The pH of sample highly affects the procedure. The optimum pH conditions depend on the particular metal and the optimal deposition potential for a particular metal. In general, the amount of hydrogen released on the cathode is increased at more negative potentials, so that the optimal pH is usually shifted to higher pH values. The deposition potential is also important because precision is deteriorated when using more negative potentials. Other variables affecting the procedure are the electrolyte concentration (the amount of deposited metals increased up to with concentration of electrolyte, for higher concentration the signal decreased slightly); the deposition time (the dependence of the signal on the deposition time is lineal for deposition times lower than 15 min), the optimum distance between electrodes and the electrodes surface area.

V.2.4.2. Electrochemical Deposition for Cadmium Pre-Concentration

Electrodeposition is a very advantageous technique for trace metal determination in sea water, where alkali metal halides form an ideal medium for the electrodeposition. Table 11 summarizes methods based on electrodeposition as pre-concentration methods for cadmium.

Table 11. Electrochemical deposition procedures for cadmium pre-concentration from natural waters

Brief description	Sample	LOD	Concentration factor	Analytical technique	Ref.
Electrochemical pre-concentration was performed at a controlled potential (-1000mV vs Ag/AgCl for 2.0 min and 0.1 M KCl as supporting electrolyte) on to a W wire filament from 20 mL of sample. After deposition, the whole system was rinsed and the potential of the cathode was set at 4000 mV for 5 s to analyte dissolution.	Seawater	0.1 µg L^{-1}	-	ETAAS	[515] [516]
Electrochemical pre-concentration was performed by electrolysis on a hanging mercury drop electrode (-1000 mV for 11 min) from 50 mL of sample. After the electrolysis, the drop mercury (1.8 mm^2 of surface) is transferred to a graphite boat and removed by evaporation, and cadmium is determined by atomization.	Seawater	-	-	ETAAS	[517]
Electrochemical pre-concentration was performed at a controlled potential (-1000 mV vs Ag/AgCl for 2.0-5.0 min) on to a Pt wire filament. After the electrolysis (25 mL of sample at pH 5.0 by using a acetate buffer), the Pt spiral filament was directly heated in a flame or placed into a quartz tube to increase the sensitivity of the method.	Seawater	0.1 µg L^{-1}	-	FAAS	[518]
Electrochemical pre-concentration was performed at a controlled potential (-800 mV SCE for 20 min) on to a graphite rod. After the electrolysis (pH 7.0, potassium chloride (0.1 M) as supporting electrolyte), the graphite rod was rinsed and inserted in the graphite furnace.	Seawater, mineral water, spring water and river water	4.0 ng L^{-1}	-	ETAAS	[520]
Electrochemical pre-concentration was performed at a controlled potential (-1000 mV vs Ag/AgCl for 2.0-6.0 min) on a graphite disk electrode. After the electrolysis (70 mL of sample at pH 4.5-5.0 with $NaNO_3$	Seawater	6.4 ng L^{-1}	-	ETAAS	[521]

(0.01 M) as supporting electrolyte), the electrode was placed by automatic sampler into a graphite tube atomizer for analysis by ETAAS.					
Electrochemical pre-concentration was performed at a controlled potential (-1000 mV (vs SCE for 2.0 min) on a three coil tungsten wire (cathode). After the electrolysis (20 mL of sample and sulphuric acid (0.1 M) as supporting electrolyte), the wire was rinsed and placed in a graphite furnace.	River water	0.01 ng mL^{-1}	-	ETAAS	[522]
Electrochemical pre-concentration was performed at a controlled potential (-1000 mV (vs Ag/AgCl)) on the graphite ridge probe from 50 mL of sample deposited in a 100 mL polyethylene cell magnetically stirred. After the electrolysis (pH 4.75 by using an acetate buffer and sodium nitrate (0.01 M) as supporting electrolyte), the electrode-probe was rinsed and inserted in the graphite furnace.	Seawater	ng L^{-1}	-	ETAAS	[523]
Electrochemical pre-concentration was performed Pt (plate of helix), Ti, Ta, W, Mo, Ni, glassy-carbon and spectrographic graphite electrodes. After electrolysis (performed at room temperature for 15 min. in 20 mL of a supporting solution on vigorous stirring at constant potential of 1000 mV), electrodes were manually inserted into the burner flame.	River and seawater	4.0 ng L^{-1}	-	FAAS	[524]
Flow-injection on line electrochemical deposition on the glassy-carbon electrode in a flow-through micro-cell of volume 3 µL under galvanostatic control (-1.0 V vs sat. Ag/AgCl). After deposition, the whole system was rinsed and the potential of the cathode was set at 0 V to analyte dissolution, which is transported to the flame by HNO_3 (0.1 M) flow.	Fresh water	80 ng L^{-1}	26	FAAS	[525]
Flow injection on line elcectrochemical deposition on the graphite electrode under conditions of controlled current (1.0 mA during 10 min) in a flow-through mode (0.35 mL min^{-1}) in a microcell of 2.6µL volume. After electrolysis (pH 4.5 by using a sodium acetate buffer and ammonium chloride (0.1 M) as supporting electrolyte), deposited metal was dissolved in 40 µL of HNO_3 (0.1 M) and the whole volume was direct injected into the atomizer.	Tap water, surface water and seawater	25 ng L^{-1}	-	ETAAS	[526]

SCE, saturated calomel electrode

V.2.5. Liquid-Liquid Extraction

Liquid-liquid extraction (LLE) is the oldest and the most widely used sample pre-treatment method for liquid samples. Extraction of low concentrations of inorganic and organic analytes from complex aqueous samples into organic solvents is the aim of this approach.

V.2.5.1. General Aspects

LLE procedures are based on the relative solubility of the target compound in two immiscible phases, which leads to an improvement on the selectivity by isolating the analyte and an increase on sensitivity by transferring the analyte to a small solvent volume. The isolation and pre-concentration is attained at the same time using the solvent more suitable for analytical signal acquisition. An ideal LLE procedure must be guarante a quantitative isolation of the analyte fom the aqueous matrix sample moreover interference species remain in the aqueous phase.

The pre-concentration of trace metals by LLE procedures is commonly achieved by the addition of a chelating agent to a buffered aqueous sample, and by extracting the formed metal complexes into an organic phase. Chelating agents are organic compounds with functional groups containing nitrogen (as amines, amides or nitriles), oxygen (carboxylic, hydroxyl or ether functional groups) or sulphur (thiols, thiocarbamides) atoms, which are capable of chelating trace metals. Therefore, the most suitable compounds with the highest affinity towards selected metals can be chosen according to the functional groups present. Chelating agents used for LLE procedures should offer two characteristics. Firstly, these substances must gurantee the complexition/extraction of most of the target metals; and secondly, the extraction of these metal-complexes must be equally well over the some fairly wide range of pH so that there would be some allowable error in adjusting the pH of the solution. This last task is not always possible because most of the chelating agents show strong pH dependence. Thus, a pH adjustment to within one pH unit or less is required which can result in serious errors in routine applications.

In order to fix the pH, the use of buffer solutions is nedeed in routine extraction work. The buffer solution must offer the following characteristics: (1) it must be a stable solution, (2) it must have a high buffering capacity, and (3) its components must not participate in any reaction. Citrate- and acetate- based buffer solutions are the most useful for fixing the pH before LLE, although acetate buffer is unfavourable for lead and silver extraction due to acetate could combine with these metals to form stable Pb- and Ag-acetates which are not readily extracted.

In addition, the solvent used to extract the metal complex must also show several desirable characteristics:

1) Solvents must extract the desired metal chelate by using one or few extraction steps, i.e. two extraction steps are generally required for quantitative recovery of most common trace metal when Freon-TF is used.
2) Solvents must be immiscible with the aqueous solution, i.e. the water solubility of 4-methyl-2-pentanone (IBMK) is relatively high, which originates salt carry-over and matrix interferences when this organic phase is used.

3) Solvents must not form emulsions because emulsion formation hinders phases separation.
4) Solvents must have good burning characteristics (some common analytical techniques for determination are FAAS, ETAAS and ICP-OES).
5) Solvents must enhance rather than suppress the atomic signal sensitivity as compared to metal sensitivity in water.
6) Finaly, the extracted metal complexes must be stable in the organic solvent.

Table 12 summarizes the physical properties of organic solvents which are commonly used in LLE of trace elements [527]. Most of these solvents offer several drawbacks, e.g. the direct analysis of IBMK extracts is limited by the poor stability of dithiocarbamate complexes [528] (dithiocarbamate complex stability is greatest in chloroform, follows IBMK and Freon-TF). In addition, IBMK solubility in water is higher than those offered by cyclohexane or tetrachloromethane and dual extractions were required for the quantitative recovery of most common trace metals when using Freon TF.

As it was commented, the efficiency of LLE process depends on the affinity of analytes with the extracting solvent, the pH of the aqueous phase and the stability of the formed chelate. Thus, several variables such as the chelating agent type, the concentration of chelating agent, the organic phase nature, the organic phase/aqueous phase ratio, the pH of aqueous phase (selection of the buffer solution), the acid and/or metal concentration in the back-extraction stage, the number of sucesive extractions, and the shaking and standing times, must be carefully selected.

Table 12. Physical properties of the organic solvents used for cadmium extraction by LLE and LLME

Solvent	Boiling point / °C	Density / g cm^{-3}	Water solubility / mg L^{-1}	Surface tension / dyn cm^{-1}	Viscosity / cP	Vapour pressure / Torr
Benzene	80.1	0.87	1791	28.2	0.60	95.2
Chloroform	61.2	1.48	8500	26.53	0.54	194.8
Cyclohexane	80.7	0.78	55	24.65	0.90	97.8
Nitrobenzene	210.8	1.20	1900 (20°C)	42.17	1.62 (30°C)	0.28
Tetrachlorometane	76.6	1.58	770	26.13	0.90	115.2
Toluene	110.6	0.86	515	27.92	0.55	28.5
Xylene	138.4	0.86	156	28.48	0.64	8.7
1-octanol	195.2	0.82	538	26.92	7.30	0.08
1,1,2 trichloro-2,2,1-trifluoroethane (Freon-TF)	47.6	1.57	200	23.9	0.68	270
2,6-dimethyl-heptan-4-one (DIBK)	166.0	0.81	500	20.7	0.89	1.7
4-Methyl-2-pentanone (IBMK)	117.4	0.80	17000	23.64 (20°C)	0.55	18.8

Although LLE procedures have proven to be reliable and efficient techniques, these methodologies are time, reagent and labor-intensive consuming, which has the tendency for emulsion formation and poor potential for automation. As LLE is a multi-step proces, it often results in analyte losses. In addition, LLE requires large amount of high purity solvents, which are expensive and toxic reagents, and the procedure generates hazardous laboratory wastes, which disposal or treatment can be harmful to the environment and to the public health. Moreover, a derivatization step previous to extraction into the organic phase is necesary for metals preconcentration; and a back-extraction of the pre-concentrate chelate into the aqueous phase is generally required for detection and quantification by several spectrometric techniques.

V.2.5.2. Liquid-Liquid Extraction for Cadmium Pre-Concentration

Analytical procedures for cadmium separation and pre-concentration by LLE are listed and commented in Table 13. Sodium diethyldithiocarbamate (NaDDC) [529-532], ammonium pyrrolidin dithiocarbamate (APDC) [455,533,534], APDC and NaDDC mixtures [499,535-543] are commonly used as chelating agents for cadmium (and for other trace metals) extraction because the low pH dependence. 8-hydroxiquinoline (8-HQ) [106] and H_2DZ [455,544] can also be used. However, APDC is the most popular because its solutions are stable in acidic conditions and it operates in a broad pH range without any decomposition. Other reported chelating agents for Cd isolation by LLE are 1-(2'-pyridylazo) naphthol (PAN) [545], capriquat [546], thiothenoyltrifluoroacetone–TOPO [547], 1,1,1-trifluoro-4-mercapto-4-(2-thienyl)but-3-en-2-one–TOPO [548] and pelargonic acid [549].

The buffer solutions used in these procedures are mainly based on citrate [532,535] and acetate [499,531,533,541,542,544,548]. Solvents such as IBMK [532,533,535,538,539], 2,6-dimethyl-heptan-4-one (DIBK) [537,541,542], trichlorofluoroethane (Freon TF) [499,529,530,534,536,539], chloroform [455, 539,540,543,545], cyclohexene [547,548] or tetrachlomethane [531,544] are commonly used. All these solvents offer a relatively low solubility in water, high complex extraction efficiencies and poor emulsion formation.

After LLE, the organic phases can be introduced into conventional atomizers [532,534,535,544-548]. However, the direct analysis of organic phases causes several solvent effects due to changes in the density, surface tension and viscosity which has a decisive effect on the droplet sizes of the aerosol (FAAS), or which affect the drop deposition into the graphite furnace by the autosampler (ETAAS). These solvent effects may cause a poor accuracy and repeatability when FAAS and ETAAS are used as quantification techniques. Finally, other solvents such xylene commonly originates turbulent and unstable flames, while chloroform tends to evaporate quickly leaving the solid complex behind, which clogs the spray chamber.

Because of direct analysis of organic extracts by atomic spectrometry instruments is not advisable due to the low repeatability and sensitivity, an organic solvent evaporation step followed to aqueous dissolution [531], or a back-extraction step into aqueous medium [499,529,530,533,536-542] are usually performed. Several acid back-extraction procedures by using nitric acid [499,529,530,533,536-539] have been reported for cadmium determination. However, acid back-extractions are slow processes and times close to 20 min are required to break the complexes. Thus, metal exchange back-extraction by using metals such as Hg or Pd [540-542], which offer greater chelate stability constants (in the range of

10^{40}) than the other analyte complexes overcome the slow kinetic of acid back-extraction. In these cases, quantitative back extraction can be completed after tmes within 1 to 3 minutes.

The implementation of LLE techniques in on-line mode coupled to atomic spectrometry instruments has been developed to minimize many of these drawbacks moreover offer the advantages of the automatic methods of analysis. Most of the drawbacks of conventional LLE procedures are also overcome by miniaturizing the whole procedure, mianly by drastically reducing the extractant phase volume. This has lead to new extractive techniques based on liquid-liquid microextractions. Some examples of these methodologies are the liquid-liquid micro-extraction, the single-drop micro-extraction, the dispersive liquid-liquid micro-extraction or the hollow fibre liquid-phase micro-extraction. These new methodologies will be described in next sections.

V.2.5.3. Flow Injection on-Line Pre-Concentration Systems by LLE

The introduction of flow injection liquid–liquid extraction (FI-LLE) as a mean for metal pre-concentration in flame atomic absorption spectrometry (FAAS) was first performed by Nord and Karlberg [550,551]. After this first application, the number of contributions in this field was rather few compared to other on-line pre-concentration systems [552]. This is due to the limited range of flow rate ratios of aqueous to organic phase of the segmented stream that on-line phase separators can handle and, in addition, the insufficient stability, versatility and robustness of them.

V.2.5.3.1. General Aspects

The extraction process in these FI modes can be divided into three steps:

1) Solvent segmentation. At this first step, the two initial streams of the two phases converging on it. Thus, regular segments of both immiscible liquids entering into the reaction coil. The commonest type of segmentor consists of a tube with three openings. The aqueous phase enters through a glass capillary and the organic phase through a platinum capillary perpendicular to the former. Two adjustable concentric PTFE tubes shut off the tube at the other opening in the same direction as the aqueous phase. The length of the inner tube is adjustable allowing the checking of the length segments.
2) Coil extraction. At the second step, the segmented phases are carried to an extraction coil made of PTFE in which the analytes are transferred from the sample solution into the organic solvent.
3) Phase separation. At this last step, the solvent extract is separated from the waste sample solution and finally it is carried into the detector. The separation is rarely quantitative and usually analytical recoveries between 80-95% are achieved. However, the phase in which the determination is to be carried out should be kept completely free from the other.

In addition to the solvent segmenter and the reaction coil, an essential component of the manifold used for on-line LLE is the phase separator in which phases are isolated. Various types of phase separators have been reported. One of them is a micro-porous PTFE membrane (0.7-0.9 µm pore diameter) which offers a selective permeability towards the phase which wets the membrane material, and allows the organic phase crossing it to be completely free from aqueous phase [553]. The main drawback offered by membrane type phase separators is the short lifetime of the membrane and the permeation of small amount of aqueous phase especially when high flow rate ratios of aqueous to organic phase and high segmented flow rate is required. Thus, it is difficult to maintain the hydrophobicity of membrane for long times and to prevent leakage of aqueous phase.

Table 13. Liquid-liquid procedures for cadmium pre-concentration from natural waters

Brief description	Sample	LOD	Concentration factor	Analytical technique	Ref.
Complexation by addition of 2.5 mL of APDC (1.1 % (m/v)) / DDC (1.1 % (m/v)) to 200 g of sample at pH 4.5 (1.5 mL of ammonium acetate buffer). After shaking with 20 mL of Freon-TF (2.0 min) and phase separation (5.0 min), the extraction was repeated and the organic extracts were combined. Addition of 400 μL of HNO_3 for back-extraction (shaking time of 1.0 min and 15 min of reaction) and 6.0 mL of Milli-Q water (shaking time of 1.0 min and 5 min phase separation). The final volume is 14 mL.	Seawater	-	40	ETAAS	[499]
Complexation by addition of 1.0 mL of APDC/DDC (0.5 % (m/v) each one) to 100 mL of sample at pH 4.5 (ammonium acetate buffer 2.0 M). After shaking with 20 mL of Freon-TF (2.0 min) and phase separation, the extraction was repeated with 10 mL of Freon-TF and the organic extracts were combined. Addition of 100 μL of conc. HNO_3 for back-extraction (5 min of reaction) and 4.9 mL of Milli-Q water (>30 min phase separation).	Estuarine and river water	-	20	ETAAS	[529]
Complexation by addition of 20 mL of APDC/DDC (2.0 % (m/v) each one) to 350 mL of sample at pH 8.0 (ammonia solution). After shaking with 35 mL of Freon-TF (6.0 min) and phase separation, the extraction was repeated and the organic extracts were combined. Addition of 100 μL of HNO_3 for back-extraction (shake time of 30 min and 5.0 min of reaction) and 2.5 mL of Milli-Q water (shaking time of 30 min). The back-extraction was repeated and the acid extracts were combined.	Seawater	0.04 nM	70	ETAAS	[530]
Complexation by addition of 4.0 mL of DDC (1.1 % (m/v)) to 500 mL of sample at pH 4.5 (10 mL of ammonium acetate buffer 0.2 M). After shaking twice with 5.0 mL of CCl_4 (2.0 min) and phase separation, organic extracts were evaporated to dryness (30°C) and the	Seawater and river water	10 pg	100-250	ETAAS	[531]

residue was mineralised with 0.1 mL of conc. HNO_3. Dilution up 2.0-5.0 mL with H_2O.					
16 mL of conc. HNO_3 was added to sample (400 mL) and the mixture was heated on a hot plate (< 500°C) until the sample was reduced to 100 mL. Complexation by addition of 5.0 mL of DDC (10 % (m/v)) and 10 mL of $(NH_4)_2SO_4$ (40%) to treated sample at pH 6.5 (10 mL of ammonium citrate solution (25%)). After shaking with 20 mL of IBMK and phase separation (1.0 min), the organic phase was aspirated and the cadmium determined.	River water	-	20	FAAS	[532]
Complexation by addition of 2.0 mL of APDC (1.0 % (m/v)) to 200 mL of sample at pH 4.0 (ammonium acetate buffer). After mechanical shaking with 7.0 mL of IBMK (25 min) and phase separation (20 min), 5.0 mL of HNO_3 (4.0 N) was added for back-extraction (shaking time of 20 min and 20 min of standing).	Seawater	3.0 ng L^{-1}	40	ETAAS	[533]
Complexation by addition of 25 µL of chelating agent / pH adjustment (APDC 2.0 % (m/v) / $NaHCO_3$ 5.0 % (m/v) to 1250 mL of sample (pH 4.5). After mechanical shaking with 250 µL of Freon-TF (10 min) and phase separation (5.0 min), organic extracts were direct analysed	Seawater	18 ng L^{-1}	-	ETAAS	[534]
Complexation by addition of 5.0 mL of APDC / DDC (1.0 % (m/v) each one) to 200 mL of sample at pH 4.0 (4.0 mL of citrate buffer). After shaking with less than 35 mL of IBMK (0.5-1.0 min) and phase separation (5.0-10 min), organic phase is direct nebulised.	Natural water	-	-	FAAS	[535]
Complexation by addition of 3.0 mL of APDC / DDC (1.0 % (m/v)) to 500 mL of sample at pH 5.0 (3.0 mL diammonium hydrogenocitrato 0.5 M). After shaking with 20 mL of Freon TF (150 s) and phase separation, the extraction was repeat using 10 mL of Freon TF. 0.2 mL of conc. HNO_3 (shaking time of 20 min and 5.0 min of standing) and 10 mL of water (shaking time of 20 min) were added for back-extraction.	Seawater	1.0 ng L^{-1}	50	ETAAS	[536]
Complexation by addition of 20 mL of APDC or DDC (5.0 % (m/v)) to sample at pH 3.0-6.0 (ammonium acetate buffer). After shaking with 8.0 mL of DIBK (2.0 min for APDC and 1.5 min for DDC) and phase separation, organic phases are directly nebulised.	River water	0.2 µg L^{-1}	-	FAAS	[537]

Table 13. Continued

Brief description	Sample	LOD	Concentration factor	Analytical technique	Ref.
Complexation with APDC / DDC (4.0 mL, 1.0 % (m/v) each) to 400 mL of sample (pH 4.0-4.5). After mechanical shaking with 20 mL of $CHCl_3$ (20 min.) and phase separation (5.0 min), 4.0 mL of Hg(II) (1000 μg L^{-1}) was added (back-extraction, shaking time, 3.0 min).	Seawater	30 ng L^{-1}	-	ETAAS	[540]
Complexation by addition of 100 μL of APDC - DDC (0.5% each) to 80 mL of sample at pH 4.5 (0.73 mL of ammonium acetate buffer). After shaking at 1000 rpm with 5.0 mL of DIBK (10 min) and phase separation (5.0 min.), 4.5 mL of the organic phase was shaken for 2.0 min with 1.0 mL Hg solution (100 mg L^{-1}) for back-extraction.	Seawater	-	72	ETAAS	[541]
Complexation by addition of 100 μL of APDC - DDC (0.5% each) to 80 mL of sample at pH 4.5 (ammonium acetate buffer). After shaking with 5.0 mL of DIBK (10 min) and phase separation (5.0 min), organic extracts (4.5 mL) were centrifuged. Addition of 1.0 mL of Hg back-extraction solution (100 mg L^{-1}) for 2.0 min.	Seawater	0.2 ng L^{-1}	16	ICP-MS	[542]
Complexation by addition of 1.0 mL of APDC / DDC (1.0 % (m/v)) to 250 mL of sample at pH 4.0 (2.0 mL of ammonium acetate buffer). After shaking (8.0 mL of $CHCl_3$, 2.0 min) and phase separation (5.0 min), the extraction was repeated (6.0 mL of $CHCl_3$). 4.0 mL of 7.5 M HNO_3 was added for back-extraction. Combined extracts were evaporated to dryness and the residue treated with 250-500 μL of conc. HNO_3.	Seawater	0.2 ng L^{-1}	200	FAAS	[543]
Complexation by addition of 3.0 mL of H_2DZ (0.78 mM) - tributylphosphine oxide (0.1 M) mixture in CCl_4 to sample at pH 4.0 (10 mL of ammonium acetate buffer). After shaking for 20 min and phase separation, the organic phase is analysed.	River and seawater	-	100	ETAAS	[544]
Monosegmented flow system solvent extraction preconcentration/separation system by using PAN (0.05 % (m/v) in chloroform. Sampling frequency of 30 h^{-1}.	Aqueous standards	85 ng L^{-1}		UV-Vis	[545]
Complexation by addition of thiothenoyltrifluoroacetone (1.0 nM) - TOPO (1.0 mM) to 100 mL sample. After shaking with cyclohexane and phase separation, the organic phase was analysed.	River water	<0.5 ng mL^{-1}	-	ETAAS	[547]
Complexation by addition of 5.0 mL of	Seawater	-	6	ETAAS	[548]

1,1,1-trifluoro-4-mercapto-4-(2-thienyl)but-3-en-2-one - TOPO in cyclohexane to 30 mL of sample at pH 5.0 (5.0 mL of ammonium acetate (0.1 M) or NaB_4O_7 (0.05 M) buffer). After shaking for 20 min and phase separation by centrifugation, the organic phase was analysed.					
Continuous flow extraction by using DDC in Freon 113. Sampling frequency of 30 h^{-1}.	Aqueous standards	<10 ng L^{-1}	50-100	ETAAS	[553]
Flow injection on-line liquid-liquid extraction preconcentration/separation system associated with a newly designed gravitational phase separator, by using APDC (0.5 % (m/v) at 1.0 mL min^{-1}) in IBMK (at 0.45 mL min^{-1}) medium at pH 2.0 by using nitric acid. Sampling frequency of 33 h^{-1}.	Tap water, river water and seawater	20 ng L^{-1}	155	FAAS	[554]
Sequential injection on-line solvent extraction-back extraction preconcentration/separation system associated with a newly designed dual conical gravitational phase separator, by using APDC (0.5 % (m/v) in IBMK. Sampling frequency of 13 h^{-1}.	Fresh water	2.7 ng L^{-1}	21.4	ETAAS	[557]
Complexation by addition of 100 µL of APDC - DDC (0.5% each) to 100 mL of sample at pH 4.0 (1.0 mL of citrate buffer). After shaking with 1.0 mL of DIBK (1.0 min) and phase separation (10 min), 750 µL portion of the organic phase was transferred to a polyethylene cup for back-extraction for 1.0min with 500 µL of $PdCl_2$ (100mg) and 10 mL of HNO_3 (65%).	Seawater and marine interstitial water	-	150	ETAAS	[564]
Complexation by addition of 3.5 mL of H_2DZ (0.01 % (m/v)) in xylene to 500 mL of sample at pH 5.0 (boric acid and sodium hydroxide). After shaking and phase separation (4.0 min), 600 µL of HNO_3 was added for back-extraction (1.0 min of shaking time and phase separation).	Tap water	8.2 – 11.1 ng L^{-1}	250 - 331	FAAS	[565] [566]

APDC, ammonium pyrrolidin dithiocarbamate; DDC, diethyldithiocarbamate; DIBK, 2,6-dimethyl-heptan-4-one; ETAAS, electrothermal atomic absorption spectrometry; FAAS, flame atomic absorption spectrometry; Freon-TF, 1,1,2 trichloro-2,2,1-trifluoroethane; H2DZ, ditizone; IBMK, isobuthyl methyl ketone; ICP-MS, inductively coupled plasma-mass spectrometry; PAN, 1-(2-pyridylazo) 2 naphthol; TOPO, 1,1,1-trifluoro-4-mercapto-4-(2-thienyl)but-3-en-2-one.

Gravitational separators are a second type [554]. These devices are T-shaped separators which use the gravity with or without a sort of phase guide made of a material wetted by one phase but not by the other. These phase separators are appealing because large range of flow rate ratios of aqueous to organic phase can be used.

The state-of-the-art of automated solvent extraction procedures exploiting flowing stream techniques is thoroughly presented and discussed in papers by Miro et al. [555] and Anthemidis et al. [556]. Although FI-LLE was originally associated with the use of a solvent extractor involving a segmentor and a phase separator, current analytical trends are focused on the design and characterization of novel strategies for improving the efficiency of phase separation as well as the repeatability, sensitivity and accuracy of the analytical chemical assays.

V.2.5.3.2. Flow Injection On-Line LLE for Cadmium Pre-Concentration

Few FI-LLE procedures for metals and inorganic compounds have been reported. Developments for cadmium pre-concentration are listed in Table 13. These applications include cadmium extraction from tap water, river water and seawater [554], from aqueous standard solutions, simulatenously extracted with copper and zinc [545], or with cobalt, copper, iron nickel and lead [553], and from sediments after dissolution of the sample [557].

V.2.5.4. Liquid-Liquid Micro-Extraction

Liquid-liquid micro-extraction (LLME) is a solvent micro-extraction technique in which the phase ratio values are higher than 100 [558]. The technique is faster and simpler than conventional LLE methods. It is also inexpensive, sensitive and effective for removing interferences. Compared with conventional LLE, LLME may provide poorer analyte recoveries, but the concentration in the organic phase is greatly enhanced. In addition, the amount of organic solvent is reduced and only one step of manipulation is necessary, which reduces problems of contamination and analyte losses.

The Murray flask and its following adaptations were firstly described for the pre-concentration of organic pollutants (organochloride pesticides [559,560] and organophosphorus pesticides [561]) using LLME from aqueous matrices. The Murray flask consists of a 1000 mL flask with a capillary tube at the top and a lateral arm near the base. The capillary tube has the function of collecting the small volume of organic solvent after agitation of the sample. The lateral arm is used to move the organic extraction solution to the capillary tube using either water or air. The use of air allows multiple extractions of the sample since the aqueous volume is maintained.

The first application of modified Murray flask for LLME of metals is attributed to Carasek [562] who developed a LLME for the determination of gold in seawater by FAAS. The method uses 800 μL of xylene after gold chelation with ammonium O,O-diethyl-dithiophosphate (DDTP). The extraction is then carried out until the aqueous to organic phase ratio achieved a 1000-fold preconcentration of metal.

Flow injection systems have been also proposed for LLME [563]. In this case, the metal (lead) is complexed by dithizone at pH 9.3 (borate/NaOH buffer) in carbon tetrachloride. The extraction process is performed in a miniaturized way by placing a small plug of the organic solvent containing dithizone at the detection point. Then, a large aqueous sample volume was passed through the organic plug, which was gradually enriched with the analyte.

Despite the advantages of this technique, few works have been reported using LLME and/or micro volume back-extraction for the pre-concentration of metals. For cadmium pre-concentration from environmental samples (Table 13), only three developments have been addressed [564-566].

V.2.6. Cloud Point Extraction

Cloud point extraction (CPE) was initially introduced for the pre-concentration of metals, in the form of their hydrophobic complexes, by Miura et al. in 1976 [567] as an alternative method for avoiding the use of organic solvents. It was then extensively exploited for inorganics [568-574] and organics isolation [568-570,575,576]. Current trends and new developments of CPE are mainly refereed to automation and to the use of ionic liquids and micellar formations [572,577,578].

V.2.6.1. General Aspects

Aqueous solutions of certain surface-active agents (such as non-ionic or amphoteric surfactants) display the so-called cloud point phenomenon in which the aqueous surfactant solution (surfactant above the critical micellar concentration, CMC) suddenly becomes turbid because of a decrease in the solubility of the surfactant in water [570]. Surfactant agents are amphiphilic molecules with distinct hydrophobic and hydrophilic moieties; a polar or ionic group connected to a long hydrocarbon tail (linear, branched or containing aromatic rings). At low concentrations, surfactant molecules are mainly as monomers. When their concentration increases above a certain threshold (CMC) surfactant monomers spontaneously accumulate to form colloidal-sized clusters, known as micelles. Depending on the specific surfactant and solution conditions, micelles can adopt a variety of shapes, ranging from roughly spherical to ellipsoidal.

CPE is based on certain properties of non-ionic or amphoteric surfactants at levels upper to their CMC. If some condition, such as temperature or pressure, is appropriately altered or, if an appropriate substance (electrolyte) is added to the solution when non-ionic surfactant are above CMC, the system composed by a unique phase is separated into two isotropic phases. One of these phases contains a surfactant at a concentration below, or equal to CMC, and the other phase is a surfactant rich phase [573]. Above 'cloud point', hydrophobic species (hydrophobic organic compounds or metal ions after reaction with a suitable hydrophobic ligand) present in sample can be entrapped in the micelles of the surfactant, thus can be separated from the bulk sample matrix and can be concentrated in the small volume of the surfactant-rich phase [579].

In the micellar structure, surfactant aggregates orientate their hydrocarbon tail towards the center of the formation, creating a non-polar core. Hydrophobic and covalent compounds initially present in the aqueous solution are favorably partitioned in the non-polar microenvironment. The whole process resembles traditional LLE, the only difference is that the 'organic' phase is generated within the aqueous phase, converting a previously homogeneous solution to heterogeneous one by simply gathering its previously scattered hydrophobic suspensions [577]. Although the exact mechanism via which this phenomenon occurs is yet to be defined, several studies have shown that these phase separations result from the competition between entropy (which favors miscibility of micelles in water) and enthalpy (which favors separation), so the clouding and phase separation procedure is reversible [577]. This behavior, showed by numerous hydrophilic groupings, is especially observed with polyoxyethylene surfactants, and it can be attributed to the two ethylene oxide segments in the micelle that repel each other at low temperature when they are hydrated, and that attract each other as the temperature increases due to dehydration. This effect causes a decrease in the effective area occupied by the polar group on the micelle surface, increasing the size of the micelle, which can be considered to become infinite at the cloud point, leading to the phase separation. Any hydrophobic species, such as metallic chelates, remain preferentially in the surfactant-rich phase, and they can be extracted or pre-concentrated. More detailed discussion on CPE phenomenon can be consulted in specialized literature [570,577,580].

When applying CPE techniques for isolating trace metals, a previous complexation stage is required. This fact usually leads to the optimization of several parameters such as pH and ionic strength. In general, the following factors must be optimized to achieve a successful CPE procedure [577]:

a) Selection of the appropriate chelating agent. The ligand is selected with the requirement that the derived complex is sufficiently hydrophobic, it possesses a high partition coefficient and it is formed quickly and quantitatively with the least possible excess. Based on their reactivity and formation constants with the target metal species, some of the most widely applicable reagents are carbamates, pyridylazo, quinoline and naphthol derivatives. These molecules are universal chelators that form hydrophobic compounds with most of metals and they can be applied when an element-specific detector is available. Other reagents, such as O,O-diethyldithiophosphate, have been used for more specific applications.
b) Concentration of chelating agent. The concentration of the chelating agent has to compensate sufficiently for any consumption of the reagent by other metals. Additionally, chelating agents with lower partition coefficients than others have to be in sufficiently large excess to extract efficiently.
c) pH. Although the chelate formation is strongly dependent on the pH, this variable offers little influence on the extraction efficiency of the complexes formed, since these complexes are bulky, uncharged and covalent.
d) Ionic strength. Although ionic strength has proved to have a negligible effect on the performance of CPE, increasing ionic strength enhances phase separation. Moreover, the addition of salts can markedly facilitate the phase-separation process. This has been demonstrated when working with some non-ionic surfactant systems, since the density of the bulk aqueous phase is changed.
e) Surfactant selection. The non-ionic surfactant must offer high extraction efficiency for the analyte. In addition, the two phases must be easy to separate after centrifugal settling and cooling stages should be omitted. Moreover, surfactant must not be expensive, it must be present a low cloud point temperature, which simplified the procedure due to incubation step is avoided. Triton X-114, Triton X-100, PONPE 7.7 and Tween 80 are the most used surfactants for metal extraction by CPE. The cloud point achieved with these reagents is low, 23, 65, 25 and 65°C, for Triton X-114, Triton X-100, PONPE 7.7 and Tween 80, respectively.
f) Surfactant concentration. There is a narrow range within which easy phase separation, maximum extraction efficiency and analytical signal are accomplished. Increasingly, outside this optimal range, the pre-concentration factor decreases. However, if the surfactant concentration is lower than the recommended value, poor repeatability is achieved.
g) Incubation temperature. Temperature seems to play an additional role in enhancing pre-concentration efficiencies and enhancement factors, and it is reported that the

application of high temperatures leads to dehydration of the micelle, and the increase of the phase-volume ratio, and thus the signal enhancement by a factor higher than 3.

h) Incubation time. Metal reaction with chelating agents and their transportation inside the micelle are kinetically controlled (although thermodynamically favored, simulating the shift of equilibrium towards precipitation). It is therefore essential to maintain the reaction time above a minimum threshold for quantitative extraction. In most studies, a reaction time of up to 10 min is reckoned to be optimum.

i) Centrifugation time. Centrifugation time hardly ever affects micelle formation but it can accelerate phase separation similar to conventional solid phase (precipitate) – liquid separations. Centrifugation times around 5–10 min have been found to be adequate in most of CPE procedures.

The use of CPE procedures offers an interesting green alternative to the conventional extraction systems because CPE avoids the use of large amounts of expensive, toxic and flammable organic solvents. Commercial surfactants are environmentally friendly and cost effective, and the amounts used for effective extraction schemes are minimal compared to the amounts of organic solvents used in conventional extractions.

Other advantages of CPE procedures are related to the quantitative recoveries and large pre-concentration factors (between 10 to 100) [571] because the analyte is transferred from a great volume of original solution to a reduced surfactant-rich phase volume, which rarely exceeds 100 μL per 10 mL of sample. Pre-concentration factors can also be modified on demand by varying the amount of surfactant. In addition, the presence of surfactant can minimize losses of analytes due to their adsorption onto the container.

CPE can be applied for extracting thermally sensitive analytes, such as molecules of biological and environmental interest because the mild conditions inherent to CPE techniques.

The surfactant-rich phase is compatible with most mobile phases used in hydrodynamic analytical systems, while it increases atomic signal in FAAS and wettability of the graphite surface in ETAAS techniques.

There are, of course, several limitations of CPE procedures, mainly attributed to the manipulation of the surfactant-rich phase obtained. As this phase is viscous, it can not be injected directly to conventional analytical instruments, so it must to be diluted with an aqueous or organic solvent to reduce its viscosity. This fact decreases the theoretical pre-concentration factor. Moreover, surfactant-bearing chromophores interfere with UV detection by overlapping with the analyte signal. This problem is not important when atomic detectors are used.

V.2.6.2. Cloud Point Extraction for Cadmium Pre-Concentration

Two different approaches have been published on the pre-concentration of cadmium and other metals from natural waters by CPE (Table 14).

In the first approach, cadmium is pre-concentrated by CPE after the formation of sparingly water-soluble complexes with a suitable chelating agent such as 1-(2-pyridylazo)-2-naphthol (PAN) [112,581-583], O,O diethyldithiophosphate (DDTP) [584,585], 1-(2-thiazolylazo)-2-naphthol (TAN) [586], H_2DZ [587], diethyldithiocarbamate (DDC) [588,589], ammonium pyrrolidin dithiocarbamate (APDC) [115,590], 1-phenyl-3-methyl-4-benzoyl-5-pyrazolone (PMBP) [591], 8-hydroxyquinoline (8-HQ) [592,593], 2-(5-bromo-2-pyridylazo)-5-(diethylamino)-phenol (Br-PADAP) [594], 1-(2-thenoyl)-3,3,3-trifluoraceton

(TTA) [595], N,N'-bis((1R)-1-ethyl-2-hydroxyethyl)ethanediamide (DAD1) and N,N'-bis((1S)-1-benzyl-2-hydroxyethyl)-ethanediamide (DAD2) [596], and glyoxal-bis (2-hydroxyanil) (GBHA) [597].

The second approach (simplified procedure) is based on direct application of CPE procedure without any ligand. By this method, cadmium has been pre-concentrated without adding chelating agents (i.e. by using iodide) [598] or by without any ligand (ligandless cloud point extraction) [599,600].

The surfactants which are used in cloud point extraction are mostly nonionic surfactants, such as (1,1,3,3-Tetramethylbutyl)phenyl-polyethylene glycol (Triton X-114) [115,116,581,583-590,592-598], 4-(1,1,3,3-Tetramethylbutyl)phenyl-polyethylene glycol (Triton X-100) [582,591], 7.5, polyethyleneglycolmonop-nonylphenylether (PONPE 7.5) [599] and polyethylene glycol sorbitan monooleate (Tween 80) [600].

Table 14. Cloud point extraction procedures for cadmium pre-concentration from natural waters

Brief description	Sample	LOD	Concentration factor	Analytical technique	Ref.
10 mL of sample (at pH 3.2) containing APDC (0.04 % w/v) and Triton X-114 (0.2 % v/v was heated (40°C for 10 min) and then centrifuged (3200 rpm for 5.0 min). After cooling (5.0 min), the surfactant-rich phase was separated off and 600 μL of ethanol (20 % v/v) and HNO_3 (10 % v/v) was added.	River water	40 ng L^{-1}	13	TS-FF-AAS	[115]
CPE using PAN and Triton X-114 .	Drinking water	25 ng L^{-1}	59	TS-FF-AAS	[116]
15 mL of sample containing PAN (0.25 % w/v), Triton X-114 (0.05 % w/v) and borax buffer solution (0.05 M, pH 8.0) was heated (40°C for 5.0 min) and then, centrifuged (3500 rpm for 5.0 min). After cooling, the surfactant-rich phase was separated off and 100 μL of a methanol containing HNO_3 (0.1 M) was added.	Seawater	< 0.4 μg L^{-1}	120	FAAS	[581]
To a 10 mL sample, 0.1 mL of PAN (0.01 M), 0.2 mL of Triton X-100 (1.0 % w/v), 0.25 mL of NaCl (30 % w/v) and 1.5 mL of phosphate buffer (pH 8.5) were added and then the mixture was heated (70°C for 15 min) and centrifuged (3500 rpm for 15 min). After cooling, the surfactant-rich phase was separated off and 200 μL of HCl (2.0 M) was added.	Tap water	5.9 ng L^{-1}	50	ETAAS	[582]
CPE using PAN (5:1 M excess) and Triton X-114 (0.25 % w/v).	Tap water and river water	-	-	ICP-OES	[583]
40 mL of refrigerated sample containing	Seawater and	6.0 ng L^{-}	29	ICP-MS	[584]

DDTP (1.0 % w/v), Triton X-114 (0.05 % w/v) and diluted nitric acid solution (pH 0.5) was heated (40°C for 15 min) and then centrifuged (3500 rpm and 30°C for 25 min). After cooling, the surfactant-rich. phase was separated off and 1.0 mL of a mixture of 60 %(v/v) methanol and 40 %(v/v) of water, also containing 1.0 %(v/v) nitric acid was added	river water				
8.0 mL of sample containing DDTP (0.05 M), Triton X-114 (0.043 % w/v), NaCl (1.0 % w/v) and dilute HCl solution (0.32 M) was hand shaken and then, centrifuged (2500 rpm for 10 min). After cooling, the surfactant-rich phase was separated off and acidified ethanol (0.1M HNO_3) was added.	Mineral water and lake water	0.9 µg L^{-1}	-	FAAS	[585]
50 mL of sample containing TAN (1.0 10^{-3} M), Triton X-114 (0.05 % w/v) and 1.0 mL of borax (0.1 M) buffer solution (pH 8.6) was heated (40°C for 15 min) and then centrifuged (5000 rpm for 15 min). After cooling, the surfactant-rich phase was separated off and treated with 200 µL of 0.1 M acidified methanol.	Tap water, river water and seawater	99 ng L^{-1}	58	FAAS	[586]
To 40 mL sample, 1.0 mL of H_2DZ (0.975 M), 200 µL of Triton X-114 (10 % w/v) and 1.0 mL of borate (0.2 M) buffer (pH 9.0) were added and then the mixture was heated and centrifuged (3000 rpm for 30 min). After cooling (for 15 min), the surfactant-rich phase was separated off and 500 µL of HNO_3 (1:1 v/v) was added.	Industrial produced water	0.1 µg L^{-1a}	21	ICP-OES	[587]
20 mL of sample at pH 9.0 containing DDC (0.01 g L^{-1}), Triton X-114 (1.5 g L^{-1}) was heated (40°C for 20 min) and then centrifuged (3000 rpm for 6.0 min). After cooling (5.0 min), the surfactant-rich phase was separated off and 100 µL of acidified methanol (HNO_3 1.0 % v/v) was added.	Seawater	2.0 ng L^{-1}	-	ETAAS	[588]
10 mL of sample containing DDC (3.0 10^{-4} M), Triton X-114 (0.05 % v/v) and phosphate (0.005 M) buffer solution (pH 7.0) was heated (40°C for 10 min) and then centrifuged (3500 rpm for 10 min). After cooling (5.0 min), the surfactant-rich phase was separated off and 200 µL of a THF was added.	Tap water, river water, seawater and waste water	0.31 µg L^{-1}	52	FAAS	[589]

Table 14. Continued

Brief description	Sample	LOD	Concentration factor	Analytical technique	Ref.
500 or 1000 mL of sample containing APDC (0.2 % w/v), Triton X-114 (0.05 % w/v) and dilute HCl or NaOH solution (pH 4.0) was heated (90°C for 15 min) and then centrifuged (3500 rpm for 15 min). After cooling, the surfactant-rich phase was separated off and acidified methanol ($HNO_3$1.0 M) was added to the surfactant-rich phase and sonicated for 5.0 min to complete dissolution.	Lake water	-	-	FAAS	[590]
To a 10 mL sample, 1.0 mL of solution containing 0.01 M of PMBP, 36 g L^{-1} of Triton X-100 and borax buffer (pH 9.0) were added and then the mixture was heated (80°C for 25 min) and then centrifuged (3000 rpm for 5.0 min). After cooling, the surfactant-rich phase was separated off and 300 µL of acidified methanol (HNO_3 0.1 M) was added.	Tap water and lake water	0.64 ng L^{-1}	23	FAAS	[591]
Sample containing 8-HQ and Triton X-114 (0.06 % v/v) was heated and then centrifuged. After cooling the surfactant-rich phase was separated off and acidified methanolic solution was added.	Tap water and river water	0.15 ng L^{-1}	-	FAAS	[592]
100 mL of sample containing 8-HQ (1.0 mM) at pH 7.0 and Triton X-114 (1.0 % v/v) was shaken and heated (65°C for 15 min) and then centrifuged (3000 rpm for 10 min). After cooling, the surfactant-rich phase was. separated off and 3.0 ml of acidified methanol (HNO_3 1.0 M) was added	Lake water	16 ng L^{-1}	-	FAAS	[593]
To 50 mL sample, 0.5 mL of Br-PADAP (0.25 % w/v) and Triton X-114 (3.2 % w/v), 0.55 mL of TRIS / HCl (0.4 M) buffer (pH 9.0) and 1.0 mL of NaCl (5.0 % v/v) were added and then, the mixture was heated in a microwave oven (10 % maximum power) and centrifuged (2500 rpm for 15 min). After cooling (for 15 min), the surfactant-rich phase was separated off and 400 µL of HNO_3 (1:1 v/v) was added.	Waste water	81 ng L^{-1}	22	ICP-OES	[594]
To 25 mL sample (pH 6 by using ammonium acetate 2.5 10^{-3} M), 50 µL of TTA (0.5 M) and 1.0 mL of Triton X-114 (2.5 % w/v) was added and then the	Tap water, well water, seawater and mineral water	0.1 µg L^{-1}	71	ICP-OES	[595]

mixture was heated (65°C). When solution is clouded, it was on.line loaded into a cotton column (at 7.5 mL min^{-1}), and then eluted at 3.0 mL min^{-1} using propanol (0.5 M) acidified with HNO_3 (75:25 v/v) and directly introduced into the nebulizer of the ICP-OES.					
40 mL sample containing DAD1 (2.75 mM) or DAD2 (1.75 mM) and Triton X-114 (0.25 % w/v) was heated (45°C for 15 min) and centrifuged (4000 rpm for 10 min). After cooling, the surfactant-rich phase was separated off and 2.0 mL of acidified methanol (HNO_3 1.0 M) was added.	Tap water	0.45 or 0.44 ng mL^{-1}	18 or 20	FAAS	[596]
10 mL of sample containing GBHA, Triton X-114, SDS and sodium hydroxide-sodium tetraborate buffer (pH 12.4) was centrifuged at room temperature (incubation step is not necessary).	Tap water, lake water and seawater	7.0 ng L^{-1}	22	ETAAS	[597]
Sample containing Triton X-114 (0.2 % v/v), iodide (30 mM) and dilute H_2SO_4 solution (1.0 M) was shaken, heated (40°C for 5.0 min) and then centrifuged (3800 rpm for 5.0 min). After cooling, the surfactant-rich phase was separated off and 0.4 mL of ethanol was added.	River water, seawater and waste water	1.0 ng L^{-1}	11	FAAS	[598]
To 46 mL of sample, 1.0 mL of PONPE 7.5 (0.06 % v/v) and 3.0 mL of boric acid buffer (0.012 M) pH 8.0 were added. The resultant solution was immediately turbid at room temperature. After centrifugation (4000 rpm for 10 min), liquid phase was separated completely by a 25 mL syringe without cooling in an ice bath.	Snow, rain, subterranean, canal water and waste water	0.56 ng L^{-1}	62	CV-AAS	[599]
To 5.0-10 mL of sample, 3.0 mL of Tween 80 (4.0 % w/v) and 5.0 mL of phosphate buffer (pH 8.5) were added and then the mixture heated (60°C for 60 min), cooled in a refrigerator (+4°C for 20 min) and then centrifuged (3500 rpm for 20 min). The surfactant-rich phase was separated off and 1.0 mL of HNO_3 (1.0 M) in methanol was added.	Tap water and mineral water	1.7 µg L^{-1}	10	FAAS	[600]

[a] LOQ (limit of quantification)

APDC, ammonium pyrrolidin dithiocarbamate; Br-PADAP, 2-(5-bromo-2-pyridylazo)-5-(diethylamino)-phenol; CV-AAS, cold vapour atomic absorption spectrometry, DAD1, N,N' bis((1R)-1-ethyl-2-hydroxyethyl)ethanediamide; DAD2, N,N'-bis((1S)-1-benzyl-2-hydroxyethyl)-ethanediamide; DDC, diethylditiocarbamate; DDTP, ammonium *O,O*-diethyl-dithiophosphate; ETAAS, electrothermal atomic absorption spectrometry; FAAS, flame atomic absorption spectrometry; GBHA, glyoxal-bis (2-hydroxyanil); H_2DZ, ditizone; ICP-OES, inductively coupled plasma-optical emission spectrometry; ICP-MS, inductively coupled plasma-mass spectrometry; PAN, 1-(2-pyridylazo)-2-naphthol; PMBP, 1-phenyl-3 methyl-4-benzoyl-5-pyrazolone; PONPE 7.5, polyethyleneglycolmonop-nonylphenylether; SDS, sodium dodecyl sulfate; TAN, 1-(2-thiazolylazo)-2-naphthol; THF, tetrahydrofuran, TRIS, tris(hydroxymethylaminomethane); Triton X-100, 4-(1,1,3,3-tetramethylbutyl)phenyl-polyethylene glycol; Triton X-114, octylphenoxypolyethoxyethanol; TS-FF-AAS, thermospray flame furnace atomic absorption spectrometry; TTA, 1-(2-thenoyl)-3,3,3-trifluoraceton; Tween 80, polyethylene glycol sorbitan monooleate; 8-HQ, 8-hydroxiquinoline.

Usually, the test tubes are cooled in an ice bath after phases separation by centrifugation. This leads to an increase on the viscosity of the micelles, which facilitates the micelles adheration to the bottom of the vial, and water removal is easily achieved by simple decantation.

Complete removal of water traces can be attained with a Pasteur pipette or by evaporation under a nitrogen, argon or helium stream. The inherent high viscosity of the surfactant-rich phase must be decreased before performing measurements. Most commonly, a methanolic solution of nitric acid, as a diluting agent, is included in analytical schemes to accommodate this need [581,584,586,588,590-592,600]. There is an optimum volume for methanol with respect to the analytical signals. This should be a compromise between the capability of methanolic solution to lower the viscosity, thus improving the homogeneity of the solution, and the amount of organic solvent that can, without problems, be tolerated by some detectors (for example, flames and plasmas). Ethanolic [115,585,598] and propanolic [595] solutions of nitric acid and diluted nitric acid [594] or hydrochloric acid [582] solutions, and tetrahydroforan (THF) [589] are also used for the same purpose.

As commented above, although higher temperatures are commonly applied for attaining the cloud point, other possibilities, such as pressure or the use of additives (including electrolytes), have been proposed. These strategies allow that the cloud point can be attained at room temperature with inherent advantages to the process. The addition of electrolytes can increase or decrease the surfactant cloud point, and these effects are respectively known as 'salting-in' and 'salting-out'. Studies on the effects of some electrolytes, as NaCl, KNO_3 and $MgCl_2$, on the cloud point behavior of non-ionic surfactants have been reported [601,602]. It was observed that the presence of electrolytes decreases the cloud point (salting-out effect), whih results in a more efficient extraction. The lower cloud point is attributed to electrolytes promoting dehydration of the polyoxyethylene chains [603]. The salting-out phenomenon is directly related to desorption of ions to the hydrophilic parts of the micelles, which increase inter-attraction between micelles and consequently leading to the precipitation of surfactant molecules [602]. Therefore, Coelho et al. [585] and Filik et al. [597] have applied a mixed micelle mediated system for Cd pre-concentration where the phase separation (at room temperature) was induced by hydrochloric acid or NaCl for lowering the cloud point.

V.2.6.3. Flow Injection on-Line Pre-Concentration Systems by CPE

For on-line coupling of CPE to FI, the flow rates must ensure the complete adsorption of the surfactant-rich phase, preventing any back pressure effects and blocking. Low flow rates are not expected to impair or to enhance the extraction efficiency but increase the time of analysis. Higher velocities cause flow disturbances and insufficient retention of the micellar phase on the reactor. The length of the column of the reactor is other important parameter. Short columns are inadequate for retaining the total amount of the surfactant phase, while, for longer columns, the signal decreases due to insufficient washing and blockage of the flow tubes. Other parameters, such as the amount of cotton additive and the elution flow rate, can affect the extraction efficiency. If large amount of the absorptive material is produced, blockage of the tubes and insufficient wash-out of the micellar phase can occured. However, if elution of the surfactant phase is too fast, incomplete desorption can be attained; while, for lower elution rates, the time of analysis can be significantly increased [577].

The combination of the FI technique with CPE has been reported in several publications for off-line or on-line pre-concentration procedures for cadmium and other trace metals.

A) Off-line coupling of CPE to FI

The first paper on the off-line coupling of FI with CPE was reported for the determination of uranium (IV) by using PAN/Triton X-114 (complexing/surfactant) in tap and river water [604]. After this pioneering work, several metals were conveniently determined by off-line coupling of FI with CPE [577]. In these papers, FI was used to introduce the surfactant rich phase into various analytical devices. In that way, the dissolution of the surfactant-rich phase can be performed in small volumes (higher pre-concentration factors) and it minimize repeatability problems that manual applications may cause.

B) On-line coupling of CPE to FI

Off-line CPE involves a series of fussy procedures including incubation, centrifugation, surfactant-rich phase separation and dilution of surfactant-rich phase. All these procedures may result in poor precision, low pre-concentration factors and time consuming procedures. To avoid these problems, Fang et al. [605] proposed the on-line incorporation of CPE to flow injection analysis. Since then, few studies have exploited the analytical advantage offered by the on-line application of this technique for metal pre-concentration [577]. This method is based on the on-line mixing of the sample with a surfactant rich solution, which allows phase separation in the flow on a high surface material, generally a glass tube packed with a suitable filtering material (cotton, nylon or glass-wool) that can quantitatively intercept the surfactant aggregates. Subsequently, the surfactant-rich phase containing the analytes is desorbed by a proper elution agent and is transported on-line towards the measuring device.

Yamini et al. [595] (Table 14) have applied this last approach in the determination of cadmium and other trace elements in water by using TTA/Triton X-114. When solution is clouded, it was on line loaded into a cotton column. Under these conditions, the surfactant rich phase was trapped inside the column and the retained chelates were eluted using acidified propanol and directly introduced into the nebulizer of an ICP-OES instrument.

V.2.7. Single-Drop Micro-Extraction

V.2.7.1. General Aspects

Single-drop micro-extraction (SDME) is a miniaturization of the traditional LLE. Classically, this pre-concentration technique is based on the use of a micro-drop water immiscible of solvent exposed to the aqueous sample donor solution (direct-SDME or continuous flow micro-extraction (CFME)). However, other variation of the technique consists of thin water immiscible organic extraction phase (of lower density than water) layered over the aqueous sample (liquid–liquid–liquid micro-extraction, LLLME). In addition, when analytes are volatile or semi-volatile, pre-concentration can be achieved by exposing the drop to the headspace above the sample, which leads to the headspace single-drop micro-extraction (HS SDME) [606]. SDME is not an exhaustive extraction method, and only a small fraction of analyte is extracted/pre-concentrated for analysis.

The first SDME application was reported by Liu and Dasgupta [607] for the determination of NH_3 and SO_2 in air samples by using an acidic aqueous micro-drop for extracting the analytes. In 1996 the same authors developed another extraction system (drop-in-drop system) [608], which consisted of a micro-drop of chloroform (13 μL) suspended in a larger aqueous drop, for extracting sodium dodecylsulphate as an ion pair with methylene blue. Once the analytical signal due to the organic drop was measured, the organic phase was pumped away, and the analytical cycle could be repeated. Jeannot and Cantwell [609] introduced other micro-drop extraction technique for extracting 4-methylacetophenone, which consisted of a micro-drop (8 μL) of n-octane suspended at the end of a Teflon rod immersed in a stirred aqueous solution. After sampling, the rod was removed from the sample solution and an aliquot of the organic drop was analysed. The same authors suggested an alternative drop-based extraction technique, which involved an improvement over the previous attempts, using an organic drop (1 μL) directly suspended from the tip of a microsyringe needle immersed in a stirred aqueous solution [610]. After extracting, the micro-drop was retracted back into the microsyringe needle and analysed.

After these first developments, SDME has been used as an extraction technique for several organic analytes [611]. The first application dealing with inorganic species dates from 2003, and it is attributed to Chamsaz et al. [612], who used the headspace mode for determining As(III) and total As by ETAAS. The number of applications for metal pre-concentration has then significantly grown, and the potential of SDME for pre-concentrating metals is nowadays well-recognized [606,613,614].

In addition to variables affecting SDME (solvent type, solvent drop size, extraction time and magnetic stirring; as well as sample flow rate when using CFME) other factors related to the formation of metal complexes (chelating agent selection/concentration and pH of the complex formation and extraction) must be taken into account when dealing with metals.

A) Solvent selection. Several solvents such as carbon tetrachloride, dichloromethane, 1,2 dichloroethane, chloroform, nitrobenzene, toluene, benzene, IBMK and DIBK can be selected according to the polarity of metal complex formed.
B) Drop size. Drop size shows an important influence on the extraction efficiency. It has been found that the use of a large organic drops can result in an increase on the analytical response of the instrument. However, larger drops are difficult to manipulate and it is easy to dissolve or displace into aqueous solutions especially suffering comparatively fast stirring and long extraction time. It was also found that when the drop volume is larger than 4 μL, the micro-drop may drop down from the needle of the microsyringe.
C) Extraction time. Extraction time is also an important factor because the mass transfer is a time dependent process. Therefore, the reaction time for the transfer of metal complex into the drop solvent increases the extraction efficiency up to reach the equilibrium (around 15 min) and leveling off at higher extraction time. Thus, SDME will attain the maximum sensitivity after the equilibrium between aqueous and organic phase has been achieved. However, large extraction times may result in drop dissolution and can produce drop losses.
D) Magnetic stirring. Magnetic stirring was used to allow the mass transfer process, enhancing extraction and reducing the time required to reach thermodynamic equilibrium. Magnetic stirring gives a dramatic increase on the analytical response of the instrument, but high stirring rates (>400-500 rpm) usually lead to drop instability, drop displacement or drop dissolution. These problems are attributed to the relatively

large vortex formed in the lower region of the drop, and the main evidence is a bad repeatability for measurements.

E) Sample flow rate. The flow rate of sample solution affect extraction dynamics remarkably since the thickness of the interfacial layer surrounding the microdroplet will vary with the change of flow rate. However, for high flow rate, although extraction efficiency was better, air bubble formation frequently occurred, which lead to quantification problems.

SDME is a simple, low-cost, fast and virtually solvent-free sample preparation technique based on a high reduction of the extractant phase-to-sample volume ratio. However, SDME is not very robust, and the droplets can be lost from the needle tip of the microsyringe during extraction. More reliable methods are obtained when working for short times and at low stirring rates. This increases precision of the SDME procedures, although decreases sensitivity.

V.2.7.1. Single-Drop Microextraction for Cadmium Pre-Concentration

Although there are four modes of SDME for extracting/pre-concentrating inorganic analytes (direct-SDME, CFME, HS-SDME and LLLME), only direct-SDME and CFME modes have been applied to cadmium extraction from waters (see Table 15).

A) Direct single-drop micro-extraction (Direct-SDME)

Direct-SDME is based on the exposure of a micro-drop of water-immiscible extractant phase suspended from the tip of a microsyringe needle to a stirred aqueous sample. After extracting for fixed time, the drop is retracted back into the microsyringe needle and finally injected into the detector to obtain the analytical signal.

Because of two liquid phases are in direct contact in absence of any porous solid support at the interface, when one of the phases is mechanically stirred (i.e. directly convected), the other phase is also subjected to a convective mixing. This occurs because the first phase transfers momentum to the second phase as a result of frictional drag at the liquid-liquid interface. Mass transfer of the analytes from the aqueous sample to the extraction micro-drop (by convective-diffusive mass transfer process) continues until thermodynamic equilibrium is attained or extraction is stopped [610]. Direct-SDME requires the use of a water-immiscible extractant phase and analytes more soluble in the extractant phase than in the sample solution.

The main drawback of Direct-SDME is the instability of the drop at high stirring rates or at high temperatures, especially when samples are not perfectly clean. Moreover, solvents with relatively high water solubility and low boiling points are not suitable for Direct-SDME because of their high rate of dissolution or evaporation. Acidic samples, such as digests obtained after acid digestion procedures, or the presence of large non-polar species that can saturate the organic phase, could be troublesome when applying this micro-extraction mode [606].

Four different Direct-SDME-ETAAS methods were proposed for the determination of Cd in natural waters (Table 15). Dithizone was used in two of them as a complexing agent by addition to the sample and subsequent extraction of the resultant cadmium dithizonate by a micro-drop of chloroform [615] or toluene [616]. A method based on the formation of a cadmium ion pair in a drop of nitrobenzene and ammonium tetraphenylborate after formation of a cationic complex in the sample solution by addition of 2-(5-bromo-2- pyridylazo)-5-diethylaminophenol (5-Br-PADAP) [617] was also developed. This last method involves an automatic SI system coupled to ETAAS, in which Cd is on line chelated with ammonium diethyldithiophosphate (DDPA) and it is on line extracted into a 60 μL micro-drop of DIBK [618]. The extraction procedure was performed into a newly designed flow-through extraction cell coupled on a SI manifold. As the complex Cd(II)-DDPA flowed continuously around the microdroplet, the analyte was extracting into the solvent micro-drop.

Table 15. Single-drop micro-extraction procedures for cadmium pre-concentration from natural waters

Brief description	Sample	LOD	Concentration factor	Analytical technique	Ref.
After the extraction system has been filled with the sample solution (at pH 6.5 by using potassium acid phthalate / potassium hydroxide), a 10 μL micro-syringe was used to push out smoothly 4.0 μL of 8-HQ (0.1 M)-chloroform to form drop above the PFA tubing outlet in the extraction chamber and sample flows around the organic drop continuously (0.05 mL min^{-1}). After extracting for 15 min, the microdrop was retracted into the micro-syringe and was then injected into the graphite furnace for further ETV-ICP-MS analysis	Tap water and lake water	4.6 ng L^{-1}	140	ETV-ICP-MS	[218]
A 10 μL micro-syringe containing 3.0 μL of H_2DZ (0.01 M)–chloroform was clamped above the vial containing 5.0 mL of sample at pH 8.3 (by using ammonia) that was stirred at 400 rpm. After extracting for 10 min, the micro-drop was retracted into the micro-syringe and injected into the ETAAS.	Tap water, river water and rain water	0.7 ng L^{-1}	-	ETAAS	[615]
A micro-syringe containing 4.0 μL of toluene was clamped above the vial containing 2.0 mL of sample at pH 5.5 and H_2DZ (1.2 mM) that was stirred at 400 rpm. After extracting for 10 min, the micro-drop was retracted into the micro-syringe and injected into the ETAAS.	Tap water, spring water, river water, pond water and lake water	2.0 ng L^{-1}	118	ETAAS	[616]
A micro-syringe containing 4.0 μL of ammonium tetraphenylborate (1.0 % w/v)- nitrobenzene was clamped above the vial containing 2.0 mL of sample at pH 6.0 and 0.5 mL of 2-Br-PADAP (0.2% w/v) that was stirred at 600 rpm. After extracting for 15 min, the micro-	Tap water and seawater	6.5 ng L^{-1}	390	ETAAS	[617]

drop was retracted into the micro-syringe and injected into the ETAAS.					
Sequential injection on-line SDME system where sample (at pH 2.0 and at 1.5 mL min^{-1}) is complexed (by using DDPA (1.2 % (w/v) at 0.3 mL min^{-1}) and on-line extracted into 60 µL micro-drop of DIBK during 10 min. Sampling frequency was 6 h^{-1}.	Tap water, river water and seawater	10 ng L^{-1}	10	ETAAS	[618]
After the extraction system has been filled with the sample solution, a 10 µL micro-syringe was used to push out smoothly 4.0 µL of BZA (0.1 M)-benzene to form drop above the PFA tubing outlet in the extraction chamber and sample flows around the organic drop continuously (at 0.2 mL min^{-1}). After extracting for 10 min, the micro-drop was retracted into the micro-syringe and was then injected into the graphite furnace for further ETV-ICP-MS analysis.	Synthetic water	0.27 ng L^{-1}	180	ETV-ICP-MS	[620]

BZA, benzoylacetone; DDPA, ammonium diethyldithiophosphate; DIBK, , 2,6-dimethyl-heptan-4-one; ETAAS, electrothermal atomic absorption spectrometry; ETV-ICP-MS, electrothermal vaporization-inductively coupled plasma-mass spectrometry; H_2DZ, ditizone; 2-Br-PADAP, 2-(bromo 2 pyridylazo)-5-diethylaminophenol; 8-HQ, 8-hydroxiquinoline.

B) Continuous-flow micro-extraction (CFME)

Continuous-flow micro-extraction is another mode of SDME, where the sample, instead of being stirred, is pumped continuously at a constant flow rate and, when the extraction chamber is full of sample, a drop is formed at the tip of a microsyringe needle. From this time, sample is flowing through the chamber in contact with the drop. In contrast to other SDME micro-extraction modes, a solvent drop makes always contact with a fresh and flowing sample solution.

In CFME the flow induces mass transfer in the drop via momentum transfer (the extraction phase indirectly experiences convection as a consequence of convection of the aqueous sample). The rate of extraction increases with increasing flow rate of the aqueous solution, consistent with a decrease in thickness of the Nernst diffusion films [619]. This mode facilitates mass transfer and high pre-concentration factors can be achieved. Sample flow rate should ensure an effective micro-extraction of analytes without drop dislodgement or bubble formation. Xia et al. [620] have reported a modification of this micro-extraction mode, called cycle-flow micro-extraction, in which the waste outlet of tubing was put into the sample reservoir, and the sample volume was set as 1 mL. The extraction was performed in the same way as CFME, and after the extraction was finished, the droplet was injected to ETV- ICPMS for further analysis.

Two different CFME-ETV-ICP-MS methods were proposed for the determination of Cd in natural waters (Table 15). In the first, Cd is simultaneously chelated and extracted by exposing an 8-HQ-chloroform drop to the aqueous sample [218]. In the second one, Cd is chelated and extracted using a micro-drop of benzoylacetone (BZA) in benzene [620].

V.2.8. Dispersive Liquid-Liquid Micro-Extraction

V.2.8.1. General Aspects

Dispersive liquid–liquid micro-extraction (DLLME) is a micro-extraction technique introduced by Rezaee et al. [621] in 2006. DLLME is based on the cloudy solution formed when an appropriate mixture of an extraction solvent (i.e., a few microlitres of an organic solvent with high density such as tetrachlorometane, chloroform, carbon disulphide, nitrobenzene, bromobenzene, chlorobenzene or 1,2-dichlorobenzene) and a disperser solvent (i.e., methanol, ethanol, acetonitrile or acetone) is quickly injected into the aqueous sample. The water-immiscible extractant solvent should have higher density than water, while the disperser solvent should be miscible in the extractant solvent and the aqueous sample.

When the mixture of extractant phase and disperser is rapidly injected into the sample, a high turbulence is produced. This turbulent regimen gives rise to the formation of small droplets, which are dispersed throughout the aqueous sample. Emulsified droplets have large interfacial area. The nature of the emulsifier (disperser solvent) can also have an influence on droplet size distribution, the mean droplet size, and also on emulsion viscosity [622]. Liquid–liquid dispersions play an essential role in separation processes and reaction systems. This is because the large interfacial area due to dispersion facilitates mass transfer and reaction rate [623]. Turbidity is formed in the aqueous phase because the fine dispersion of the extractant distributed throughout the aqueous sample, which is facilitated by the disperser solvent. After formation of the cloudy solution, the surface area between extraction solvent and aqueous sample is very large, so the equilibrium state is achieved quickly and the extraction time is very short. After centrifuging the cloudy solution, the sedimented phase at the bottom of a conical tube is recovered and analyzed by the most appropriate analytical technique.

As it has been commented for SDME, parameters such as the nature of the chelating agent, the concentration of chelating agent and the pH, play a unique role on metal–chelate formation and subsequent extraction. On the other hand, several variables, such as the extraction solvent and disperser types, the volumes for both extraction solvent and disperser, the extraction time and the ionic strength, must be carefully optimized.

A) Solvent extraction selection. Special attention deserves the selection of the extraction solvent. This solvent must higher density than water, high extraction capabilities for the target analytes, and low solubility in water. In addition, a stable cloudy solution must be formed and it must be easily removed by the auto-sampler for injection into the atomizer (ETAAS).
B) Extraction solvent volume. Extraction solvent volume is an important factor becuase the use of low volumes decreases the volume of the sedimented phase. This can offer problems when removing the sedimented phase for injection into the atomizer ETAAS. On the other hand, the enrichment factor decreases for high volumes of solvent because the volume of the sedimented phase increases. Subsequently, high enrichment factors can be obtained when using low volumes of the extraction solvent.
C) Disperser selection. The main criterion for selecting the disperser solvent (acetone, acetonitrile, methanol or ethanol) is the miscibility of the disperser in the extraction solvent and in the aqueous sample.

D) Disperser volume. It is clear that the solubility of complexes in water increases when increasing the volume of methanol. Therefore, the extraction recovery decreases.

E) Extraction time. Extraction time is defined as the time between the injection mixture of disperser and extraction solvent and the starting to centrifuge. This variable is not significant on the extraction efficiency. An explanation can be the infinitely large surface area between extraction solvent and aqueous phase after formation of cloudy solution. Thereby, complex formation/transfer from aqueous phase to extraction solvent is fast. Subsequently, equilibrium state is achieved quickly therefore, the extraction time is very short.

F) Ionic strength. Ionic strength has no significant effect on the enrichment factor. It is maybe because of two opposite effects of salt addition; one of them is the increase on the volume of sedimented phase that decreases the enrichment factor, and another is the salting-out effect that increases the enrichment factor. Therefore, the enrichment factor is nearly constant by changing this parameter.

As advantage, DLLME does not require a syringe as a drop holder during the extraction process, and problems such as drop dislodgment are avoided. Nevertheless, this technique is limited to a small number of extractants, which should efficiently extract the analytes and which must have higher density than water, must form a stable cloudy solution and must be easily removed. In addition, this micro-extraction technique appears to be difficult to automate [606].

V.2.8.2. Dispersive Liquid-Liquid Micro-Extraction for Cadmium Pre-Concentration

Due to DLLME has been recently described, few methods for the determination of metals have been developed [606]. In all cases, a complexing agent is added to the sample before the mixture of the extractant and the disperser solvent. DLLME has been mainly used in combination with ETAAS for quantitation. For instance, Au, Cd, Co, Ni and Pb have been determined in different water samples by ETAAS after extraction of the corresponding complex into a chlorinated organic solvent, generally tetrachlorometane. DLLME procedures applied to Cd extraction [624,625] are summarized in Table 16.

V.2.9. Hollow Fibre Liquid Phase Micro-Extraction

V.2.9.1. General Aspects

To overcome drawbacks associated to SDME, Bjergaard and Rasmussen [626], introduced an alternative concept for liquid phase micro-extraction (LPME) based on the use of single, low-cost, disposable, porous, hollow fibers made of polypropylene. In this concept, the analytes of interest are extracted from aqueous samples through a thin layer of organic solvent immobilized within the pores of a porous hollow fiber, and into an acceptor solution inside the lumen of the hollow fiber. Hollow fibre liquid-phase micro-extraction (HF-LPME) is a technique which allows extraction and pre-concentration of analytes from complex samples in both a simple and inexpensive way. Two reviews describe this appealing technique in detail [627,628].

In HF-LPME the organic phase is protected by the fiber, and it appears that the hollow fiber decelerates the process of organic solvent dissolution into the bulk solution. Another factor which contributes the improvement on sensitivity is that the surface area for the rod-like configuration of the two-phase HF-LPME system is larger than the spherical surface adopted by the drop-based SDME methods. The disposable nature of the hollow fiber totally eliminates the possibility of sample cross-contamination and ensures precision. In addition, the small pore size prevents large molecules and particles present in the donor solution from entering the accepting phase and at the same time, most of macromolecules do not enter the hollow fiber because they are not soluble in the organic phase present in the pores, thus yielding very clean extracts.

Table 16. Dispersive liquid-liquid micro-extraction procedures for cadmium pre-concentration from natural waters

Brief description	Sample	LOD	Concentration factor	Analytical technique	Ref.
500 µL methanol containing 34 µL carbon tetrachloride and 0.10 mg APDC was rapidly injected by syringe into the water sample (5.0 mL) placed in a 10 mL screw cap glass test tube with conic bottom. After the cloudy solution formation (complex extracted into the fine droplets of tetrachloromethane) and centrifugation (5000 rpm for 2.0 min), droplets were sedimented at the bottom of the test tube. Then a 20 µL of sedimented phase was determined by ETAAS.	Tap water, river water and seawater	0.6 ng L^{-1}	125	ETAAS	[624]
500 µL methanol containing 34 µL carbon tetrachloride and 0.10 mg Salen(N,N'-bis(salicylidene)ethylenediamine) was rapidly injected by syringe into the water sample (5.0 mL). After the cloudy solution formation and centrifugation (5000 rpm for 2.0 min), droplets were sedimented at the bottom of the test tube. Then a 20 µL of sedimented phase was determined by ETAAS.	Tap water, rain water and seawater	0.5 ng L^{-1}	122	ETAAS	[625]

APDC, ammonium pyrrolidin dithiocarbamate; ETAAS, electrothermal atomic absorption spectrometry.

Solvent impregnation of the fibre is essential since extraction occurs on the surface of the immobilized solvent. The pores of a porous hydrophobic polymer membrane are filled with an organic liquid, which is held by capillary forces. The extractant must have a polarity matching that of the hollow fibre to be easily immobilized within the pores.

Two sampling modes are used in LPME: two phases and three phases.

A) In the two phases LPME sampling mode (HF-LPME), the analyte is extracted from an aqueous sample to a water-immiscible extractant immobilized in the pores of a hydrophobic hollow fibre, typically made of polypropylene and supported by a microsyringe. In this sampling mode, the acceptor phase is organic and the extraction process occurs in the pores of the hollow fibre, where the solvent is immobilized.

B) In the three-phases sampling mode, hollow fibre liquid–liquid–liquid micro-extraction (HF-LLLME), the analyte is extracted from an aqueous sample through the water-immiscible extractant immobilized in the pores of the hollow fibre and then back-extracted into an aqueous phase inside the lumen of the hollow fibre. The acceptor phase is aqueous in this micro-extraction mode. This operation mode is only applied to analytes with ionisable functionalities.

Moreover the nature and concentration of the chelating agent and the pH, several factors, such as the nature of the extraction solvent, the sample stirring speed and the extraction time, must be optimized in LPME procedures. In addition, the stripping solution concentration must be also considered in the three-phase sampling mode.

A) Organic solvent selection. The final choice of the organic solvent must be based on several considerations. Firstly, it should have a low solubility in water to prevent its dissolution into the aqueous phase, and a low volatility, which will restrict solvent evaporation during extraction. Secondly, it should be strongly immobilized in the pores of the hollow fiber to prevent leakage. Finally, it should provide high extraction recoveries.
B) Sample stirring speed. Magnetic stirring enhances extraction and reduces the time required to reach the thermodynamic equilibrium because stirring allows continuous exposure of the extraction surface to fresh aqueous sample. In LPME, the acceptor solution is confined within the fiber, and it can tolerate very high stirring speeds (around 1000 rpm). However, high stirring rates can cause bubbles which can be attached on the surface of the hollow fiber and can decrease analyte transfer moreover promote solvent evaporation. This leads to bad precision in the measurements.
C) Extraction time. The extraction efficiency of HF-LPME is based on the analyte's partitioning between the aqueous sample and the organic solvent. This equilibrium is attained only after exposing the acceptor solution to the sample for a 'long' period of time for which solvent loss due to dissolution could occur.
D) Stripping solution concentration. In the three-phase sampling mode (HF-LLLME) the effect of the stripping solution concentration must be considered. In these cases, strong acidic solutions are commonly used for stripping of the metal ions from supported liquid membrane phase.

In general, the extraction efficiency achieved with HF-LPME is higher than with Direct-SDME because hydrophobic hollow fibres allow the use of vigorous stirring rates to accelerate the extraction kinetics. Moreover, the contact area between the aqueous sample and the extractant phase is higher than in the case of SDME, which invreases the mass transfer rate. The use of the hollow fibre provides protection of the extractant phase and hence, the analysis of dirty samples is feasible. The small pore size allows micro-filtration of the sample, thus yielding very clean extracts. However, it should be noted that this technique suffers from some drawbacks such as the manipulation of the hollow fibre at the time of placing it at the tip of the needle of the microsyringe. This operation is carried out before the micro-extraction process and it could be a source of contamination.

Table 17. Hollow fibre liquid phase micro-extraction procedures for cadmium pre-concentration from natural waters

Brief description	Sample	LOD	Concentration factor	Analytical technique	Ref.
1.5 cm segments of the HF (4.0 µL of internal volume) were sonicated in acetone (for 2.0min) to remove contaminants then, acetone was evaporated. 4.0 µL CCl_4 was drawn into the microsyringe. The needle tip was inserted into the HF and the assembly was immersed in the organic solvent (for 5.0 s). Then, the organic solvent in the syringe was injected completely into the HF. The HF was then removed from the organic solvent and immediately immersed into the stirred aqueous sample (at pH 8.0) containing DDC (0.5 % w/v) during 15min at 1000 rpm. After extracting, the extract in the HF was retracted into syringe. The HF was detached, discarded and the extract was injected into the graphite furnace for analysis.	Lake water and river water	4.5 ng L^{-1}	29	ETV-ICP-MS	[217]
4.5 cm segments of the HF was immersed in 1-octanol solution containing H_2DZ (1.6 mg L^{-1}), and then was taken out and washed outside and inside with water. ~20 µL of HNO_3 0.05 M (stripping solution) was drawn into the microsyringe. The needle tip was inserted into the HF and the stripping solution in the syringe was injected to wash and fill the lumen of the HF. Then, the two ends of the HF were sealed with heated tweezers. The extraction was carried out by placing the HF in 100 mL of aqueous solution (at pH 5.2). After stirring for 30 min, the HF was taken out from the solution and one of the sealed ends was cut to allow its connection with the microsyringe needle; the other sealed end was then cut to permit the flushing of the stripping solution to a glass vial. The HF was detached and discarded. The solution was directly analysed.	Seawater	0.8 ng L^{-1}	387	ETAAS	[629]

DDC, Diethyldithiocarbamate; ETAAS, Electrothermal atomic absorption spectrometry; ETV-ICP-MS, electrothermal vaporization-inductively coupled plasma-mass spectrometry; HF, hollow fibre; H_2DZ, ditizone.

V.2.9.2. Hollow Fibre Liquid Phase Micro-Extraction for Cadmium Pre-Concentration

Although HF-LPME has been widely applied to the determination of organic compounds, especially in biological and environmental samples, the use of this technique for extracting inorganic species is still scarce [606]. HF-LPME [217] and HF-LLLME [629] procedures applied to Cd extraction are summarized in Table 17.

V.2.10. Solid Phase Extraction

V.2.10.1. General Aspects

Solid phase extraction (SPE) is based on the partitioning of compounds between a liquid phase (sample) and a solid phase (ion exchange resin, polymeric resin, activated carbon, silica, etc.) whereby the intermolecular forces between the phases influences retention and elution.

There are two SPE modes, the batch and column methods. The batch method consists of mixing the sample, at a fixed pH, with few grams of the resin, and then, the mixture is shaken or stirred for several minutes to facilitate analyte adsorption onto the solid phase. After filtration or centrifugation, desorption is achieved by stirring the solid phase with few milliliters of a dilute acid. Metals are finally determined in this eluate solution. In the column method, a PTFE or glass column or a syringe is packed with few milligrams of solid phase. A small amount of glasswool, or a frit made of suitable materials, is placed at both ends to prevent loss of the particles during sample loading. The use of commercial cartridges, disks or SEP-PAK devices (various adsorbents within a plastic syringe) has also been used in this pre-concentration mode. Before using these commercial supports, acid solutions and doubly distilled de-ionized water were successively passed through the column in order to equilibrate, clean and neutralize it. Sample, at the desired pH value, is then passed through the column at a fixed flow rate by using a peristaltic pump. Afterwards, the metal ions retained on column are eluted with the appropriate solvent and the analytes in the elute solution are determined.

Retention on the solid phases may involve non-polar, polar or ionic interactions and Van der Waals forces, or hydrophobic interactions, which are not very strong. Other times, strong interactions can be attained. In this case, the addition of chelating agents to the sample prior to the extraction or the use of solid phases with immobilized chelating groups are required.

The wide range of sorbents available for trace metal SPE pre-concentration, including synthetic (divinylbenzene polymers, zeolites, fullerenes and polyurethane foam with or without chelating ligands, etc.) and natural materials (microbial cells, microoragnisms, dead biomass, humic and fulvic acids, seaweed, clay, fly ash, moss, iron-oxide-coated sand, modified wool and cotton, etc.), provides a wide range of interactions. The current researches in SPE are mainly focused on the development of new sorbents such as nanometer-sized materials, egg-shell membranes, modified silica beads, ion imprinted polymers, or mesoporous materials. These new solid supports are mainly prepared to improve selectivity.

Optimization of the SPE process involves the study of several factors, such as: (1) the choice of sorbent; (2) the volume of the sample flushed through the column (dynamic mode); (3) the chelating agent (batch mode); (4) the mass of the resin and breakthrough volume of the resin (volume of sample per gram of adsorbent resin which causes the analyte molecules to migrate from the front of the adsorbent bed to the back of the adsorbent bed); (5) the elution profile; (6) the sample pH; and, (7) the retention (adsorption) and elution flow rates. The choice of sorbent is the key point in SPE because it can control parameters such as selectivity, affinity and capacity. This choice depends strongly on the analyte and the interactions of the chosen sorbent through the functional groups of the analyte. However, it also depends on the kind of sample matrix and its interactions with both the sorbent and the analyte.

Another important consideration is related to the washing and conditioning of the column before the pre-concentration step. After sorbent material packing into the column or micro-column (and plugged with a small portion of glass wool or frits at both ends), methanol, dilute acids or high purity de-ionized water must be passed through the column in sequence to clean the solid support. The column must then be conditioned to the desired pH by passing trough the column the same buffer solution used for fix the pH of samples.

SPE has gradually replaced classical LLE and it becomes the most common sample pre-treatment technique in environmental areas. SPE overcomes the problems arising from the use of organic solvents and offers several major advantages that include (1) higher enrichment factors, (2) simple operation, (3) lower cost and less time, (4) improved selectivity, specificity and repeatability, (5) safety with respect to hazardous reagents, (6) less volume requirement and the use of environmentally friendly reagents, (7) absence of foaming and emulsions, (8) the ability for coupling with different modern detection techniques in on-line and off-line modes, and (10) ease of automation.

However, SPE offers as main drawback the fact that organic substances presents in natural waters must be removed previous to the pre-concentration step. Studies undertaken during the last two decades have shown that organic complexation is important for a number of trace metals in natural waters. An important fraction (30–99.9%) of metals, including cadmium, are complexed by natural (humic acids, fulvic acids, glycollic acid, peptides, proteins, amino acids, lipids, polysaccharides) and anthropogenic (ethylenediaminetetraacetic acid (EDTA), nitrilotriacetate, phosphonates, citric acid, tartaric acid and surfactants) organic ligands in aquatic environments. These complexes are not retained on chelating or polymeric resins. Thus, it is necessary to release the trace metal from the metal–organic complex prior to the pre-concentration step.

V.2.10.2. Solid Phase Extraction for Cadmium Pre-Concentration

The characteristics (composition, structure and functional groups) of the different solid supports used for cadmium SPE pre-concentration are summarized in Table 18. There are different requirements for solid phase materials so that they can be used as solid-phase extraction adsorbents. Firstly, these supports must allow the extraction of several metals in a wide pH range, and they must offer a high sorption capacity, selectivity and enrichment factor. Moreover, a fast and quantitative adsorption of the target metals and an easy desorption of the adsorbed target ions is desirable. Finally, the solid supports should have a long life time so that repeated adsorption/desorption cycles could be carried out with the same sorbent.

Different substances (Table 19) have been proposed and applied as solid-phase extraction sorbents for cadmium. These materials can be splitted in two general groups as synthethic (ion exchange resins, polymeric resins and chelating resins) or natural-biological (microorganisms and vermicompost) sorbents. In addition, new materials such as carbon nanotubes, nanoparticles, ordered mesoporous silica and ionic imprinted polymers, have also been proposed for cadmium solid phase extraction.

Table 18. Composition, structure and functional groups of several supports used for Cd pre-concentration by SPE

Sorbent	
Amberlite XAD	Hydrophobic polyaromatic resin. Macroreticular styrene-divinylbenzene copolymer.
Amberlite IT-120	Cation exchange resin registered trademark.
Amberlist® 36	Cation exchange resin registered trademark.
Capcell-NH_2	Chelating resin. Silicone coated silica gel with alkyl amino groups.
Chelamine	Chelating resin. Organic polymer containing pentamine ligand (1,4,7,10,13, pentaazatridecane) immobilised.
Chelex-100	Cation exchange resin. Styrene-divinylbenzene resin containing iminodiacetic acid groups ($HN(CH_2CO_2H)_2$).
Cellex P	Cation exchange resin. Cellulose sorbent with phosphonic acid groups ($R\text{-}PO(O^-H^+)_2$).
Chitosan modified ordered mesoporous silica	Natural polysaccharide containing hydroxyl and amino groups immobilized on aminopropyl modified ordered mesoporous silica.
Chromosorb	Hydrophilic cross-linked polystyrene resin. Styrene-divinylbenzene copolymer.
Dowex 50Wx4	Strongly acidic cation exchange resin. Styrene-divinylbenzene (gel) containing sulfonic acid functional groups ($R\text{-}SO_2O^-H^+$).
Dowex Optipore V-493	Hydrophobic resin. Macroporous styrene-divinylbenzene copolymer.
Polyorgs VII	Chelating resin. Poltacrylonitrile fibrous matrix containing amide and oxime complex forming groups.
SPE disk 3M™ Empore™	Cation exchange resin. Poly-styrene-divinylbenzene copolymer containing iminodiacetic acid groups ($HN(CH_2CO_2H)_2$).
Sepabeads	Chelating resin. Polyvinyl polymer with alkyl amino groups.
Sephadex	Strong anion exchange resin registered trademark.
SiO_2-TPP	Chelating resin. Aminopropyl silica gel containing porphyrin ligand (carbonyl phenyl-10,15,20-triphrnyl porphyrin) covalently attached.
TIOPAN-13	Chelating resin. Fibrous sorbent containing PAN with mercaptobenzothiazole functional groups.
Toyopearl AF Chelate 650M	Cation exchange resin. Macroporous methacrylate backbone containing iminodiacetic acid groups ($HN(CH_2CO_2H)_2$).
Zeolite A 4	Cation exchange resin. Highly crystalline aluminosilicate framework comprising SiO_2 and Al_2O_3 tetrahedral units.

Table 19. Solid-phase extraction procedures for cadmium pre-concentration from natural waters

Brief description	Sample	LOD	Concentration factor	Analytical technique	Ref.
180 mL of sample at pH 2.5 (10 mL of 0.1 M citric acid plus 0.1M HCl/0.1 M NaOH, as buffer solution) was mixed with 200 mg of chlorosulfonated polymer functionalized with thiourea. The mixture is stirred for 30 min and after centrifugation, Cd is eluted with 3 mL of 1.5 M HCl.	Tap water, lake water, mineral water	0.021 ng mL^{-1}	20	STAT-FAAS	[79]
An injection loop is filled with 2.0 mL of sample (pH 2.0 with 10 %(v/v) HCl), and then sample is displaced by a solution containing 0.5 %(m/v) DDTP at a flow rate of 2.4 mL min^{-1}. The Cd-DDPT complex that formed is adsorbed onto 20 mg of PUF into a mini-column for 2.0 min. Elution is performed with an etanol/water (80/20) mixture pumped at a flow rate of 1.0 mL min^{-1}.	Lake water and mineral water	0.62 µg L^{-1}	5.2	TS-FFAAS	[112]
Sample (volume of 5 to 20 mL) at pH 2.0 was pumped to C18 column functionalized with APDC at flow rate of 15 mL min^{-1}. Elution was carried out by pumping 500 µL of a solution methanol/water (60/40) containing 0.1 %(m/v) APDC at a flow rate of 0.9 mL min^{-1}.	Well water	0.03 to 0.007 µg L^{-1}	36 to 107	TS-FFAAS	[113]
Sample is pumped at a flow rate of 2.3 mL min^{-1} for 60 s and it is on-line mixed with a solution of 0.6 %(m/v) DDTP (1.1 mL). The complex is then adsorbed onto C18 mini-column previously rinsed with DDPT. Before elution, 0.5 %(v/v) HNO_3 is pumped to remove samples solution retained in the interstitial volume, and the mini-column is washed with acif¡dified DDPT solution. Elution is performed with 25 µL of methanol.	Seawater	0.0002 µg L^{-1}	45	ETV-ICP-MS	[215]
190 mL of sample acidified with HCl (final HCl concentration of 2.0 M) was contacted with the membrane (PVDF, average pore-size of 0.2 mm, porosity of 75%, and average thickness of 125 mm) activated with Aliquat 336 under continuous stirring (800 rpm) for 150 min. At the end of	Seawater	0.075 µg L^{-1}	--	HE-P-EDXRF	[316]

the preconcentration, the loaded membranes were rinsed with deionized water and allowed to dry at room temperature. Finally, the cadmium-loaded membranes were directly placed into the sample holder device for XRF analysis					
190 mL of sample acidified with HCl (final HCl concentration of 2.0 M) was contacted with the membrane CTA-NPOE with physical inclusion of Aliquat 336 under continuous stirring (800 rpm) for 150 min. At the end of the preconcentration, the loaded membranes were rinsed with deionized water and allowed to dry at room temperature. Finally, the cadmium-loaded membranes were directly placed into the sample holder device for XRF analysis.	Seawater	0.71 μg L^{-1}	--	HE-P-EDXRF	[317]
1000 mL of sample at pH 6.0 was passed through a chelating SPE disk 3M™ Empore™ (cation exchange). Elution with HNO_3.	Tap water	3.8 ng mL^{-1}	-	XRF	[325]
Same procedure than reference 316	Seawater	0.17 mg L^{-1}	--	WDXRF	[326]
50 mL of sample was stirred (5.0 min) with 0.5 g of Amberlite IT-120. After drying at 100°C, the resin was grinded and direct analysed.	Seawater	-	-	ETAAS	[455]
Sample at pH 5.0 was passed through the 50 mL syringe filled with Chelex-100 at 7.0 mL min^{-1}. 1.8 mL of HNO_3 (2.0 M) was used for elution step.	Lake water	0.2 ng L^{-1}	25	ICP-MS	[469]
Sample at pH 7.0 was passed through the column filled with Dowex 1x2 loaded with 8-HQ-5-sulfonic acid. HNO_3 (7.0 M) was used for elution step.	Natural water	-	-	ETAAS	[485]
50 mL of sample at pH 8.0 (ammonium acetate buffer) containing 0.25 mL of HEDC (0.5 M) solution was passed at 8.0-10 mL min^{-1} through the column filled with 1.0 g of Dowex Optipore V-493. After elution with 5.0 mL of HNO_3 (1.0 M) in methanol, the eluate was evaporated to near dryness and made up to 1.0-2.0 mL of HNO_3 (1.0 %).	Seawater	-	25	ETAAS	[499]

Table 19. Continued

Brief description	Sample	LOD	Concentration factor	Analytical technique	Ref.
Sample was on-line preconcentrated at 1.0 mL min^{-1} into a Toyopearl AF-Chelate 650 M mini-column using a flow injection system. After elution with HNO_3 at 0.5 mL min^{-1}, eluate was transported directly into the nebulizer.	Seawater	-	-	ICP-MS	[543]
Sample was lowered at a constant rate from the zone of a circular resistance heater to a cooling thermostat chamber (−15°C) filled with liquid antifreeze and cooled by means of a compression-type refrigerator. During this process sample was mixed by reverse rotation of the vessel about its vertical axis (the period of rotation in one direction and the duration of a stop being 3 and 1 s, respectively). On completion of crystallization, 1.5 mL of hot water was poured onto the upper part of the ingot where the impurity was located.	Lake water	-	-	ICP-OES	[590]
50 mL of sample containing 1.0 mL of ammonium acetate buffer was passed through column filled with 5.9 mL of Chelex-100 slurry at 1.0-2.0 mL min^{-1}. 5.0 mL of HNO_3 (5.0 M) was used for elution step.	Seawater	2.0 ng L^{-1}	10	ETAAS	[630]
250 mL of sample at pH 4.0 was stirred for 30 min with 0.05 g of Chelex-100. After resin filtration and suspended in 5.0 mL of water, Cd was directly analysed in the suspension.	River water	0.1 µg L^{-1}	-	ETAAS	[632]
500 mL of sample at pH 5.4 or 8.0 (20 mL of ammonium acetate buffer) was passed at 4.0 mL min^{-1} through the column filled with 0.4 g of Chelex-100. Elution with 4 x 2.0 ml of HNO_3 2.0 M and dilution to 10 mL.	Seawater	1.0 ng L^{-1}	50	FAAS	[633]
Sample at pH 7.0 was on-line preconcentrated at into a Chelex-100 mini-column using a flow injection system. After elution with HNO_3, the eluate was transported directly into the nebulizer.	Seawater	17 ng L^{-1}	-	FAAS	[636]
50 mL of sample at pH 5.0-6.0 (acetate buffer) was passed at 4.0 mL	Tap water	1.0 µg L^{-1}	10	ETAAS	[638]

min^{-1} through 0.2 g of ethylcellulose with iminodiacetic groups in a polyethylene funnel at 7.0-8.0 mL min^{-1}. Elution with 2 x 2.5 mL of HNO_3 1.0 M and dilution to 10 mL.					
2000 mL of sample at pH 7.1 was passed at 8.0 mL min^{-1} through a column filled with 2.0 g of Chellex-P. 25 mL of HNO_3 (1.0 M) was used for elution step.	Tap water	-	80	ETAAS	[639]
100 mL of sample at pH 6.0 was on-line preconcentrated at 1.5 mL min^{-1} into a column filled with 0.08 g Dowex 50W by using a flow cell connected to a FIA. HCl (3.0M) at 1.5 mL min^{-1} was used for elution.	Tap water	0.3 µg	-	EDXRF	[640]
800 mL of sample at pH 6.0 (diluted sodium hydroxide or nitric acid) was stirred with 100 mg of Zeolite A-4. After filtration (membrane filter), resin was dissolved in 2.0 mL of HNO_3 (2.0 M).	Natural water and waste water	20 µg mL^{-3}	400	ETAAS	[641]
Sample was passed through the column filled with Amberlyst 36. 5.0 mL of HCl (3.0 M) at 1.0 mL min^{-1} was used for elution step.	Tap water	0.51 µg L^{-1}	200	FAAS	[642]
3.0 mL of sample at pH 5.6 was on-line preconcentrated at 1.0 mL min^{-1} into a column filled with Toyopearl AF-Chelate 650M resin. HNO_3 (1.5 M) at 1.0 mL min^{-1} was used for elution.	Estuarine water	1.4 ng L^{-1}	-	ICP-MS	[643]
100 mL of sample at pH 8.0 (dilute HCl or ammonia) was passed at 1.0 mL min^{-1} through the column filled with 0.5 g of AXAD-7. 5.0 mL of HCl (1.0M) at 1.0 mL min^{-1} was used for elution step.	Lake water and seawater	-	-	ETAAS	[644]
100 mL of sample at pH 8.0 (dilute ammonia) was passed at 1.0 mL min^{-1} through the column filled with AXAD-7. 5.0 mL of HNO_3 (1.0%) was used for elution step.	Tap water	-	-	ETAAS	[645]
500 mL of sample at pH 6.0 containing 10 mL of NN (10 mM) was passed through 500 mg of AXAD 4column at 5.0 mL min^{-1}. After elution with 7.0-8.0 mL HNO_3 (1.0 M), eluate was evaporate near to dryness (hot plate at 35°C) and then dilute to 5.0 mL with HNO_3 (1.0 M)	Tap water	16 ng L^{-1}	-	ETAAS	[646]

Table 19. Continued

Brief description	Sample	LOD	Concentration factor	Analytical technique	Ref.
100 mL of sample at pH 5.0 (dilute HCl 0.1 M) was passed at 2.0 mL min^{-1} through a column filled with 1.5 g of In-treated AXAD-2. After column disassembled, desorption (3.0 mL of HNO_3 (0.5 M), ultrasound (1.0 min)), filtration on Nucleopore filter (0.4 µm), evaporation to dryness and dissolution in 1.0 mL of HNO_3 (0.5 M) were carried out.	River water	-	100	ETAAS	[647]
100 mL of sample at pH 5.0 (dilute HCl0.1 M) was passed at 2.0 mL min^{-1} through a DEAE-Sephadex A-25 column. After column disassembled, desorption (3.0 mL of HNO_3 (4.0 M), ultrasound (1.0 min)), filtration on Nucleopore filter (0.4 µm), evaporation to dryness and dissolution in 1.0 mL of HNO_3 (0.5 M), were carried out.	River water	-	100	ETAAS	[647]
Sample at pH 9.0 (NH_3/NH_4Cl buffer) containing HMA-HMDTC was stirred with AXAD-16.	Seawater	-	150	FAAS	[648]
Sample at pH 8.0 containing PAR (50-fold excess) was passed through 50 mg of C-18 column at 3.0 mL min^{-1}. After elution with 3.0mL HNO_3 (1.0 M) at 0.1 mL min^{-1}, eluate was directly determined.	Natural water	30 µg L^{-1}	20	FAAS	[651]
600 mL of sample at pH 8.5 (maleic acid/ammonium hydroxide) containing HEDC (1.2 mM) was passed through C-18 cartridge at 10 mL min^{-1}. After elution with 3.0 mL HNO_3 (1.0 M) in methanol at 10 mL min^{-1}, eluate was evaporated to near dryness and then dilute to 1.0 mL with HNO_3 (0.1 M).	Seawater	-	600	ICP-MS	[652]
500 mL of sample containing 0.5 mL of 8-HQ (0.5 M) was passed through a column filled with 1.0 g of C18 at 4.0mL min^{-1}. 10 mL HNO_3 (2.0 M), was used for elution step.	Seawater	-	50	ICP-OES	[653]
100-1000 µL of sample at pH 6.0-8.0 containing 40µL APDC (5.0 % m/v) was passed through a mini-column filled 5 mg of C-18 at 0.1 mL min^{-1}. 20 µL of methanol was used for elution.	Seawater	0.178 ng L^{-1}	5-50	ETAAS	[656]
100 mL of sample at pH 8.0 containing 8-HQ (5.0 10^{-4} M) was passed through	Seawater	0.04 µg L^{-1}	40	ICP-OES	[657]

C-18 cartridges at 6 mL min^{-1}. 2.5 mL of nitric acid (2.0 M) was used for elution at a flow rate of 4.5 mL min^{-1}.					
150 mL of sample at pH 10 (2.0 mL of ammonia/ammonium acetate) containing 2.0 mL of nitroso-S (0.1 % w/v) and 2.0 mL of TBDA was treated with 2.0 mL of naphthalene in acetone (20%) under continuous shaking. After filtration, the solid was shaken with 9.0 mL of HCl (1.0 M) and 1.0 mL KCl (1.0 M) and directly measured.	Tap water	90 ng mL^{-1}	15	DPASV	[658]
200 mL of sample at pH 8.5 was mechanical stirred (15 min) with 1.5 g of H_2DZ co-crystalline with NAP. After filtration with paper filter, solid was washed with 15 mL of HNO_3 (8.0 M).	Saline water	5.0 µg L^{-1}	13	ICP-OES	[659]
Sample containing 1-BPzDC was extracted with microcrystalline NAP. After eluting with KCN solution, Cd was directly determined.	Natural water	34 µg L^{-1}	50	FAAS	[660]
Sample at pH 7.0-10.5 containing TAN-TPB was stirred with NAP adsorbent. After filtration, the solid mass was shaken with 8.0 mL of HCl (2.0 M) and 2.0 mL of $NaBH_4$ (4.0%).	Natural water	66 ng mL^{-1}	-	DPP	[661]
Sample at pH 8.2 -11 was passed through a PAN co-crystalline with NAP packed column. After eluting with 9.0 mL of HCl (1.0 M), Cd was directly determined.	Natural water	70 µg L^{-1}	-	DPP	[662]
600 mL of sample pH 6.0 containing 3.0 mL of APDC (0.05 % w/v) was passed through a mini-column filled with 500 mg of Chromosorb-102 at 2.0 mL min^{-1}. After elution wit 10 mL of acetone, elute was evaporated to near dryness and dilute to 2.0-10 mL with HNO_3 (1.0 M).	Tap water and seawater	0.10 µg L^{-1}	-	FAAS	[663]
250 mL of sample pH 9.0-9.5 (NH_3/NH_4Cl buffer) was passed through a column filled with 500 mg of Chromosorb-102 at 5.0 mL min^{-1}. After elution with10 mL HNO_3 (1.0 M), the eluate was evaporate to dryness (hot plate at 30-40°C) and then dissolved in 2.0 mL of HNO_3 (1.0 M).	Tap water and seawater	0.8 µg L^{-1}	-	FAAS	[664]

Table 19. Continued

Brief description	Sample	LOD	Concentration factor	Analytical technique	Ref.
5 x 5 mL of sample at pH 6.0 (HCl 0.1 M or NaOH 0.1 M) treated with 1.0mL APDC (0.1 % w/v) was drawn at 20 mL min^{-1} into a syringe filled with 0.3 g Chromosorb-107. Elution was carried out with 2.5 mL HNO_3 (3.0 M) at 20 mL min^{-1}.	Seawater	1.2 μg L^{-1}	10	ETAAS	[665]
500 mL of sample at pH 9.0 (ammonia/ammonium chloride buffer) containing 3.0 mL of PAN (0.01 M) solution was passed at 5.0 mL min^{-1} through the column filled with 600 mg of Chromosorb-106. After elution with 8.0-10 mL of HNO_3 (1.0 M) in acetone at 5.0 mL min^{-1}, the eluate was evaporated to near dryness and made up to 2.0 mL with HNO_3 (1.0 M).	Tap water, river water, mineral water and seawater	0.19 μg L^{-1}	-	FAAS	[666]
250 mL of sample at pH 2.0 containing 5.0 mL of DBDTC (0.01 M) was passed at 4.0 mL min^{-1} through the column filled with 600 mg of Dowex Optipore V-493. After elution with 5.0-10 mL of HNO_3 (1.0 M) at 4.0 mL min^{-1}, eluate was evaporated to 0.5-1.0 mL and dilute to 2.5mL with HNO_3 (1.0 M).	Tap water, river water, mineral water and seawater	0.43 μg L^{-1}	100	FAAS	[667]
1.0 L of sample at pH 7.0 (sodium citrate 0.7 mM and HCl) was stirred for 1.0 h with 0.5 g of Polyorgs VII M sorbent. After filtration, Cd was eluted with 8.0-9.0 mL of HNO_3 (1.0 M).	Mineral water and spring water	-	-	ETAAS	[668]
0.5 L of sample at pH 7.0 (sodium citrate 0.7 mM and HCl) was stirred for 1.0h with 0.2 g of Polyorgs VII M sorbent. After resin filtration and suspended in 5.0-10 mL of water, Cd was directly analysed in the suspension.	River water, groundwater and seawater	1.0 ng mL^{-1}	-	ETAAS	[669]
10 mL of sample at pH 6.0-8.0 was stirred ultrasonically for 2.0 min with 20 mg of Polyorgs VII M sorbent. After centrifugation and suspended in 1.0 mL of water, Cd was directly analysed in the suspension.	Seawater	-	-	ETAAS	[670]
10-20 mL of sample at pH 6.0-8.0 was ultrasonically stirred (5.0 min) with 0.02 g of Polyorgs YIIM. After	Seawater	10 ng L^{-1}	-	ETAAS	[672]

centrifugation (6000 rpm) and rinse with 10 mL of water, the solid mass, as slurry, was directly analysed.					
500 mL of sample at pH 7.0 (ammonia) was stirred for 45 min with 0.5 g of TIOPAN-13 sorbent. After elution with 2 x 5 mL of HNO_3 (1.0 M), the combined extract was evaporated to 5.0 mL and analysed.	River water	8.0 ng L^{-1}	-	ETAAS	[673]
400 mL of sample at pH 8.0 (TRIS 1.0 M and ammonia 2.0 M) was passed through the Chelamine column at 0.3 mL min^{-1}. 4.0 mL of HNO_3 (2.0 M) was used for elution.	Seawater	2.3 ng L^{-1}	100	ETAAS	[674]
1000 mL of sample at pH 4.5-5.0 (20 mL of acetate buffer) was shaken with 0.5 g of AC loaded with ZnPDC for 10 min. After filtration, the solid phase was treated with 2.0-5.0 mL of Hg(II) (0.1 M) and stirred 10 min.	Tap water and seawater	11.8 ng mL^{-1}	-	FAAS	[675]
50 mL of sample at pH 5.0 was passed at 1.0 mL min^{-1} through the column filled with 500 mg of APDC-AC. HNO_3 (1.0 M) in acetone at 0.6 mL min^{-1} was used for elution step.	Tap water and pom water	20 ng L^{-1}	-	FAAS	[676]
Sample at pH 4.5 was passed through the column filled with AC impregnated with DDC. After elution at 1.5 mL min^{-1} with 10mL of HNO_3 (1.0 M) in acetone, the eluate was direct analyzed.	Natural water	-	-	FAAS	[677]
Sample was on-line preconcentrated at 1.1 mL min^{-1} into a Dowex 1 X 10 column loaded with dithizone using a flow injection system. After elution with NaOH (1.0mM) at 1.0 mL min^{-1}, eluate was measured at 544 nm. Sampling frequency of 11.4 h^{-1}.	Surface water, well water and tap water	5.4 µg L^{-1}	6.5	UV-Vis	[678]
800 mL of sample at pH 7.0 was passed at 2.0 mL min^{-1} through the column filled with 500 mg of AXAD-4 modified with AMPDAA. 8.0 mL of HCl (1.0 M) – NaCl (1.0 M) and 2.0 mL of water was used for elution step.	Tap water and river water	28 ng L^{-1}	-	FAAS	[679]
Sample was on-line pre-concentrated at 3 mL min^{-1} into a AXAD-2 column loaded with TAM using a flow injection system. After elution with HCl (1.0 M) at 1.0 mL min^{-1}, eluate was transported directly into the nebulizer.	Fresh water with low saline conc.	22 ng L^{-1}	548	FAAS	[680]

Table 19. Continued

Brief description	Sample	LOD	Concentration factor	Analytical technique	Ref.
1000 mL of sample at pH 8.0 was passed at 5.0 mL min^{-1} through the mini-column filled with 200 mg of AXAD-2 loaded with DNDAA. After elution with 10 mL of HCl (0.1 M), the eluate was direct analyzed.	Tap water and river water	62 ng L^{-1}	100	FAAS	[681]
50mL of sample at pH 5.0 (ammonium hydroxide) was passed at 1.0 mL min^{-1} through a micro-column packed with 80 mg of 8-quinol immobilized on SG. 1.7 mL of HCl (2.0 M)-HNO_3 (0.1 M) was used for elution step.	Seawater	42 ng L^{-1}	-	ETAAS	[682]
20 mL of sample at pH 5.0 (acetate buffer) was passed through a SG modified with DDC or HMA-HMDTC column at 1.0 mL min^{-1}. 2.0 mL IBMK at 1.0 mL min^{-1} was used for elution.	River water	0.04 μg L^{-1}	-	ETAAS	[683]
Sample at pH 9.0 was stirred during 30 min with 20 mg of porous SiO_2 modified with IE11. HNO_3 (2.0 M) was used for elution.	River water, tap water and seawater	-	-	FAAS	[684]
Sample at pH 5.0-10 was passed at 2.0 mL min^{-1} through the column filled with 1.0 g of TAA immobilized on SG. 10.0 mL of HCl/HNO_3 (0.1/0.5 M) at 2.0 mL min^{-1} was used for elution step.	Tap water and river water	0.96 μg L^{-1}	200	FAAS	[686]
Sample was passed through a column filled with of functionalised SG-ofloxacin. After elution with HCl (0.5 M), the eluate was directly measured.	River water and seawater	0.29 ng mL^{-1}	100	ICP-OES	[687]
500 mL of sample at pH 8.0 was passed at 4.0 mL min^{-1} through the column filled with 300 mg of SG-TREN. After elution with 5.0 mL of HCl (0.1 M), the eluate was direct analyzed.	River water	0.14 ng mL^{-1}	100	ICP-OES	[688]
1000 mL of sample pH 4.0-4.5 was passed through a column filled with 1.0 g of TSA-AXAD-2 at 2.0–5.0 mL min^{-1}. 5.0 mL of HNO_3 (2.0M) at 1.5-2.5 mL min^{-1} was used for elution.	Tap water and river water	0.48 μg L^{-1}	200	FAAS	[689]
1000 mL of sample at pH 9.0 containing 5.0 mL of PAN (0.75 mM) was passed at 5.0-10 mL min^{-1} through a column filled with 0.5 g AXAD-4. After elution with 8.0-10	River water, tap water and lake water	0.16 μg L^{-1}	200	FAAS	[690]

mL of HNO_3 (1.0 M) in acetone, eluate was evaporated to near dryness and dissolved in 5.0 mL of HNO_3 (1.0 M).					
Sample at pH 8.0 (borate buffer) was on-line preconcentrated at 5.0 mL min^{-1} into a AXAD-2 mini-column functionalized with AT using a flow injection system. After elution with HCl (0.5 M) at 5 mL min^{-1}, eluate was transported directly into the nebulizer.	River water, well water, tap water and mineral water	0.89 μg L^{-1}	28	FAAS	[691]
Sample at pH 5.0 was passed through a 2,2'-dithiobisaniline modified AXAD-2 column. After elution with HCl (2.0 M), eluate was transported directly into the nebulizer.	Tap water and river water	0.1 μg L^{-1}	80-200	FAAS	[692]
Sample at pH 7.5 (borate buffer) was on-line pre-concentrated at 7 mL min^{-1} into a PUF column loaded with Me-BTANC using a flow injection system. After elution with HCl (1.0 M) at 5.0 mL min^{-1}, eluate was transported directly into the nebulizer. Sampling frequency of 19 h^{-1}.	River water, tap water and well water	0.8 μg L^{-1}	37	FAAS	[694]
100 mL of sample at pH 7.0 (sodium citrate 0.7 mM and HCl) was mixed with 10 mg of freeze-dried Stichococcus bacillaris. After centrifugation and suspended in 1.0 mL of water, Cd was directly analysed in the suspension.	River water and seawater	-	-	ETAAS	[695]
100 mL of sample at pH 7.5 (dilute HCl or ammonia) was passed at 3.0 mL min^{-1} through the column filled with 0.2 g of sepiolite immobilized with Saccharomyces cerevisiae. 10 mL of HCl (1.0 M) was used for elution step.	River water and seawater	44 ng mL^{-1}	75	FAAS	[696]
500 mL of sample at pH 6.0 (dilute HCl or ammonia) was passed at 3.0 mL min^{-1} through the column filled with 0.3 g of SG 60 loaded with Aspergillus niger. 10 mL of HCl (1.0 M) at 1.0 mL min^{-1} was used for elution step.	Tap water and lake water	1.4 ng mL^{-1}	50	FAAS	[697]
Sample was passed through a column filled with SG-60 loaded with Saccharomyces carlsbergensis. HCl was used for elution step.	Dam water, lake water and tap water	1.48 ng mL^{-1}	705	FAAS	[698]

Table 19. Continued

Brief description	Sample	LOD	Concentration factor	Analytical technique	Ref.
Sample was on-line pre-concentrated at 4 mL min^{-1} into a mini-column packed with humic acids or vermicompost using a flow injection system. After elution with HNO_3 (1.0 M) at 4.0 mL min^{-1}, eluate was transported directly into the nebulizer.	Mineral water and tap water	0.8 μg L^{-1} (0.4 μg L^{-1})[a]	27 (46)[a]	FAAS	[699]
50 mL of sample at pH 8.0 was passed at 2.0 mL min^{-1} through a column filled with 30 mg MWCNT. 2.0 mL of HNO_3 (0.5 M) was used for elution step.	Lake water	48 ng L^{-1}	25	ICP-OES	[703]
200 mL of sample at pH 9.0 was passed through a 0.2 g MWCNT packed cartridge. 10 mL of HNO_3 (1.0 M) at 1.5 mL min^{-1} was used for elution.	Tap water, reservoir and stream water	-	-	ETAAS	[704]
Sample at pH 4.9 was on-line preconcentrated at 5.0 mL min^{-1} into a microcolumn filled with 30 mg of MWNTs. HNO_3 (0.5 M) was used for elution.	Mineral water, tap water and river water	11.4 ng L^{-1}	51	TS-FFAAS	[705]
Sample at pH 5.0 was passed through column filled with DDC-TiO_2 nanoparticles. After elution, eluate was transported directly into the nebulizer.	Natural water	0.52 ng mL^{-1}	-	ICP-OES	[707]
Sample at pH 9.0 (ammonia 0.01 M) was flow injection on-line preconcentrated at 2.0 ml min^{-1} into a column filled with Al_2O_3 nanoparticles. 0.4 mL HNO_3 (1.0 M) at was used for elution. Sampling frequency of 15 h^{-1}.	Lake water	79 ng L^{-1}	5	ICP-MS	[708]
6.0 mL of sample at pH 6.5 was on-line preconcentrated at 2.0 ml min^{-1} into a PTFE micro-column filled with 50 mg of chitosan modified ordered mesoporous silica. 0.3 mL of HCl (1.0 M) was used for elution. Sampling frequency of 10 h^{-1}.	Lake water and river water	50 ng L^{-1}	20	ICP-OES	[710]
100 mL of sample adjusted at pH 6.0 was passed through the column (a glass tube of 100 mm in length and 10 mm in diameter) containing 300 mg of IIP at a flow rate of 3 mL min^{-1}. Elution was performed with 5 mL of 0.2 M HCl (or HNO_3) at a flow rate of 1 mL min^{-1}.	River water, lake water, tap water	0.14 μg L^{-1}	20	ICP-OES	[714]

100 mL of sample adjusted at pH 5.0 was passed through the column (a glass tube of 100 mm in length and 10 mm in diameter) containing 500 mg of IIP at a flow rate of 0.5 mL min^{-1}. Elution was performed with 2 mL of 0.2 M HCl at a flow rate of 0.5 mL min^{-1}.	River water, tap water, river sediment	0.093 μg L^{-1}	50	FAAS	[715]
25 mL of sample was stirred at constant potential (20.5 V) for 1000 s for extraction of Cd on polypyrrole-coated electrode. After the washing step, desorption was made into a vial containing 200 μL of HNO_3 (0.01 M) by scanning the potential to +0.8 V, at a scan rate of 100 mV s^{-1}. The latter potential was maintained for 2 min.	Natural water	-	125	ICP-MS	[718]

[a] value in brackets corresponds to vermicompost

AC, activated carbon; AMPDAA, 2-acetylmercaptophenyl diazoamino- azobenzene; APDC, ammonium pyrrolidin dithiocarbamate; AT, aminothiophenol; AXAD, Amberlite XAD; CTA, cellulose triacetate; CZE, capillary zone electrophoresis; DBDTC, dibenzyldithiocarbamate; DDC, diethyldithiocarbamate; DDTP, O,O-diethyldithiophosphate; DNDAA, 2,4-dinitrophenyldiazoaminoazobenzene; DPP, Differential pulse polarography; EDXRF, energy dispersive x-ray fluorescence; ETAAS, electrothermal atomic absorption spectrometry; FAAS, flame atomic absorption spectrometry; FIA, flow injection analysis; HEDC; bis (2-hydroxyethyl)dithiocarbamate; HE-P-EDXRF, high-energy polarized- energy dispersive x-ray fluorescence; HMA-HMDTC, hexamethyleneammonium hexamethylenedithiocarbamate; IBMK, 4-methyl-2-pentanone; ICP-MS, inductively coupled plasma-mass spectrometry; ICP-OES, inductively coupled plasma-optical emission spectrometry; IE11, N-propylsalicylaldimine; Me-BTANC, 2-(6'-methyl-2'-benzothiazolylazo)chromotropic acid; MWCNT, multi-walled carbon nanotubes; NAP, microcrystalline naphthalene; NN, 1-nitroso-2-naphtho; Nitroso-S, 2-nitroso-1-naphthol-4-sulfonic acid; NPOE, 2-nitrophenyl octyl ether; PAN, 1-(2-pyridylazo)-2-naphthol; PAR, 4-(2-pyridylazo)resorcinol; PUF, polyurethane foam; PVDF, polyvinylidene difluoride; SG, functionalized silica gel; STAT-FAAS, slotted tube atom trap - flame atomic absorption spectrometry; TAA, thioacetamide; TAM, 2(2-thiazolylazo)-5-dimethylaminophenol; TAN, 1-(2-thiazolylazo)-2-naphthol, TDBA, tetradecyldimethylbenzylammonium; THF, tetrahydroforan; Tm-APP, tetra(m-aminophenyl)porphyrin; TPB, tetraphenylborate; TREN, tris(2-aminoethyl) amine; TSA, thiosalicylic acid; TS-FFAAS, thermospray flame furnace atomic absorption spectrometry; UV-Vis, ultraviolet- visible; WDXRF, wavelength dispersive x-ray fluorescence spectrometry; XRF, x-ray fluorescence spectrometry; ZnPDC, Zn-piperazinedithiocarbamate; 1-BPzDC, 1-benzylpiperazinedithiocarbamate; 8-HQ, 8-hydroxiquinoline.

V.2.10.2.1. Synthetic Sorbents

V.2.10.2.1.1. Ion Exchage Resins

Cation exchange resins such as Chelex-100 [466,469,630-637], ethylcellulose with iminodiacetic groups [638], Cellex-P [635,637,639], Dowex-50 W [485,635,637,640], Zeolite [641], Amberlite IT-120 [455], Amberlist-36 [642], Toyopearl AF-Chelate 650M [543,643] and 3M™ Empore™ SPE disks [325], have been extensively used as solid phase extractors for cadmium (and other metals).

V.2.10.2.1.2. Polymeric Resins

Polymeric resins such as activated carbon (AC), silica gel (SG), Amberlite XAD (AXAD), cellulose, Chromosorb, polyurethane foam, silica, etc., have also been proposed for cadmium pre-concentration. There are two ways for trace metals pre-concentration by using these solid supports. In the first method, chelates of trace metals must be formed by the addition to the sample of a specific chelating reagent, and then the sample solution is passed through the column filled with the resin (column mode) or the sample is stirred with the resin (batch mode) to retain the chelates. In the second method, chelating agents are firstly immobilized on the solid phase (chelating resins) and then, the sample is passed through the column containing this modified solid phase, or sample is stirred with the chelating resin.

Therefore, Amberlite XAD (AXAD) [644-648], Sephadex [647,649], nitrocellulose membrane [650], C-18 [113,215,651-657], co-crystallized NAP adsorbent [658-662], Chromosorb [663-666]Dowex Optipore [499,667] or polyurethane foam (PUF) [112], have been used for cadmium sorption on the solid phases previous chelating forming step. The chelating agents used in these applications were O,O-diethyldithiophosphate (DDTP) [112,215], 1-nitroso-2-naphtol (NN) [646], hexamethyleneammonium hexamethylenedithiocarbamate (HMA-HMDTC) [648], 1-(2-pyridylazo)-2-naphthol (PAN) [650,662 ,666], or 4-(2-pyridylazo)resorcinol (PAR) [649,651], bis (2-hydroxyethyl)dithiocarbamate (HEDC) [652], 8-hydroxiquinoline (8-HQ) [653,657], tetra(m-aminophenyl) porphyrin (Tm-APP) [654], ammonium pyrrolidin dithiocarbamate (APDC) [113,656,663,665], 2-nitroso-1-naphthol-4-sulfonic acid (Nitroso-S) and tetradecyldimethylbenzylammonium (TDBA) [658], H_2DZ [659], 1-benzylpiperazinedithiocarbamate (1-BPzDC) [660], and 1-(2-thiazolylazo)-2-naphthol (TAN) [661].

V.2.10.2.1.3. Chelating Resins

The use of chelating resins is especially appealing bcause the high selectivity offered. Therefore, the use of chelating resins such as Polyorgs VII [668-671], Polyorgs YIIM [672], TIOPAN-13 [673], SiO_2-TPP [637], and Chelamine [654,674], and sorbents with functional chelating groups introduced into the resin structure, have extensively been used for trace metal pre-concentration.

Chelating resins can be obtained after synthesis of the support containing the chelating groups, although other times, the chelating groups can be bound to conventional materials. In this last case, the complexing reagents can be chemically introduced into the conventional supports (functionalized sorbents), or they can be physical bounded to the sorbent by impregnating the solid matrix with a solution containing specific molecules (loaded sorbents). Physical binding is the simplest method to be used in practice. However, the chemical binding approach allows elevated lifetime for the resin due to covalent bonds between the ligand and the support. This property avoids the possible flush of the ligand molecule from the column during sample percolation or elution steps. Several complexing agents have been loaded or immobilized on a variety of solid matrices and they have successfully been used for the pre-concentration and determination of cadmium. Some examples include Zn-piperazinedithiocarbamate (ZnPDC) [675], APDC [676], and DDC [677] which can be loaded or impregnated on AC support. Simmilarly, H_2DZ [678] can be loaded on Dowex, 2-acetylmercaptophenyl -diazoaminoazobenzene (AMPDAA) [679], while 2(2-thiazolylazo)-5-dimethylaminophenol (TAM) [680] and 2,4-dinitrophenyldiazoaminoazobenzene (DNDAA) [681] have been loaded on AXAD. In addition, a wide range of new sorbents for trace cadmium extraction have been developed by immobilizing several chelating groups on silica or ion-exchange resins. Agents such as 8-quinol [682], DDC [683], HMA-HMDTC [683], N-propylsalicylaldimine (IE11) [684], calcon [685], thioacetamide (TAA) [686], ofloxacin [687], and tris(2-aminoethyl) amine (TREN) [688] have been immobilized on silica; while applications of AXAD impregnated with thiosalicylic acid (TSA) [689], PAN [690], aminothiophenol (AT) [691] or 2,2'-dithiobisaniline[692] have been also reported. Finally, 8-quinol [693] has also been impregnated on Capcell-NH_2 and Sepabeads supports.

Applications of the impregnation of 2-(6'-methyl-2'-benzothiazolylazo) chromotropic acid (Me-BTANC) on PUF [694], thiourea on chlorosulfonated polymers [79] or tricaprylmethylammonium chloride (Aliquat 336) on polyvinylidene difluoride (PVDF) films

[316,326] or Aliquat 336 physically included in cellulose triacetate – 2-nitrophenyl octyl ether (CTA-NPOE) membranes [317], have also been reported.

Chemically modified silica gel (SG) is one of the most successful adsorbents (Table 19), because the SG supports does not swell or shrink like the other polymeric resins. The modified SG may be employed for aqueous and organic solvents medium, moreover offer good thermal stability and appropriate accessibility of ions to the adsorbent groups. In addition, the modified SG exhibits higher sorption capacities than other polymeric resins because the number of organic molecules immobilized on the support surface is larger. However, silica sorbents are unstable in a broader pH range.

V.2.10.2.2. Natural/Biological Sorbents

V.2.10.2.2.1. Microorganisms

Living or non-living microorganisms can accumulate metals from aqueous solutions by a different chemical and biological mechanism without pre-concentrating the matrix. This ability is attributed to stable complexes between metals and any functional groups present in the cell walls of the microorganisms. In addition, the microbial cell products such as metabolites (polysaccharides) are effective in metal accumulation. Thus, the use of microorganisms for biosorption of trace metals has been proposed. In this technique, microorganisms are immobilized on a solid support instead of the chelating agent. Microorganisms have been used as freely suspended cells or as loaded cells on a support. The use of loaded cell systems on a support has many advantages over the use of freely suspended cells. These include better capability of reusing the biomass, better precision for recovery, easy separation of cells from the reaction mixture, high biomass loadings, and minimal plugging in continuous flow systems. For cadmium pre-concentration, freeze-dried *Stichococcus bacillaris* [695] has been used as suspended cells. In addition, immobilization of *Saccharomyces cerevisiae* [696], *Aspergillus niger* [697], and *Saccharomyces carlsbergensis* [698] on materials such as sepiolite [696] and SG [697,698] has been used to design solid phase extractors.

V.2.10.2.2.2. Vermicompost

Vermicompost is a humic substance recently characterized which has also been applied to cadmium pre-concentration by SPE [699]. This material offers a high cationic exchange capacity (CEC) for different species due to the presence of carbonyl groups and phenolic and alcoholic hydroxyls, among others functional groups, in its structure. This very stable and complex product is formed when earthworms (*Lumbricus rubellus*, mainly) metabolise organic residues from different sources [699]. Fulvic and humic acids are present in vermicompost are responsible for the high CEC. In this application, humic substances (vermicompost and purified humic acid), are packed in small columns which can be coupled to FAAS.

The use of natural sorbents for metals has many advantages over other sorbents. Some of them are the higher recoveries obtained and the simplicity of the procedure, economical advantages and environmental protection. However, these sorbents offer poor selectivity due to the predominance of electrostatic forces among sites (placed on their surfaces) and analytes. In this sense, unspecific retentions are usually observed. In addition, these sorbents present low uniformity composition and, obviously, this aspect can damage the precision and accuracy of the results.

V.2.10.2.3. New Sorbents

Conventional sorbents used in typical SPE procedures for both organic compounds and trace elements pre-concentration can be affected from interfering concomitants from the matrix which can be co-extracted with the target analytes. Therefore, more selective systems for separation are required and the development or synthesis of new sorbents has increased in recent years.

High selectivity can be expected from sorbents such as inmunosorbents (ISs) and molecularly imprinted polymers (MIPs), although MIPs are advantageous over ISs because the fast and less expensive synthesis and the high degree of molecular recognition [700]. Carbon nanotubes, nanometer-sized materials, mesoporous materials, biomaterials, C60–C70 and their derivatives have been extensively explored as new adsorbent in SPE technique. Some of then, have been applied to cadmium pre-concentration and will be commented in this section.

V.2.10.2.3.1. Carbon Nanotubes

Since carbon nanotubes (CNTs) were first prepared in 1991 [701], they have become appealing materials for their novel structure characteristics and unique properties. Novel mechanical and electronic properties, a large specific surface area and a high thermal stability indicate the tremendous potential of CNTs for engineering applications such as hydrogen storage, field emission, quantum nanowires, catalyst supports, chemical sensors and packing material for gas chromatography [702]. CNTs can be visualized as a graphite sheet rolled up into a cylinder, and they are classified into single-walled carbon nanotubes (SWNTs) and multi-walled carbon nanotubes (MWNTs) according to the carbon atom layers in the wall of the nanotubes. The hexagonal arrays of carbon atoms in graphite sheets of CNTs surface have a strong interaction with other molecules or atoms, which make CNTs a promising adsorbent material substituted for activated carbon in many ways. All of the facts mentioned above reveal that CNTs may have great analytical potential as a solid phase extraction adsorbent for metal ions [703,704]. This novel sorbent has been also applied for cadmium pre-concentration from natural water using a micro-column packed with MWNTs [703-705], and also MWNTs packed cartridges [704].

V.2.10.2.3.2. Nanoparticles

Nanometer-sized materials have attracted substantial interest in the scientific community because of their special properties [706]. The size range of nanoparticles is from 1.0 to almost 100 nm, which falls between the classical fields of chemistry and solid-state physics. The relatively large surface area and highly active surface sites of nanoparticles enable them to have a wide range of potential applications, including shape-selective catalysis, chromatographic separations, enzyme encapsulation, DNA transfection, drug delivery and sorption of metal ions.

One of nanoparticle's properties is that most of the atoms are on the surface. The surface atoms are unsaturated and therefore can bind with other atoms that possess highly chemical activity. Consequently, the nanometer material can adsorb metal ions with high capacity. Investigations of the surface chemistry of highly dispersed oxides, e.g. TiO_2, Al_2O_3, ZrO_2,

CeO_2 and ZnO, show that these materials have a very high sorptive capacity and give promising results for pre-concentrating trace metals from different water samples and solutions of high-purity alkali salts. The sorption properties of many oxides, including nanometer TiO_2, mainly depend on the characteristics of the solid, e.g. crystal structure, morphology, defects, specific surface area, hydroxyl coverage, surface impurities, and modifiers. Due to these nanometer-sized metal oxides are not target-selective, the change of the characteristics of the solid by using a suitable coating has been proven to be one of the most efficient ways to improve the selectivity. Thus, TiO_2 nanoparticles immobilized on DDC were synthesized and used as an SPE adsorbent for separating and pre-concentrating cadmium and other metals from natural and biological samples [707]. Simmilarly, nanometer Al_2O_3 packed micro-columns coupled to ICP-MS were described for simultaneous determination of Cd and V, Cr, Mn, Co, Ni, Cu, Zn and Pb in environmental liquid samples [708].

V.2.10.2.3.3. Ordered Mesoporus Silica

Mesoporous materials have been quickly developed during the last decade and they have attracted much attention in various scientific areas of physics, chemistry and material science. These materials offer performances, such as large surface area, high pore porosity, well-defined pore size and an ordered pore arrangement, excellent mechanical resistance properties, none swelling, good chemical stability and well modified surface properties [709]. Therefore, ordered mesoporous silica has become an ideal adsorbing material for SPE procedures, although for trace metals pre-concentration modified mesoporous silica is preferred. Modified mesoporous adsorbents are synthesized in the presence of surfactants, which used as template and can be later functionalized with different organic groups. The development of functionalized mesoporous materials for adsorption applications has generated a considerable interest. Among the variety of adsorption applications, the preparation of highly effective adsorbents for heavy metal ions trapped by the grafting or by incorporatind the ligands into mesoporous materials is clearly one of the most promising methods for environmental clean-up. In particular, materials whose surfaces have been functionalized with groups containing sulfur and nitrogen as active donor atoms offer high selectivity. For this purpose, chitosan (TS), a natural polysaccharide, can react with silica through the–NH_2 functional group found on its surface, and it could be used as a modifier for mesoporous silica, support which pre-concentrate cadmium and other trace metals [710].

V.2.10.2.3.4. Ionic Imprinted Polymers

Addressing molecurlarly imprinted polymerization in the presence of ions as templates, instead of molecules, new supports for recognizing ions (ionic imprinted polymers, IIPs) can be synthesized. These polymers offer all the benefits derived from MIPs and a high capacity for recognizing ions. In addition to bulk polymerization, different approaches for synthesizing MIPs and IIPs (suspension, emulsion, dispersion and precipitation methods) have been proposed by several authors. Detailed information on these methodologies can be found elsewhere [711].

In contrast to MIP for which the organic compound target (template) can react with different vynilated monomers (4-vinylpyridine, 4-VP, or 2-(diethylamino) ethyl methacrylate, DEM), when using ions as templates stable ion-vynilated monomer complexes are not formed. Therefore, only certain complexing agents which can react with the ion and which can polymerize through vynil groups can be used. As it has recently be reviewed by Rao et al. [712,713], bifunctional reagents, showing vynil groups and functional groups to interact with the dissolved ions, can be used following three main approaches: (i) linear chain polymers carrying metal-binding groups being cross-linked with a bifunctional reagent; (ii) chemical immobilization by preparation of binary complexes of metal ions with ligands having vinyl groups, isolation and then polymerization with matrix-forming monomers; and (iii) surface imprinting conducted on an aqueous-organic interface. Polymerization under these conditions leads to complexing ligands to be chemically immobilized in the polymeric matrix. However, the main drawback of these approaches is that vynilated ligands, which act as a complexing reagent for the target ion and as a monomer for polymerizing, are not commercially available and they must previously be synthesized.

Additionally, a different approach consisting of trapping a non-vinylated chelating ligand via imprinting of binary/ternary mixed ligand complexes of metal ions with non-vinylated chelating agent and vinylated ligand is also possible. In this case, the use of special synthesized vinylated ligands is avoided. After polymerization, the non-vinylated ligand is not chemically bonded to the polymer chains, but instead it is trapped inside the polymeric matrix.

For cadmium, IIPs synthesized by using the dual-ligand reagent (2Z)-N,N'-bis(2-aminoethylic)but-2-enediamide has been applied for the SPE of cadmium from river water, lake water and tap water (Table 19) [714]. By mean of the trapping approach, the non-vynilated ligand diazoaminobenzene (DAAB) in the presence of 4-VP as a monomer, has been used for synthesizing IIPs for cadmium pre-concentration from river and tap waters and acid digests from river sediments (Table 19) [715]. Similarly, the synthesis and evaluation for cadmium retention of a hierarchically imprinted sol-gel sorbent based on the use of cetyltrimethyl ammonium bromide (CTAB) as a non-vynilated reagent has carried out for cadmium aqueous standards [716].

V.2.10.3. Final considerations for SPE methods to assess cadmium

As it was commented, SPE can lead to low recoveries when water samples containing high naturally organic matter are treated [543]. In these cases, a previuos stage to release the trace metal from the metal–organic is needed. For this task, several procedures have been proposed. Humic substances removal from tap water can be performed by passing the sample at a pH lower than 2.0 through an AXAD-7 or Bio Beads SM-7 support [645]. Other times, organic matter can be oxidized by using sulphuric acid combined with ammonium or potassium peroxodisulphate and heat [673,684]. The organic matter digestion can also be assisted or not with hydrogen peroxide or UV-photolysis [543,679,681,715]. The use of UV digestion is highly recommended becuase the addition oxidant reagents can be minimized. Furthermore, UV digestion is effective and can be readily incorporated in flow injection manifolds, allowing stand-alone trace metal analysis [543].

Finally, as it was summarized in Table 19, after pre-concentration of target analyte onto the solid phase, metals are eluted by using an appropriate solvent (usually dilute acid solutions) in which the target metal is directly determined (in some cases an evaporation step can be use). However, other applications lead to the direct determination of metals adsorbed on the solid phase (without elution stage). In these cases, the support material is directly sampled (slurry sampling or solid sampling techniques) into the detector.

V.2.10.4. Flow Injection on-Line Pre-Concentration Systems by SPE

As commented in preceding sections, it is evident that the coupling of flow injection on-line SPE techniques provides an improvement on the detection limits and precision of measurements, and it reduces the interference from the matrix sample. In comparison with the off-line batch modes, these systems offer a number of significant advantages for trace metals determination, such as higher efficiency, lower consumption of sample and reagents, possibility of working in a closed system with a significant reduction of airborne contamination, and increased sampling frequency.

On-line column pre-concentration systems are based on the retention of the analytes in micro-column packed with adsorbents that determines the sensitivity and the selectivity of the analytical method. After the first paper by Olsen et al. [717] on FI sorbent extraction by using an ion exchanger (Chelex-100) as column packing material for Cd, Cu, Pb, and Zn pre-concentration and determination by FAAS, this promising technique has been extensively applied using several packing materials for the pre-concentration of several trace metals from environmental and biological samples. For cadmium, methods based on FI on-line pre-concentration with micro-columns packed with C18 [113,215], Chelex-100 [636], Dowex-50 [640], AXAD [680,691], PUF [112,694], MWNTs [705], ordered mesoporous silica and Toyopearl AF-Chelate 650M [543,643] have been proposed. The FI systems have been coupled to both spectrometric and electroanalytical detectors.

FI sorbent extraction consists in two steps. In the first stage, the sample is aspirated by a pump (A) through a valve (load position) into the column and then to waste (this allow the sample volume taken to be adjusted according to the analyte concentration). In the second step, the valve is switched to another position (elute position) and a pump (B) propels the eluting solvent through the column. Analytes are rapidly released from the column and led to the detector. Detailed description of FI manifold components has been shown in section IV.11.

V.2.11. Electrochemical Solid Phase Micro-Extraction

Electrochemical solid phase micro-extraction (SPME) combines the use of electrochemistry or electronically conducting polymers with SPME. In this technique, the analyte extraction/desorption on a conducting polymer coated on an electrode is controlled simply by changing the potential of an electrode. The main advantage of using conducting polymers in SPME is that the charge of the coatings can readily be controlled by oxidation and reduction of the polymers. These potential controlled ion exchangers can therefore be used to extract ions. Moreover, it should be possible to obtain pre-concentration of ions present in small sample volumes since electrochemically controlled desorption can be carried out in small volumes of solutions (~ 100 μL). By electrodeposition of a layer of conducting polymer, the surface area of the electrode can be increased to provide sufficiently high ion exchange capacities for the use of these electrodes in miniaturized analytical systems. The properties of the polymer (e.g. the conductivity, pore size and the mechanical characteristics) can be changed by using different counter ions, surfactants or different solvents during the electropolymerisation step. If an anion with a high mobility in the polymer film is incorporated during the electropolymerisation, the film becomes an anion exchanger. Polymers doped with anions of low mobility, such as large organic anions, operate as cation exchangers. As a result of the dependence of the polymer morphology on the size of the dopant ions and interactions between irreversibly incorporated dopants and ions, it is also possible to control the selectivity in the micro-extraction process. The extraction of Cd using electrochemical SPME has so far only been described by Liljegren et al. [718] using a polypyrrole-coated electrode. These authors also determined Co and Zn, as well as chloride, nitrite, bromide, nitrate, sulfate and phosphate in natural waters.

V.2.12. Low Temperature Directed Crystallization

Low-temperature directed crystallization (LTDC) is based on the crystallization of the aqueous sample during mixing so that impurities of different nature (organic, inorganic, particles) are displaced by the crystallization front into the liquid phase. On completion of the crystallization process, the top of the ice ingot is separated and analyzed by conventional instrumentation. The mineral composition of water, the chemical forms of elements and the crystallization conditions offer a great influence on the process. This procedure has also been applied to Cd pre-concentration from lake waters [590].

V.3. Sample Pre-Treatments for Solid Environmental Samples

As it was commented in previous sections, sample pre-treatments for elemental analysis in solid materials can be minimized or even omitted when using spectrometric methods based on laser ablation or electrothermal vaporization coupled to plasma techniques [206-208]. However, these analytical methods can only be applied to limited kinds of samples [719] and therefore, sample pre-treatment for spectrometric analysis is an active research field in analytical chemistry. This is especially important after the establishment of the Twelve Principles of Green Chemistry [720] and the application of such fundamentals to analytical procedures, which has led to the concept of GAC [449]. Therefore, aspects involving none or simple sample preparation, or little or no sample destruction, have been also introduced as principles of GAC. By this way, the sample pre-treatment stage has received special attention to make it greener [719,721,722] and several sample pre-treatments have been developed and applied to biological and environmental samples as green and alternative methods to the conventional acid digestion procedures. These procedures consist of direct approaches such as those based on the slurry sampling or solid sampling techniques, lixiviation or acid leaching procedures and enzymatic hydrolysis [723].

V.3.1. Decomposition Sample Pre-Treatments

Acid digestion procedures are the most used methodologies as sample pre-treatments for the determination of major and trace elements in biological and environmental materials. These procedures require the use of mineral/oxidizing acids and an external heat source to decompose the sample matrix.

The nature of the sample matrix determines the type of acid and also its individual use or a certain combination of them. Therefore, hydrochloric acid (boiling point at 110°C) can be used to decompose matrices containing salts of carbonates and phosphates, as well as some oxides and sulfides. In addition to the strong acidity provided by hydrochloric acid, this reagent offers complexing properties which facilitates the dissolution of transition elements. The oxidizing properties of sulfuric acid (boiling point at 338°C) make this acid as an useful reagent for dissolving inorganic matrices such as ores, alloys, and most of oxides and hydroxides. Nitric acid (boiling point of 122°C) is commonly used to dissolve samples containing high organic matter content, such as biological and clinical samples, and biota. This acid also dissolves carbonates, phosphates, oxides and sulfides that are not well attacked by hydrochloric acid. Perchloric acid (boiling point of 203°C) is also a strong oxidizing agent for organic matter, but it must be used with caution because the violent reaction. In these cases, samples are firstly treated with nitric acid before using perchloric acid. Hydrofluoric acid (boiling point at 112°C) is the unique agent capable of dissolving matrices containing silica (geological materials, soils, sediments, atmospheric particulate matter) by forming SiF_6^{2-}. This acid is very dangerous (its use is forbidden in some laboratories) and it must be handled with extreme caution. In addition to the individual use of these mineral acids or oxidizing agents, certain combinations of acids are useful. The most used mixtures are sulfuric acid/nitric acid and nitric acid/hydrochloric acid. When using the first mixture (5/20 sulfuric acid/nitric acid) the nitric acid attacks most of the organic matter but it is not hot enough (122°C as boiling point) to destroys the last traces. As it boils off during the digestion process, only sulfuric acid remains at the end of the digestion and it attacks the remaining organic matter. The nitric acid/hydrochloric acid mixture, at the proportion 1/3, called as aqua regia, is also quite popular. Nitric acid provides an oxidizing media while hydrochloric acid acts as a complexing agent and allows the dissolution of noble metals (gold, palladium and platinum) and many alloys. Sometimes, the addition of bromine or hydrogen peroxide to mineral acid increases their solvent action, especially for attacking samples with a high organic matter content.

Instead of mineral acids, certain alkaline solvents, such as tetramethylammonium hydroxide (TMAH or TMAOH), a quaternary ammonium salt with the molecular formula $(CH_3)_4NOH$, have been used to solubilize samples with high organic matter contents. The use of this reagent does not totally decompose the sample, and a centrifugation step is needed to obtain the aqueous digest. TMAH is mainly used when extracting elements which can be lost in oxidizing medium such as iodine, although it has also be proposed for extracting other elements [163,174].

In addition to the mineral acid, the other parameter to be controlled in an acid digestion is the heat source to be applied and the vessels containing the oxidizing mixture and the sample. The type (material) of the vessels is dependent on the type of heat source and also the nature of acid. Therefore; PTFE, exhibiting a melting point around 320°C, must not be used when using sulfuric acid (boiling point of 338°C); or glass must not be used when using hydrofluoric acid because this material is attacked by this acid. In addition, the source of heating commonly requires certain properties for the reaction vessels. For example, materials transparent to microwave radiation, such as PTFE or quartz, must be selected when using microwave energy to assist the sample decomposition. Either acid decomposition or TMAH solubilization can be carried out in open glass (sometimes PTFE or quartz) vessels. However, the use of open vessels offers as main drawbacks the risk of sample contamination and analyte losses, especially for volatile elements, such as cadmium, which can form volatile halides or oxyhalides or even hydrides.

Concerning the heat sources, conventional sample decomposition can be carried out by using hot plate or multi-sampler digestor devices. These heating systems allow the simultaneous treatment of several samples at once and they can result convenient for several applications. However, sample pre-treatments involving these heating systems (conductive heating) lead to time consuming procedures (typically from 1 to 2 hours, but it be much longer). This long time is because the poor heat conductivity properties of vessels which require a certain time to be heated and then to transfer the heating energy to the sample-acid mixture. In addition, the vaporization at the surface of the liquid generates a thermal gradient, which makes that a small portion of the sample-acid mixture is at the programmed temperature. Therefore, it can be assumed that most of the acid is above its boiling point. The poor heat transfer is the responsible of the high volumes of mineral acids required to complete the digestion. In addition, another drawback associated with these systems is the special safety considerations because the vapor generation during the digestion process as consequence of the sample destruction.

V.3.1.1. Microwave Assisted Sample Decomposition

The use of microwave energy for assisting sample preparation procedures is a current practice in analytical chemistry [724,725]. Regarding elemental analysis, after the first reported application by Abu-Samra et al. [726] to acid digestion of biological materials, microwave assisted acid digestion is nowadays a well-established methodology for treating biological, environmental and geological samples and it is by far the chosen sample pre-treatment for routine analysis [727,728].

V.3.1.1.1. Microwave Energy

Microwave energy is a no ionizing radiation exhibiting frequencies within the 300 – 300000 MHz (wavelengths from 0.005 to 1 m) that does not cause changes in the molecular structure. When microwave interacts with a material, the energy can be absorbed at a rate dependent upon de dissipation factor (tan δ) [729], parameter defined as a ratio of the sample's dielectric loss (ε'') to its dielectric constant (ε') as

$$\tan \delta = \varepsilon'' / \varepsilon'$$

The dielectric constant (ε') can be considered as a measure of the material to minimize the penetration of microwave energy, while the dielectric loss (ε'') offers an opposite effect and it is though as a measure of material's ability to dissipate microwave energy as heat. Materials which exhibit low dissipation factors, such as quartz (0.6) or PTFE (1.5), can be considered as transparent to microwave radiation, and they are commonly known as insulators. However, materials showing high dissipation factors, such as water (1570), offer absorptive properties to microwave radiation and they are referred as dielectric materials. A third type of materials, called as conductors, offer reflective properties to the microwave energy. This is the case of most of metals which reflect microwaves.

The rapid heating achieved by microwave irradiation (microwave energy lost) is caused by two different mechanisms; the molecular motion by migration of ions (dissolved salts in the solvent) and rotation of dipoles (molecular solvents). Migration of ions, also referred as ionic conduction, is the conductive migration of dissolved ions in the applied electromagnetic field of microwave radiation. Because of the resistance of the solution to this ionic movement, energy dissipation as heat is produced. Rotation of dipoles is a consequence of the alignment of molecules (permanent or induced dipole moments) in the electric field supplied by microwaves. As the electric field increases, it aligns the polarized molecules; however, the thermally induced disorder is restored when the electric field decreases or disappears and thermal energy is released. Both mechanisms occur simultaneously and the dominance of one or another process is highly dependent on the temperature. Therefore, for solvents such as water, the dielectric loss to a sample due to the dipole rotation decreases as the temperature increases. In contrast, dielectric loss due to ionic migration increases as sample temperature increases.

V.3.1.1.2. Microwave Instrumentation

In general, there are two types of commercially available microwave ovens for laboratory uses, i.e., an open-focused and closed-vessel type. Open-focused microwave ovens can be used to treat simultaneously up six samples, and vessels are commonly large test tubes made of glass or quartz. Usually, a condenser is fitted to the test tube to prevent loss of volatiles. In these systems the microwave radiation is focused to the vessel containing the sample and the acid, and a controlled and precise heating is obtained.

A closed-vessel microwave oven system also allows the simultaneous treatment of several samples but their main characteristic is that closed-vessels must be equipped with a safety relief valve to control the internal pressure. A closed-vessel microwave instrument for laboratory purposes is formed by the microwave generator (magnetron), the wave guide, the microwave cavity, the mode stirrer, a circulator and a turntable. The magnetron generates the microwave radiation which is propagated down the wave guide, and injected directly into the microwave cavity where the mode stirrer distributes the incoming energy. Most of the commercial microwave ovens have a fixed-tuned magnetron, giving an output frequency of 2450 MHz. These instruments use approximately 1200 W of electrical line power that is transformed to 600 W of electromagnetic energy. The turntable is needed in microwave instruments which use more than one digestion vessels so that all vessels can uniformly be exposed to the microwave energy. The turntable (carousel) is available that rotates 360° continuously, or that alternates back and forth 180°. In addition, other features include a function for monitoring both pressure and temperature. Typically, the pressure (up to 800 psi) is continuously measured at a rate of 200 s^{-1}, while temperature (up to 300°C) is monitored every 7 s.

Digestion vessels must be made of materials transparent to microwave radiation (low dissipation factors) so that the energy will not be absorbed by the vessel but will pass through the vessel to the solution inside. As commented, PTFE (dissipation factor of 1.5), polystyrene (3.3) and quartz (0.6) are adequate materials for microwave accessories. It must be taken into account that although quartz offer good transparency to microwave irradiation, this material have not adequate chemical resistance when using certain acids such as hydrofluoric acid, and vessels made of PTFE are preferred (good thermal and chemical resistance). Nowadays, most of the digestion vessels are closed-vessels, which give as main advantage the higher temperatures reached because the boiling point of the acid is raised by the pressure produced in the vessel. This implies a shorter digestion time. In addition, closed-vessels avoid loss of volatile elements during digestion and also minimize the possibility of airborne contamination. Less volume of acids are commonly required when working with closed-vessels because acid evaporation is not occurred. Typically, closed-vessels used for microwave digestion are designed to operate at internal pressures up to 120 psi, and in addition to the vessel body and the vessel cap, closed vessels have a safety relief valve which is open when the internal pressure exceeds the target pressure. In these cases, the safety relief valve works by lifting of the vessel cap to release excess pressure and then immediately releasing to prevent loss of sample.

V.3.1.2. Sample Decomposition for Cadmium Determination

Nowadays, acid digestion procedures based on the use of microwave energy are the most used methods to decompose biological and geological materials. Nevertheless, recent applications of ultrasound energy for digesting biological samples [730,731] and sediments [732] have been also proposed. However, because total sample decomposition is not reached, these methods are commonly referred as ultrasonic-assisted acid pseudodigestion procedures [730,731].

Table 20. Selected applications of acid digestion procedures

Sample	Processor	Acid composition	Analytical technique	Ref.
Molluscs / fish	Microwave-assisted	HNO_3/H_2O_2	ETAAS	[149]
Molluscs	Microwave-assisted	HNO_3/H_2O_2	STAT-FAAS	[72]
Molluscs	Heating block	HNO_3/H_2O_2	ICP-MS	[733]
Molluscs	Microwave-assisted	HNO_3	ETAAS	[734]
Molluscs	Heating block	2-stage digestion with HNO_3 and H_2O_2	ICP-OES	[735]
Molluscs	Microwave-assisted	$HNO_3/H_2O_2/HF$	ICP-MS	[736]
Molluscs / fish / plants / sediments	Microwave-assisted	HNO_3/H_2O_2	ICP-OES	[737]
Molluscs / fish	Microwave-assisted	HNO_3/H_2O_2	ICP-MS	[738]
Lake fish lake sediment	Microwave-assisted	HNO_3/H_2O_2 HNO_3/HCl	ETAAS	[155]
Fish	Ultrasounds-assisted Heating block	HNO_3/H_2O_2 HNO_3/H_2O_2	ETAAS	[730]
Fish	Ultrasounds-assisted Heating block	HNO_3/H_2O_2 HNO_3/H_2O_2	ETAAS	[731]
Fish	Microwave-assisted	HNO_3/H_2O_2	ETAAS	[739]
Fish	Heating at 100C for 4 h + reflux for 2 h Heating at 100C for 4 h + reflux for 2 h Heating at room temperature overnight, 100C for 4 h + reflux for 2 h Heating at 100C for 4 h + reflux for 2 Heating at 50C for 2 h + heating 70C for 2 h	HNO_3 (1.0M) + HCl HNO_3 + HCl HNO_3 + HCl HNO_3 + $HClO_4$ H_2O_2 + HCl	ASV	[740]
Antarctic krill	Microwave-assisted	HNO_3/H_2O_2	ETAAS	[151]

		HNO_3/HF		
Plants / soils	Heating block	HNO_3/HF	STAT-FAAS	[83]
Plants	Dry ashing / microwave-assisted	HNO_3/H_2O_2	STAT-FAAS	[84]
Plants	Pressurized digestion system	HNO_3	TSFF-AAS	[117]
Plants	Microwave-assisted	HNO_3/H_2O_2	TSFF-AAS	[119]
Plants	Microwave-assisted	HNO_3/H_2O_2	TSFF-AAS	[120]
Plants	Heating block	HNO_3	ETAAS	[142]
Plants	Heating block	HNO_3/H_2O_2	ETAAS	[143]
Plants	Microwave-assisted	HNO_3/H_2O_2	ETAAS	[176]
Soil / sediment / coal fly ashes		$HNO_3/HCl/HF$		
Plants / lichens / sediments	Microwave-assisted performed at atmospheric pressure by a multi-samples rotor and screw-capped disposable polystyrene line	-----	ICP-MS	[741]
Plants	Microwave-assisted	HNO_3/H_2O_2 $HNO_3/H_2O_2/HF/H_3BO_3$ $HNO_3/H_2O_2/HBF_4$	ICP-MS	[742]
Lichens	Oven closed vessel acid digestion	$HNO_3/H_2O_2/HF$	ICP-OES	[743]
	Hot plate open vessel acid digestion	2-stage HNO_3/H_2O_2 and HF		
	Microwave-assisted	$HNO_3/H_2O_2/HF$		
Lichens	Digestion, room temperature overnight	$HNO_3/HClO_4$	ICP-MS	[744]
	Microwave-assisted	$HNO_3/HClO_4$ + HF + H_3BO_3		
Soil / sediment	Microwave-assisted	$HNO_3/HCl/HF$	ETAAS	[146]
Soil / sediment	Microwave-assisted	$HNO_3/HCl/HF$	ETAAS	[167]
Soil	Microwave-assisted	$HNO_3/HCl/HF$	IDA-ICP-MS	[745]
Soil	Microwave-assisted	HCl/HNO_3	ICP-MS	[746]
	DigiPrep system	HCl/HNO_3		
Soil	Microwave-assisted	$HNO_3/HCl/HF/H_2O_2$	ICP-OES	[747]
	Microwave-assisted	$HNO_3/HCl/H_2O_2$		
	Microwave-assisted	$HNO_3/HCl/HF$		
	Microwave-assisted	HNO_3/HCl		
Soil	Microwave-asisted	HCl/HNO_3	ICP-OES	[748]
Soil	Microwave-assisted + Heating block (HF removal)	$HNO_3/HCl/HF$ H_2SO_4	ICP-MS	[749]
Soil	Microwave-assisted	$HF/HCl/HNO_3$	ICP-MS	[750]
Soil / dust	Microwave-assisted	HNO_3/HCl	ETAAS	[751]
Sediments	Heating block	$HNO_3/HClO_4/HF$	TSFF-AAS	[114]
Sediment / fly ash	Microwave-assisted + Heating block (HF removal)	$HNO_3/HCl/HF$ H_2SO_4	ETAAS	[169]
Soil / sediment	Microwave-assisted	$HNO_3/HCl/HF$	ETAAS	[170]
Fish	Microwave-assisted	HNO_3	ETAAS	[172]
Sediment		$HNO_3/HCl/HF$		

Table 20. Continued

Sample	Processor	Acid composition	Analytical technique	Ref.
Fish /mollusc	Microwave-assisted	HNO_3	ETAAS	[173]
Sediment		$HNO_3/HCl/HF$		
Sediment	Ultrasound-assisted Microwave-assisted	$HNO_3/H_2O_2/HF$ $HNO_3/H_2O_2/HF$	ICP-MS	[732]
Sediment	Open-focused microwave-assisted	HCl/HNO_3	ICP-OES	[752]
Sediment	Microwave-assisted Heating block	$HNO_3/HCl/H_2SO_4$ HNO_3/HCl	FAAS	[753]
Sediment	Hot plate acid digestion evaporation to dryness ultrasounds-assisted + oven heating evaporation to dryness dissolution in 2.0M HCl	H_2O_2/HNO_3 + HF/HNO_3	Anion exchange chromatography – IDA-ICP-MS	[754]
Sediment	Microwave-assisted	2-stage microwave-assisted acid digestion procedure using HNO_3 and HCl	ETAAS	[755]
Sewage sludge	Heating block	HNO_3/H_2O_2	STAT-FAAS	[81]
Fly ashes	Open system digestion Microwave-assisted Microwave-assisted	$HF/HNO_3/HClO_4$ HF/HNO_3 $HF/HCl/HNO_3$	ICP-OES	[367]
Atmospheric aerosol	Oven heating acid digestion Oven heating acid digestion Oven heating acid digestion Oven heating acid digestion Microwave-assisted	$HF/HNO_3/HClO_4$ $HF/HNO_3/HCl$ $HF/HNO_3/H_2O_2$ HF/HNO_3 HF/HNO_3	ETAAS	[756]
Atmospheric aerosol	Microwave-assisted	$HNO_3/HF/H_2O_2$	ICP-MS	[757]
Atmospheric particulate matter	Microwave-assisted	$HNO_3/HClO_4$	ICP-OES	[758]
Atmospheric particulate matter	Microwave-assisted	2-stage microwave-assisted acid digestion procedure using HNO_3/HF, and H_3BO_3	ICP-MS	[759]
Atmospheric particulate matter	Microwave-assisted	$HNO_3/HF/H_2O_2$	ICP-MS	[760]
Dust	Acid digestion	HNO_3/HCl	FAAS	[761]
Suspended particulate matter in seawater	Microwave-assisted	----	FI-ICP-MS	[762]

ASV, anodic stripping voltametry; BIFF-AAS, beam injection flame furnace – atomic absorption spectrometry; ETAAS, electrothermal atomic absorption spectrometry; FAAS, flame atomic absorption spectrometry; FI-ICP-MS, flow injection - inductively couple plasma – mass spectrometry; ICP-OES, inductively couple plasma – optical emission spectrometry; ICP-MS, inductively couple plasma – mass spectrometry; IDA-ICP-MS, isotopic dilution analysis - inductively couple plasma – mass spectrometry; STAT-FAAS, slotted tube atom trap - flame atomic absorption spectrometry; TSFF-AAS, thermospray flame furnace – atomic absorption spectrometry

Table 20 summarises selected applications of acid digestion procedures for several environmental materials; including biological samples, such as biota, mainly molluscs [72,149,733-738] and fish [149,730,731,737-740] soft tissues, Antarctic krill [151], plants [83,84,117,119,120,142,143,166,176,737,741,742], lichens [741,743,744], soils [146,167,745-750], sediments [146,155, 167,169,170,172,173,732,737,752-755], sewage sludge [81], fly ashes [169,367], atmospheric aerosol [756,757], atmospheric particulate matter [758-760], dust [751,761] and suspended particulate matter in seawater [762].

V.3.2. Leaching (Extraction) Procedures

Leaching (extraction or lixiviation) methods consist of using diluted reagents (commonly acids) to extract or leach the target trace elements from solid samples. These procedures do not involve the total matrix sample decomposition but the breakdown of certain chemical bonds between the trace elements and the matrix sample constituents [763]. Nevertheless, partial dissolution of the solid sample is unavoidable especially when working with samples containing high organic matter and with diluted oxidizing acids such as nitric acid. In addition, the sample dissolution ratio is expected to be increased when assisting the leaching process with an external source of energy (ultrasounds, microwaves or pressurization) [723]. Therefore, as proposed by Lorentzen and Kingston [764], the mechanisms through which the targets reach the solution can be considered as partial sample decomposition. This fact can lead to non quantitative extractions, mainly for elements associated to silicate fractions in certain environmental matrices because common diluted (or even concentrated) mineral acids do not partially decompose silica [723].

In general, because total sample matrix destruction is not required, leaching methods are considered as environmentally friendly procedures and they have received especial attention. After the first applications on using a leaching method for total metal extraction based on concentrated acids (9/1 HCl/HNO_3) and heating [765], different procedures using diluted mineral or organic acids and chelating agents, combined with different sources of energy to speed up the leaching process (microwaves, ultrasounds and pressurization), have been developed as alternative sample pre-treatments to conventional microwave assisted acid digestion procedures.

V.3.2.1. Ultrasounds Assisted Acid Leaching Procedures

Ultrasound energy has been commonly applied for assisting solid-liquid extractions [766]. Concerning trace elements, most of these procedures have involved the use of diluted reagents (mineral acids), although some applications involving concentrated mineral acids have been performed to assist total sample decomposition process [767-770].

V.3.2.1.1. Ultrasound Energy

Ultrasound is cyclic sound pressure with a frequency greater than the upper limit of human hearing. Although this limit varies from person to person, it is approximately 20 kHz in healthy, young adults and thus, 20 kHz serves as a useful lower limit in describing ultrasound. Because of the cyclic pressure variations when ultrasound passes through a material (liquid), successive expansion and compression cycles occur, and the expansion cycles are responsible of producing negative pressures in the liquid. These negative pressures generate bubble nucleation because molecules of vapor and gases can migrate to the cavities. In contrast, a fraction of gases and vapor molecules can be expelled from the bubbles during the compression cycles. The successive expansion and compression cycles originate the cavitation phenomenon which leads to bubbles collapses and the generation of high local temperatures and pressure inside the liquid [771]. When a solid is suspended in a liquid subjected to ultrasound irradiation, the cavitation process occurs by asymmetric collapses, which originates high-speed microjets toward the solid surface, leading to the solid erosion and cleavages [766,772]. This fact increases the solid surface area and under these conditions analyte transport from the solid particles to the liquid phase is more efficient and quantitative extractions in short times can be expected. In addition, due to a best interaction between the solid sample and the extracting solution, chemical bonds between the analytes and the matrix components can be broken by the action of diluted reagents. This is especially advantageous when extracting inorganic species (major and trace elements) because ultrasound energy does not modify/decompose them during the extraction process. However, partial or complete mineralization of organic compounds, as well as oxidation process, can be occurred and certain limitations have been found when extracting organic substances and organometallic species [766].

V.3.2.1.2. Ultrasound Instrumentation

As recently reviewed by Santos Júnior et al. [766], two ultrasound generator devices are available: ultrasounds water-baths (ultrasonic baths, ultrasonic cleaning devices) and ultrasonic probes.

Ultrasounds water-baths consist of a tank, with or without temperature control, in which the ultrasound transducer (normally placed at the bottom of the tank) transfers the ultrasound energy to the liquid, normally ultrapure water, contained in the tank. In these devices, the ultrasound energy is partially dissipated into both the reaction vessels placed in the bath and also through the water contained into the tank. Therefore, the ultrasound energy which is dissipated into the reaction vessel (where the sample and the extracting solution are placed) is lower than the ultrasound intensity supplied by the transducer (commonly between 1 and 5 Wcm^{-2}). In addition, it has been established that the ultrasound energy transfer depends on several factors, such as the position in which the sample vessel is situated inside the bath (vertical and horizontal position) and the number of vessels as well as wall thickness and material, the bath dimensions, the water volume contained in the tank and the transducer position. Therefore, the intensity of the ultrasound energy is not constant in all points of the bath neither inside all reaction vessels [773]. This fact together with the different ultrasounds frequencies and powers depending on the specific transducers used in the different commercial ultrasonic baths makes the attempts to compare experimental conditions difficult [766].

Ultrasounds probes consist of an ultrasound transducer coupled to a detachable metal probe or tip, commonly made of a titanium alloy (Ti-6Al-4V). The tip transmits the ultrasound energy to the sample-extracting liquid mixture, and there are different tip sizes in function of the volume of the sample-extracting liquid mixture to be sonicated. Although titanium alloys tips are the most used probes, silica glass or pure titanium tips are available to minimize metal contamination [774,775]. Conventional ultrasonic probes generate higher ultrasound intensities than ultrasonic water-baths, within the 50–750 Wcm^{-2}. In addition, because of the tip is directly immersed into the reaction vessels the ultrasound energy is completely absorbed by the reaction medium, increasing the sonication efficiency. Therefore, lower extraction times are attained and better extraction efficiencies are obtained when using ultrasounds probes. Another important consideration is related to the dimension and shape of the sample vessel to be irradiated. Conical-type vessels with the diameter as small as possible in order to rise up the liquid level are recommended to avoid the formation of aerosols and foams [774].

V.3.2.1.3. Ultrasound Assisted Leaching Methods for Cadmium Extraction

After the first work by Kumina et al. [776], several applications involving the use of diluted acids/reagents, and leaching at room temperature or at moderate high temperatures have been carried out for biological, clinical and environmental samples. Different reported methods for cadmium extraction from environmental samples are summarized in Table 21.

Quantitative cadmium recoveries were achieved by using either ultrasounds water-baths or probes working at room temperature; although sonication at 65°C was needed to quantitatively extract cadmium from seaweed [777], and a temperature of 56°C was also found necessary when pre-treating TORT-2 certified reference material [778]. Terrestrial plants [779-781], moss [782], animal tissues [783], plankton [784], aquatic plants [148,782], algae [173,784], molluscs [111,150,763,784-788], fish [111,150,173,788], sediments [173], sludge and ash [294] have been subjected to ultrasound irradiation to extract cadmium.

Ultrasound assisted acid leaching methods have shown great potential for extracting trace elements from samples containing a high organic matter content. However, these methodologies lead to non quantitative recoveries when treating samples containing silica (geological materials and sediments and soils). In such cases, concentrated acids are required to digest or partially digest the sample matrix [789].

V.3.2.1.4. Continuous on Line Ultrasound Assisted Leaching

As commented before, the implementation of flow injection analysis systems (FIA) in analytical laboratories is a current practice, leading to on-line sample pre-treatment just before analyte determination. This fact considerably shortens the pre-treatment time and mainly decreases analyte losses and sample contamination. Most of the on line ultrasound assisted extraction methods by means of a FIA imply the use of a single leaching solution to extract a certain number of analytes, but applications using different extracting solutions which are sequentially passed through the solid sample enclosed in a chamber (column or cell) have also been described [790].

Both ultrasound water bath and ultrasound probe devices have been used for continuous ultrasound assisted leaching procedures. In the first case, the solid sample enclosed in the extraction cell is immersed in the ultrasonic water bath and after loading the leaching solution in the circuit, the system is subjected to ultrasound irradiation for a certain time. When using an ultrasonic probe, similar operations regarding extraction chamber filling and leaching solution loading are applied but, the extraction chamber is immersed in a water bath and the ultrasound irradiation is applied with the probe placed 1 mm from the top surface of the extraction cell [723]. Extraction cells (chambers) containing the solid sample consist of glass mini-columns (50 mm×3 mm i.d.) or of stainless-steel cylinders (10 cm×10 mm i.d.), closed with screws at either end. The screw caps are plugged with filter paper or cellulose filters to ensure that the sample remains in the extraction unit [791,792].

Table 21. Selected applications of ultrasound assisted leaching procedures (ultrasonic bath and probe)

Sample	Processor	Leaching solution (atomic spectrometric technique)	Temperature	Leaching time	Ref.
Aquatic plant, mussel	Probe	3.0 %(v/v) HNO_3 (ETAAS)	Room temperature	1.0 min	[148]
Mussels, clams, tuna	Bath	(0.5–4.5) M HNO_3 + (2.0–4.0) M HCl + 1.5 M H_2O_2 (ETAAS)	Room temperature	10.0 min	[150]
Sludge, ash	Bath	6.0 M HCl (CVAAS)	Room temperature	20.0 min	[294]
Molluscs	Bath	(0.5–4.5) M HNO_3 + (2.0–4.0) M HCl + 1.5 M H_2O_2 (ETAAS)	Room temperature	10.0 min	[764]
Seaweed	Bath	Step1: 6.0 M HCl Step 2: 3.7 M HNO_3 + 3.0 M HCl + 3.0 M H_2O_2 (ICP-OES)	Room temperature 65°C	10.0 min 35.0 min	[777]
Seafood (TORT-2)	Bath	80% MeOH 2.0 %(v/v) HNO_3 (ICP-MS)	Room temperature	60.0 min	[778]
Plants	Bath	Diluted HNO_3 and HCl (ICP-OES)[a]	Room temperature	40.0 min	[779]
Plants	Bath	0.5 M HNO_3 (ETAAS)	Room temperature	5.0 min	[780]
Plants	Bath	1.0 M HCl + surfactants (ICP-OES)	Room temperature	40.0 min	[781]
Plants, aquatic moss	Probe	0.1 M EDTA pH 10 (ETAAS)	Room temperature	3.0 min	[782]
Animal tissues	Probe[b]	15.0 %(v/v) HNO_3 (ETAAS)	Room temperature	10.0 min	[783]
Plankton, algae, oyster	Probe	1.0 M HNO_3 (ETAAS)	Room temperature	1.0 – 5.0 min	[784]
Mussel, fish	Bath Probe	2.8 M HNO_3 (BIFF-AAS) 0.7 M HNO_3 (BIFF-AAS)	Room temperature Room temperature	10 min 5 min	[111]
Fish, algae, sediments	Probe	0.5 – 5.0 %(v/v) HNO_3 (ETAAS)	Room temperature	2.0 – 5.0 min	[173]
Mussel	Probe	3.0 %(v/v) HNO_3 (ETAAS)	Room temperature	15 s	[785]
Mussels	Bath	1.6 M HNO_3 + 1.2 M HCl + 1.5 M H_2O_2 (ETAAS)	Room temperature	120.0 min	[786]
Molluscs	Bath	(0.5–4.5) M HNO_3 + (2.0–4.0) M HCl + 1.5M H_2O_2 (ETAAS)	Room temperature	10.0 min	[787]
Fish and mussels	Bath	4.0 M HNO_3 + 4.0 M HCl + 0.5 M H_2O_2 (ETAAS)	56°C	30.0 min	[788]

[a] Concentrations not available

[b] Sample into the extraction chamber is immersed into a water bath at room temperature and the ultrasound probe is placed 1 mm from the top surface of the extraction cell

BIFF-AAS, beam injection flame furnace – atomic absorption spectrometry; CVAAS, cold vapor atomic absorption spectrometry; ETAAS, electrothermal atomic absorption spectrometry; ICP-OES, inductively couple plasma – optical emission spectrometry

Two main approaches for the continuous on line ultrasound assisted acid leaching of trace elements from solid samples are described in literature. In the first one, after immersing the extraction cell containing the sample into the ultrasonic bath, the system is loaded with a certain volume of the leaching solution, the circuit is then closed and this volume of leaching solution is pumped through the extraction cell for a certain time under sonication conditions. After this extraction time, the leachate passes to a mixing coil and a small volume (around 250 µL) is injected by means of an injection valve into a carrier (ultrapure water) stream through the detection system [791]. In the second operation mode, after the extraction chamber containing the sample is immersed in the water bath, the leaching solution is continuously circulated through the solid sample for a certain time under ultrasound irradiation. Finally, after extraction was completed, the extract is collected and conveniently diluted before analyte determination [792]. For either operation modes, the direction of the flow of the leaching solution during the extraction is changed after a fixed period of time (commonly between 30 and 80 s) to avoid compactness of the sample at the extraction chamber ends that could cause overpressure in the system.

Several applications of continuous ultrasound acid leaching coupled to atomic spectrometric detectors have been recently reviewed [723]. Most of these applications use ultrasounds water-baths and sonication at room temperature. Concerning cadmium continuous extraction, 3.0 M nitric acid has been used as an extracting solution at room temperature and for 2.0 min to pre-treat molluscs [793]. However, sonication at 50°C was needed for the continuous leaching of cadmium from welding fumes [794].

V.3.2.2. Microwave Assisted Acid Leaching Procedures

As commented in preceding sections, the use of microwave energy for sample preparation has been widely accepted for assessing inorganic and organic compounds in different biological and environmental samples. Although microwave ovens are mainly used for total sample decomposition when assessing metals, there are several applications in which diluted mineral acids are used, and metals are leached from the solid sample without implying the total sample decomposition. Similarly to ultrasounds assisted acid leaching procedures, the mechanisms through which the elements reach the solution must be considered as partial sample decomposition due to diluted oxidizing and relatively high temperature and pressure are used.

Both off-line and on-line microwave assisted acid leaching procedures have been described in literature [723]. Diluted nitric acid, hydrochloric acid and hydrogen peroxide have commonly been used as extracting solutions. In few cases, assisted acid leaching have been carried out in household microwave ovens, and high pressure microwave instruments have been mainly used and recommended.

Table 22 summarizes the application of microwave assisted acid leaching procedures for extracting cadmium from environmental samples. Molluscs [149,778,795], fish [149] and plant [796,798] samples have been efficiently treated to release cadmium before an atomic spectrometric determination. In addition, Table 22 also lists an application involving a continuous microwave assisted acid leaching procedure [797]. In this approach, the sample (plant material) is placed into an extraction cell connected to an extraction loop which is filled with the acid leaching solution (2.0 mL of 1 %(v/v) HNO_3). The extraction cell is then placed in the vessel of a focused microwave oven which contains water and it is irradiated at 300 W for 10–15 min.

Table 22. Selected applications of microwave assisted leaching procedures

Sample	Microwave oven	Leaching solution (atomic spectrometric technique)	Power / Temperature	Leaching time	Ref.
Molluscs, fish	Household microwave oven	4.5 M HNO_3 + 2.8 M HCl + 0.5 M H_2O_2 (ETAAS)	64 W	2.0 min	[149]
Seafood (TORT-2)	High pressure microwave oven	2.0 %(v/v) HNO_3 (ICP-MS)	From room temperature to 75°C	6.0 min	[778]
Mussels	High pressure microwave oven	2.5 M HNO_3 + 3.0 M HCl + 0.5 %(m/v) H_2O_2 (ICP-OES)	From room temperature to 65°C	2.5 min	[795]
Plants	High pressure microwave oven	0.02 M EDTA or 1.0 M HCl (ICP-OES)	250 W + 400 W + 550 W + 250 W	5.0 min + 10.0 min + 5.0 min + 5.0 min	[796]
Plants	Focused microwave oven[a]	1 %(v/v) HNO_3 (ETAAS)	300 W	10-15 min	[797]

[a] Continuous microwave assisted acid leaching

ETAAS, electrothermal atomic absorption spectrometry; ICP-OES, inductively couple plasma – optical emission spectrometry; ICP-MS, inductively couple plasma – mass spectrometry.

V.3.3. Pressurized Liquid Extraction. Pressurized Hot Water Extraction

Pressurized liquid extraction (PLE), also called pressurized fluid extraction (PFE), pressurized solvent extraction (PSE) or accelerated solvent extraction (ASE), is a reference extraction methodology for organic compounds, mainly organic pollutants [798,799] and it has also been popular for extracting organometallic compounds [800]. PLE consists of using solvents at a high pressure and/or high temperature without reaching the critical point. Parameters, such as the dielectric constant and the surface tension, of a solvent subjected to high pressure and temperatures are changed, and the solubility of analytes in the extracting solution has been found to be increased. When using water as a solvent, the technique is commonly called as pressurized hot water extraction (PHWE) [801], but the instrumentation as well as critical parameters controlling the extraction process is similar to PLE.

PLE takes place into stainless-steel extraction cell, in which the sample, previously dispersed in an inert support, is placed between two cellulose filters, and which is subjected to controlled temperature and pressure for a certain static extraction time, or for several extraction cycles. Dispersion of the sample in an inert support (functionalized C18, diatomaceous earth, alumina, Florisil, among other) is required to assure good solvent sample contact within the extraction cell.

High efficiencies for analyte transfer from the solid sample to the extracting solution are obtained because the high temperature and pressure used. Therefore, extraction can be completed using small amounts of solvents. When extracting inorganic species, the enhanced contact between sample and the extracting solution attributed to high pressure and temperature allows the use of diluted reagents (acids), and diluted organic carboxylic acids can be used to breakdown bounds between trace elements and sample matrix constituents. This makes the technique environmentally friendly [723].

V.3.3.1. Pressurized Hot Water Extraction / Pressurized Liquid Extraction Methods for Cadmium Extraction

First applications for extracting metals can be attributed to Luque de Castro et al., who used sub-critical water [802] or acidified sub-critical water [803,804] in a prototype extractor made of stainless steel [805]. Similar to the extraction of several major and trace elements from coal [804,805], acidified sub-critical water extraction (4.0 %(v/v) nitric acid) was firstly used to extract cadmium from plants [806] and industrial oils [807] (Table 23).

Table 23. Selected applications of pressurized hot water extraction and pressurized liquid extraction procedures

Sample	Pressurized device	PHWE/PLE conditions	Atomic spectrometry technique	Leaching time	Ref.
Plants	Laboratory-made extractor	Acidified water (1.0 %(v/v) HNO_3), 100-250°C	ETAAS	5.0 min	[806]
Industrial oils	Laboratory-made extractor	Acidified water (4.0 %(v/v) HNO_3 + 0.1 M KCl), 150°C	ETAAS		[807]
Particulate matter	Laboratory-made extractor	Acidified water (0.1 %(v/v) HNO_3), 150°C, 1500 psi	ICP-OES/ICP-MS	15.0 min (30.0 min)[a]	[808]
Squid waste	Laboratory-made extractor	Water, 170-380°C, 0.79-30 MPa	FAAS	1.0-40.0 min	[809]
Sediments	Supercritical fluid extractor	Water or water+CO_2, 80°C, 27 MPa	ICP-OES/ICP-MS	5.0 h	[810]
Particulate matter	Dionex ASE-200	40.0 mM EDTA[b], 100°C, 1000 psi, 1 cycle	ICP-OES	10.0 min	[812]
Plants	Dionex ASE-200	0.01 M CDTA[c], 75°C, 1500 psi, 1 cycle	ETAAS	5.0 min	[813]
Seafood	Dionex ASE-200	1.0 M formic acid, 125°C, 500 psi, 1 cycle	ICP-OES	10.0 min	[814]
Marine sediments, soils	Dionex ASE-200	8.0 M acetic acid, 100°C, 1500 psi, 2 cycles	ICP-OES	15.0 min	[815]
Seaweed	Dionex ASE-200	0.75 M acetic acid, room temperature, 10.3 MPa, 1 cycle	ICP-OES	5.0 min	[816]

[a] 30 min dynamic extraction time
[b] EDTA, ethylendiaminotetraacetic acid
[c] CDTA, 1,2-diaminocyclohexane-N,N,N'N'-tetraacetic acid
FAAS, flame atomic absorption spectrometry; ETAAS, electrothermal atomic absorption spectrometry, ICP-OES, inductively couple plasma – optical emission spectrometry; ICP-MS, inductively couple plasma – mass spectrometry

Morales-Riffo and Richter [808] have also used a laboratory-made extractor to extract cadmium from atmospheric particulate matter using 0.1 %(v/v) nitric acid as a extracting solution; while a similar prototype was also used by Tavakoli and Yoshida [809] to release cadmium from solid (squid) wastes. Finally, cadmium has been also extracted from sediments by pressurization using a commercial supercritical extractor (supercritical carbon dioxide extraction, SFE) [810] which involved three sequential steps: a first supercritical carbon dioxide extraction, followed by a second sub-critical water extraction, and a final sub-critical water (90%) and carbon dioxide (10%) extraction stage.

However, the first application of commercial PLE piece of equipment (Dionex ASE-200 system) for leaching metals was developed by Wanekaya et al. [811] who investigated the use of EDTA aqueous solution as extracting under pressurized conditions. Concerning cadmium extraction by using commercial Dionex ASE-200 system and chelating agents as extracting solutions, atmospheric particulate matter (PM10), previously collected on quartz fiber filters [812], and plant samples [813] (Table 23) have been successfully pre-treated. Table 23 also lists other applications of PLE using diluted formic acid to extract cadmium from molluscs [814], and diluted acetic acid for pre-treating sediments and soils [815], and seaweed [816].

V.3.4. Enzymatic Hydrolysis Methods

Enzymatic hydrolysis or enzymic hydrolysis is a group of procedures commonly considered as environmentally friendly methods because digest biological samples under mild temperature and pH conditions and in absence of polluting or toxic reagents [723]. These procedures consist of hydrolyzing biomolecules by breaking down certain bonds, allowing the selective release of metals (and/or organometallic species) from the sample matrix. Enzymes are mainly of proteolytic nature (proteases), and they attack the peptidic bonds of proteins and peptides (Pronase E, pepsin, pancreatin or trypsin). Other enzymes, such as lipases, are able to hydrolyze fats into long-chain fatty acids and glycerol, while amylases hydrolyze starch and glycogen to maltose and to residual polysaccharides. The enzyme must be carefully selected in base on the nature of sample (high protein, fat or carbohydrate content). In some cases, a mixture of different enzymes can be simultaneously used but this approach is not always possible because the optimum pH value for a certain enzyme can not be same for the other one (i.e., pepsin normally operates at pH around 2 while lipase or α-amylase work at a pH around 7) [817]. Complete information about enzymatic methods for the extraction of metals and organometals species can be found in the review by Bermejo et al. [818].

The first work involving the use of enzymes to extract metals (cadmium, copper, lead and thallium) was carried out in 1981 by Carpenter [819], who used subtilisin Carlsberg proteolytic enzyme for hydrolyzing human liver and kidney tissues. The use of enzymes for extracting metals and organometallic compounds (mainly arsenic, selenium and tin species) from biological and environmental samples was then quite popular [818,820]. Concerning cadmium extraction from environmental materials (Table 24), proteolytic enzymes such as Pronase E [158], trypsin [817] and pepsin [821] have been successfully used to digest mussel soft tissue [158,817] and seaweed samples [821]. As reported, temperature, pH and ionic strength were the most important parameters affecting the enzymatic hydrolysis of mussel soft tissue. Other times, as those obtained for seaweed material [821], the high ionic strength in enzyme/sample mixture is the factor controlling the enzymatic hydrolysis process. This can be explained by assuming the high salts content in the seaweed [821].

Table 24. Selected applications of enzymatic hydrolysis procedures

Sample	Enzyme	Enzymatic hydrolysis conditions	Atomic spectrometry technique	Hydrolysis time	Ref.
Mussels	Pronase E	37°C, buffer 0.1 M/0.1 M TRIS[a]/HCl, pH 7.4	FAAS/ETAAS	5.0 h	[158]
Mussels	Trypsin	37°C, buffer 0.1 M/0.1 M PDHP/PHP[b], pH 9.0	ICP-OES	24 h	[817]
Seaweed	Pepsin	37°C, 1.0 %(m/v) NaCl, pH 1.0 (with HCl)	ICP-OES	6.0 h	[821]
Mussels	Trypsin[c]	37°C, buffer 0.2 M/0.2 M PDHP/PHP, pH 8.0	ICP-OES	30 min	[824]
	Pancreatin[c]	37°C, buffer 0.5 M/0.5 M			
Seaweed	Pepsin[c]	37°C, 1.0 %(m/v) NaCl, pH 1.0 (with HCl)	ICP-OES	30 min	[821]
Mussel, plants	Pepsin[d]	50°C, 1500 psi, 3 PLE cycles, Milli-Q water, pH 1.0 (with HCl)	ICP-OES	6.0 min	[822]

[a] TRIS, Tris(hydroxymethyl)aminomethane
[b] PDHP/PHP, potassium dihydrogen phosphate / potassium hydrogen phosphate
[c] Ultrasound bath assisted enzymatic hydrolysis
[d] Pressurized assisted enzymatic hydrolysis
FAAS, flame atomic absorption spectrometry; ETAAS, electrothermal atomic absorption spectrometry; ICP-OES, inductively couple plasma – optical emission spectrometry

Although enzymatic hydrolysis procedures are appealing methodologies especially in organometallic speciation studies, their general application is limited because the long time required for completing the hydrolysis process, around 5-24 h [818]. Several attempts based on the use of ultrasound energy (both ultrasonic water-baths and ultrasonic probes) [818], and more recently pressurization [822] or microwave energy [823], have been developed to speed up the enzymatic hydrolysis process. The reduction on the enzymatic hydrolysis time when using ultrasounds can be attributed to cell membranes disruptions by the action of ultrasounds and the direct cytosolic content attack by enzymes [824]. By this way, enzymatic hydrolysis time can be shortened from hours to 30 min when treating mussel soft tissue with trypsin and pancreatin [824], or seaweed with pepsin [821], by using an ultrasonic water bath. Similarly, pressurization also disrupts cell membranes before enzymatic digestion of marine biota (mussel), fatty biological tissues and plants [822]. In this case, the pre-treatment time for the pressurized enzymatic hydrolysis procedure was shortened to 6 min (3 PLE cycles of 2 min each one at 50°C and 1500 psi) using pepsin or pancreatin as proteases.

V.3.5. Slurry Sampling Technique

The slurry sampling technique is a well-established methodology for direct determination of trace metals in solid samples by atomic spectrometry. As reported in several reviews on this topic, the sampling of slurries or suspensions in conventional sample introduction systems for atomic spectrometry offers several advantages. Firstly, the time for sample prepratation is shortened; the possibility of analyte losses and/or sample contamination is reduced because the minimum sample manipulation; dangerous or corrosive reagents commonly required for other sample pre-treatment are avoided, and this technique can be considered as an environmentally friendly approach; and the same atomizers and autosampler devices than those used for liquid sampling can be employed. When comparing with the solid sampling technique, mainly used for ETV coupled to ICP-OES/MS (section IV.5.5.1), the slurry sampling approach overcomes some of the problems associated with direct solid sampling in most of conventional atomizers.

Most of the applications of slurry sampling technique make use of ETAAS. In this way, several reviews have been published reporting application of the slurry sampling and solid sampling for atomic spectrometry [825] and paricularly for ETAAS [826-828].

Among the different factors affecting the slurry sampling, one of the most critical variables are the particle size distribution and the need for maintaining stable the slurry during the time required for sample introduction. The first factor is mainly important when nebulizing slurries in FAAS and ICP-OES/MS, while the second one widely affects determinations by ETAAS. Therefore, different approaches have been adopted for homogenizing the slurry just before sampling. These techniques include manual shaking, mechanical agitation by magnetic stirring or vortex mixing, gas bubbling, ultrasonic homogenization by ultrasonic baths or ultrasonic probes either hand-held or assembled to the autosampler tray (first reported by Carnrick et al. [829]) [825-828]. Concerning particle size, large sizes can lead to rapid sedimentation of the solid particles, fact which is associated to errors when sampling high density materials [830,831]. In these cases, sedimentation problems are avoided by an effective grinding (particle size reduction of the sample) and mainly with the addition of special agents (thixotropic or surfactant agents) to the liquid medium. The main problem when dealing with low-density materials is the flotation of the finely ground material on the surface of the slurry. Therefore, wetting and dispersing agents must be used and ethanol, Triton X-100 and glycerol are the most commonly reported agents [827,829]. It is well-established that a minimum number of particles must be introduced into the graphite tube to ensure the representativity of the aliquot of sample. In this sense, Miller-Ihli [832] developed an equation to calculate the representative sample mass. In this equation the percentage of analyte extracted from the solid particles to the liquid media during the slurry preparation is taken into account.

Another important factor is the liquid medium used to obtain the slurry, which acts as a suspension medium but also as an extracting solution, improving accuracy and precision. For several applications, the use of H_2O was good enough to obtain satisfactory results [827]. Other researchers have used concentrated acids, including HNO_3, HF, or combination of HNO_3 and H_2O_2 [833], or H_2O_2 [827,828]. As commented in section V.3.2.1 regarding ultrasound probes, metal extraction efficiencies from the solid particles can be increased when using ultrasound irradiation, and under these conditions dilute acids can be used. As most of the autosampler devices for ETAAS are equipped with ultrasonic probes, adequate slurry homogenization but also analyte extraction is achieved just before sample introduction into the graphite furnace.

Continuous slurry nebulization for FAAS and ICP-OES/MS is also a well-established practice. In these cases, the slurry introduction is more limited because of the restriction of the particle size. The developed methods require an acid medium to prepare the slurry and sometimes the use of ultrasound energy for increasing metal releasing from the solid particles.

As firstly reviewed by Ebdon et al. [834], and recently by Santos and Nóbrega [835], there are several applications of slurry sampling in plasma detectors mainly because the inherent high sensitivity for cadmium determination by ICP-OES/MS. A critical factor in direct analysis of slurries by ICP-OES/MS is the stability and homogeneity of the slurry. High stable and homogeneous slurries is necessary to achieve an homogeneous and reliable aerosol for introduction into the plasma, and to obtain precise and accurate results [835]. Therefore, different suspending or dispersing agents such as Triton X-100, poly(acrylate amine), NH_3, HNO_3, HCl, have been proposed [835]. Usually, the slurry is stirred before the introduction in the plasma using the same systems reported above [835]; however, the use of surfactant agent does not avoid problems related to particle size distribution, parameter which affects the efficiency on transport and atomization-excitation. In general, a mean particle size of 5-10 μm is essential to ensure that the slurry has similar transport properties than an aqueous solution and therefore, an efficient grinding process is needed.

Several approaches have been developed for metal vapour generation, mainly for hydride forming elements, from solid particles (slurries). As reviewed by Matusiewicz [836], the main problem associated to this technique is the low vapour generation efficiency from the metals trapped in the solid particles. To avoid this problem, several authors have prepared the slurry sample in acid medium in combination with a mechanical agitation or sonication processes. The use of these acids improves the vapour generation efficiency due to the analyte´s mobilization from the solid particle to the liquid phase. The disadvantages of these procedures are the time required to complete the acid slurry preparation (10 min-hours) and in some cases, the high acid concentrations used, which are not environmentally friendly. For certain matrices such as coal fly ash and urban dust, it was possible the determination of these elements using aqueous slurry sampling [837,838]. For cadmium, volatile cadmium species formation (HCl and $NaBH_4$) has been obtained from seawage sludge and Antartic krill slurries under sonication conditions [294].

Chapter VI

CADMIUM SPECIATION

According to the International Union for Pure and Applied Chemistry (IUPAC) chemical species of an element is referred to a specific form (isotopic composition, electronic or oxidation state, and/or complex or molecular structure) of that element [839]. By this way, speciation analysis is understood as all those analytical activities of identifying and/or measuring the concentrations of one or more individual chemical species in a sample. In some cases, there are not specific molecular structures or oxidation states which can define a certain compound or ion as a chemical species of an element. In these cases, the term fractionation is preferred than speciation. Fractionation procedures consist of classifying an element in a sample according to physical (size, solubility) or chemical (bonding reactivity) properties.

In contrast to other heavy metals such as mercury, arsenic, antimony, selenium, lead or tin, there are few identified chemical species for cadmium in function of the oxidation state or defined simple organometallic structures. DPASV studies have demonstrated that cadmium is mainly complexed with organic chelates (around 80%) in estuarine water [840], while most of the cadmium is present as free ion (Cd^{2+}) in lake waters [841]. Electrochemical methods have also revealed the presence of monomethyl-cadmium ion (CH_3Cd^+) in seawater (concentration around 0.5 µg L^{-1}) [842] which is biosynthesized by the plankton by a biomethylation process [842,843].

However, most of the speciation studies concerning cadmium are related to the fact that this element is one of the most important inductor for metallothioneins (MTs) biosynthesis in animals. Therefore, cadmium is present in the structures of these metal-binding-proteins. In addition, cadmium is also complexed to the phytochelatin peptides in plants cultivated in highly cadmium-contaminated soils.

Therefore, cadmium speciation studies discussed in this section will be focused to different sequential extraction methods for cadmium fractionation in environmental samples (soils, sediments, sludge, atmospheric particulate matter); and different hyphenated chromatography/electrophoresis and atomic spectrometric methods for the determination of cadmium bound to MTs.

VI.1. Sequential Extraction Schemes for Metal Fractionation

In general, metal fractionation studies can be carried out by using single or sequential extractions (different extractants with increasing extraction capacity), and both often address operationally defined fractions which identify certain groups of elements without clear identification. Although in same cases a particular single extraction can be related to certain chemical form of an element, these extraction procedures are mainly developed for isolating of a particular metal-containing matrix phase (e.g. water soluble fraction or carbonate bound fraction). This information is useful to characterize a sample, commonly soils and marine and river sediments, in base of metals mobility or availability. It is possible the identification of different defined phases (fractions) for soil and sediment samples, as well as for other similar environmental solid materials (sludge, atmospheric particulate matter), as follows:

1) Water-soluble, phase which contains the most mobile and potentially available metal species.
2) Exchangeable fraction, phase which contains weakly bound metals species that can be released by ion-exchange (mainly NH_4^+, Ca^{2+} and Mg^{2+}). Solutions containing ammonium (NH_4^+), especially ammonium acetate, are used as extractants.
3) Carbonate bound, phase which contains metals easily released at acidic pHs (sodium acetate or acetic acid, pHs approximately 5).
4) Oxides of manganese and iron, phase which contains metals easily released (easily reducible) after minimal attack of iron/manganese oxide phases. Extractants based of hydroxylamine hydrochloride at acid pH are commonly used.
5) Organically bound, phase which contains metals bound to easily extractable organic matter, mainly humic substances. Extractants commonly consist of an ammonium acetate solution after an organic matter oxidation stage, hydrogen peroxide or sodium hypochlorite solutions.
6) Moderately reducible oxides, phase which contains metals bound to moderately reducible oxides. Extractions are mainly based on hydroxylamine hydrochloride/acetic acid mixtures.
7) Oxide and sulfide fraction, phase which contains elements bound to oxides and mainly sulfides. Hydrogen peroxide and ammonium acetate are commonly used.
8) Silicate bound (residual fraction), phase which mainly contains crystalline bound trace elements. This fraction is released after silicate decomposition with concentrated acid, specially hydrofluoric acid, perchloric acid and nitric acid.

VI.1.1. BCR Sequential Extraction Scheme

The lack of uniformity of the different sequential extraction schemes to assess the fraction of elements belonging to some of the defined phases listed above, offers as a main disadvantage that the significance of the analytical results is dependent to the extraction scheme used. Therefore, comparison between different laboratories in assessing metal mobility or availability is difficult. This fact has led to the development of standardized single and sequential extraction schemes by the Standard, Measurements and Testing Programme (SMandT, formerly BCR – The Community Bureau of Reference) of the European Union to harmonize sequential extraction methods for metals in sediments [844]. In addition, efforts have been made to prepare soil and sediment certified reference materials with certified extractable single or sequential extraction fractions [845]. The BCR protocol proposes a standardized three-stage extraction procedure (BCR EUR 14763 EN), originally developed for sediments [846], which consists of the following steps:

Step 1. Extraction with 0.11 M acetic acid, 40 mL for each gram of dry sample (sediment). The sample and the extractant are shacked in a sealed container for 16 hours at room temperature (20°C) on an end-over-end mechanical shaker at a speed of 30 rpm. Elements extracted during this stage are those related to water-soluble, exchangeable, and carbonate bound phases, above coded as (1), (2) and (3), respectively. Therefore, elements weakly absorbed on the sample surface, elements that can be released by ion-exchange processes and elements co-precipitated with carbonates are extracted during this step.

Step 2. Extraction with 0.11 M hydroxylammonium chloride (adjusted pH 2 with nitric acid), 40 mL for each gram of dry sample (sediment). After centrifugation at 2500 rpm for 25 min and removal of the supernatant (0.11 M acetic acid), the residue is washed with 20 mL of ultrapure water and after centrifugation and washing solution discarding, the hydroxylammonium chloride solution is added and the mixture is again shacked in a sealed container for 16 hours at room temperature (20°C) on an end-over-end mechanical shaker at a speed of 30 rpm. Because of the reducing properties of hydroxylammonium chloride, elements bound to iron/manganese oxides are released. Therefore elements extracted during this stage are those related to easily or moderately reducible phases, above coded as (4) and (6-7), respectively.

Step 3. Organic matter decomposition (oxidation) with 8.8 M hydrogen peroxide, 10 mL plus 10 mL for each gram of sample (sediment); and extraction with 1.0 M ammonium acetate (adjusted pH 2 with nitric acid), 50 mL for each gram of dry sample (sediment). After separation of the solution containing the reducible fraction and washing stages, the hydrogen peroxide (10 mL) is added to residue and it is digested at room temperature for 1 hour with occasional manual stirring. Organic matter oxidation is then continued by heating the container at 85°C on a hot plate for 1 hour. The volume of the liquid is then reduced to 2-3 mL by further heating. After this, another 10 mL of hydrogen peroxide is added, and the mixture is again heated at 85°C for 1 hour. The liquid is also reduced to a few mL by heating, and after cooling the ammonium acetate solution is added. The mixture is then shacked for 16 hours at room temperature (20°C) on an end-over-end mechanical shaker at a speed of 30 rpm. Because of the oxidation of the organic matter, elements bound to organic matter are released. In addition, metals bound to sulfides must be also extracted during this stage. Therefore elements extracted are those related to organically bound and sulfide fraction phases, above coded as (5) and (7), respectively.

The simplicity of the BCR sequential extraction scheme based on three stages has made that this protocol overcome the early sequential extraction scheme based on five sequential stages proposed by Tessier et al. [847]. This scheme uses a mechanical stirring of the sample with 1.0 M magnesium chloride solution (pH 7.0) for 1 hour to isolate the water-soluble and exchangeable fractions (first step), followed to an extraction with 1.0 M sodium acetate (pH

5.0) for 5 hours to release metals bound to carbonates (second step). The third extraction is carried out with 0.04 M hydroxylammonium chloride and acetic acid (25 %(v/v)) for 6 hours at 96°C to release metals bound to the iron/manganese oxides fraction (third step). 8.8 M hydrogen peroxide (adjusted pH 2.0 with nitric acid) is also used to decompose the organic matter fraction, an metals associated with it are then released with 3.2 M ammonium acetate in 20% nitric acid (mechanical shacking for 30 min). During this fourth stage, elements bound to organic matter and sulfides are extracted. The fifth stage consists of an acid digestion with hydrofluoric acid and perchloric acid to decompose the silicate fraction. Therefore, metals bound to silicate are extracted.

Vi.1.2. Assisted – BCR Sequential Extraction

Because the long time required for BCR sequential extraction methods, several attempts have been adopted to speed up the whole process. Ultrasound energy (ultrasonic water baths and ultrasonic probes) and microwave energy have been commonly used for assisting the three stages of the BCR scheme. According with recent reviews [848,849], good agreement between extractability of certain fractions, mainly exchangeable and carbonate-bound) after ultrasound assistance and conventional BCR procedures has been obtained [850]. However, acceleration with microwave energy has been reported to give worse performances. This can be attributed to the efficient heating caused by microwaves which increases the extractability of certain fractions.

VI.1.3. Cadmium Fractionation in Environmental Samples by BCR Sequential Extraction Scheme

BCR scheme has been used to assess cadmium fractionation in several solid environmental samples. Some examples include soils [157,165,851-860], sediments [845,861-872], sewage sludge [165,851,873-880 and atmospheric particulate matter [154,881-884]. Assisted BCR sequential extraction scheme for cadmium fractionation in sediments has been also proposed. Both ultrasound and microwave sequential extraction energy have given extractable cadmium contents for the three stages of the BCR procedure (excluding the H_2O_2 digestion in step 3) statistically similar than conventional BCR; but the over-all procedure could be completed in 15-30 min using ultrasounds, or after 60-120 s for microwave assistance [885,886]. Ultrasound-assisted BCR extraction procedures were also optimised for the evaluation of cadmium mobility in bottom sediments [885, 887].

VI.2. Metallothioneins / Metallothioneins-Like Proteins

There are certain products biosynthesised in an organism as consequence of a toxic event that can be used to measure exposure to contaminants and to study its possible chronic toxic effects. These molecular features, named biomarkers, can be defined as biological responses to an environmental chemical which may be found at the molecular level but also at the cellular or even the whole organism level. Metallothioneins (MTs) are considered as biomarkers of heavy metal impregnation [888-892]. MTs are non–enzymic metal–binding proteins characterized by low molecular weight, high content of cysteine residues, lack of aromatic amino acids and high resistance to heat. Their sulphydrryl–rich primary structure confers them with a high capacity for metal binding, mainly copper, cadmium and zinc [893]. This type of metal–binding proteins has been also isolated from microorganisms and marine invertebrates, but they usually contain less cysteine residues and lower metal concentrations [894]. Therefore, MTs isolated from invertebrates are preferably called as metallothioneins–like proteins (MLPs). As biomarkers, both MTs and MLPs have been described as physiologically multifunctional and they are involved in the transport, storage and detoxification of metals [895].

Isolation procedures based on ion exchange properties and electrophoresis has led to the recognition of a variety of forms of MTs which were subsequently shown to differ in amino acid composition and sequence [894]. In mammalians there are at least two isometallothioneins (metallothioneins isoforms) numbered according to the order in which they elute from anion exchanger as metallothionein–1 (MT–1) and metallothionein–2 (MT–2) [896]. Advances in HPLC combined with sensitive spectrometric detectors have aided the resolution of not only differently charged MT isoforms but also other polymorphic variants not separated by classical chromatographic procedures. Thus, several MT subisoforms have been reported, mainly MTs subisoforms isolated from mammalians [897-900]. Electrospray mass spectrometry (ESI-MS) and electrospray tamdem mass spectrometry (ESI-MS-MS) have shown excellent capabilities for this characterization. These detection methods have allowed the determination of the molecular mass for the different MTs and their subisorforms, as well as structural information [897-905

In plants, MTs have been also found and they play major roles in the hyperaccumulation of toxic metals. In fact, different studies have pointed out that certain plants can be useful in the cleanup of metal-contaminated soils and waters [906]. Plant MTs differ from animal MTs by a peculiar sequence organization consisting of two short cysteine-rich terminal domains linked by a long cysteine-devoid spacer. The role of the plant MT domains in the protein structure and functionality is largely unknown [907]. Other studies have proved the existence of other plants metallo-proteins (metallohistins), such as multimeric histidine-rich proteins, coded as AgNt84 and Ag164, which are encoded by two nodule-specific cDNAs isolated from the actinorhizal host plant *Alnus glutinosa*. Characterization studies have revealed that these polypeptides are capable of binding metals such as cadmium [908].

VI.2.1. Sample Preparation For MTs/MLPs Extraction

Sample preparation for the extraction of MTs and MLPs (commonly known as cytosol preparation) are based on modifications of the procedure proposed by Roesijadi and Fowler [909]. This method consists of triturating (cutting)/homogenizing the wet sample with an extracting TRIS/HCl buffer solution at a physiological pH of 7.4. Different substances such as phenylmethylsulfonyl fluoride (PMSF) and 2–mercaptoethanol (2–MEC) are commonly added to the buffer solution to prevent protein hydrolysis and/or protein oxidation during the extraction process [910].

Cutting mixers with cutting edges made of steel [909], are commonly used to prepare the cytosol. In order to avoid or minimize metal contamination, metal free cutting devices have been also applied in the recent literature [910]. However, cutting mixers offer other drawbacks such as the strong air uptake (the vessel containing the wet sample and the extracting solution is open) and a large amount of sample needed to perform the cutting/homogenization. To overcome these problems, some authors have proposed the use of rotating pestles or sonication in closed vessels under inert Ar or N_2 atmosphere [910]. Contamination, as well as oxidation by air, is minimized by using these methods, which also allow the treatment of small samples. Other methods for cytosol preparation have consisted of freeze thaw cycles in liquid nitrogen [911] or blending the sample and the extractant inside a free-metals disposable closure bag [152]. The last procedure has been demonstrated to be free of metal contamination and it has been found adequate for small samples (lower than 5 g) [152,912]. Recently PLE has also been proposed for mussel cytosolic preparation [912].

VI.2.2. Separation/Determination of MTs/MLPs

Although electroanalytical techniques such as DPASV [913,914] are used for monitoring complexation of metals such as cadmium by α- and β-domains of MTs, and also for MTs quantification, most of the research efforts have been led to separating the different MTs/MLPs isoforms. For this purpose, poweful separation techniques such as HPLC and capillary electrophoresis (CE) have been widely tested and optimized in combination to sensitive element-specific or molecule-specific detectors, such as ICP-MS or ESI-MS [915], respectively. These detection systems offer more adequate capabilities that have overcome those offerred by early UV detectors. By this way, hyphenate techniques (HPLC or CE) with ICP-MS detection are nowadays used for the detection and quantification of metals bound to MTs/MLPs, while detection by ESI-MS is used to characterize and quantify the different isoforms and subisoforms [916,917]. CE offers great potential for the separation of the different MTs/MLPs isoforms and sub-isoforms, as well as allows the use of small volumes of sample [918]. Advances on the interfaces for hyphenating CE to ICP-MS have led to promissing developments [916]. HPLC techniques also offer excellent abilities to separate MTs/MLPs isoforms, with the advantage that HPLC and ICP-MS coupling can be easy realized. HPLC separation based on size exclusion, anion-exchange and reversed-phase have been widely applied. As reported, size exclusion does not allow the separation of MT-1 and MT-2 isoforms although the isolation of MT-3 from cytosols from human tissues is possible. Despite of this limitation, size exclusion is commonly used for isolating MTs from other metallo-proteins as a preliminary step (preparative size-exclusion chromatography) [919,920]. Anion exchange allows the resolution of the two main MTs/MLPs isoforms (MT-1 and MT-2), and recent advances have shown capabilities for distinguishing between sub-isoforms [920]. Finally, reversed-phase chromatography with C4, C8, or C18 columns and also capillary C8 reversed-phase liquid chromatography (cLC) [921] can be used for separating MT isoforms, although the long chromatogragrahic time for reversed-phase MTs separation is a drawback of this type chromatography.

Recently, multidimensional nuclear magnetic resonance spectroscopy (NMR) has also been used for characterizing the metal-thiolate connectivity of recombinant Cd7-MT10 from the sea mussel *Mytilus galloprovincialis*. The ^{113}Cd NMR spectrum of mussel MT10 shows unique features, with a remarkably wide dispersion (210 ppm) of ^{113}Cd NMR signals [922].

VI.2.3. Separation/Determination of Cadmium Bound to MTs/MLPs

Table 25 lists the different developments for MTs/MLPs separation by size-exclusion HPLC hyphenated with, mainly, ICP-MS detection. The methods have been applied for isolating MTs from mammalians tissues (bovine liver [923], rat liver [905,924], porcine liver [925], human liver [926], human thyroid [927], human tissues [928] and human brain [929]), fish tissues (eel liver [930,931], catfish liver [932], pearl cichlid liver [932], carp liver [931,933-935]), molluscs soft tissues (mussel hepatopancreas [919,920,936]), and plant extracts [937] Similarly, HPLC separations based on anion exchange are also given in Table 25. MTs/MLPs separation in cytosols prepared from rabbit liver [938], rat liver [924,939], human liver [927], mussels [920,936], carp liver [933,940], eel liver [941], were developed to assess cadmium associated with. Reversed-phase HPLC methods are also included in Table 25. In this case, most of the developments have been performed for MTs from rabbit liver [897,898,935,942-944], although some applications for isolating MTs from porcine liver [925] and rat liver [905] can also be found. Studies on fish [935,943], and mussel [936] soft tissues have been also carried out. Applications by CE are listed in Table 26. Electrophoretic MTs separations from rabbit liver [945-956], horse kidney [957], rat tissue [902], and human tissues [951,958] have been successfully addressed. Separations in cytosols from fish tissues [959-962 and mussels [962] have also been reported.

Table 25. Cadmium bound to MTs/MLPs separation by HPLC

Sample type	Size-exclusion HPLC Column / conditions	Detection	Reference
Bovine liver	TSK-GEL G3000 SW_{XL} / 50 mM TRIS/HCl, pH 7.3	ICP-MS	[923]
Rat liver	Superdex-75, Sephadex G-75 / 30 mM TRIS/HCl, 10 mM TRIS/HCl, pH 8.5	SE preparative for RP-HPLC	[905]
Rat liver	TSK-GEL G3000 SW / 25 mM TRIS/HCl, pH 8.0	SE preparative for anion exchange HPLC	[924]
Porcine liver	TSK HW 55S / 10 mM ammonium acetate + 10 mM sodium chloride, pH 7.5	ICP-MS / ESI-MS	[925]
Human liver	Superdex-75 HR (16/60) / 20 mM TRIS/HCl, pH 7.4	ICP-OES	[926]
Human thyroid	BioSEp-SEC-S3000 / 20 mM TRIS/HCl, pH 7.4	ICP-MS	[927]
Human tissues	Superdex 75 PG / 20 mM TRIS/HCl, pH 7.4	ICP-MS	[928]
Human brain	Superdex 75 PG / 20 mM TRIS/HCl, pH 7.4	ICP-MS	[929]
Eel liver	TSK-GEL G3000 PW_{XL} / 30 mM TRIS/HCl, pH 7.4	ICP-TOFMS	[930]
Eel liver	TSK-GEL G3000 PW_{XL} / 30 mM TRIS/HCl, pH 7.4	ICP-TOFMS	[931]
Catfish liver	Superdex HR 10/30 / 10 mM TRIS/HCl, pH 7.4	ICP-MS	[932]
Pearl cichlid liver	Superdex HR 10/30 / 10 mM TRIS/HCl, pH 7.4	ICP-MS	[933]

Table 25. Continued

Sample type	Column / conditions	Detection	Reference
	Size-exclusion HPLC		
Carp liver	TSK-GEL G3000 PW_{XL} / 30 mM TRIS/HCl, pH 7.4	ICP-TOFMS	[931]
Carp liver	TSK-GEL G3000 PW_{XL} / 30 mM TRIS/HCl, pH 7.4	ICP-TOFMS	[933]
Carp liver	TSK-GEL G3000 PW_{XL} / 30 mM TRIS/HCl, pH 7.4	ICP-TOFMS	[934]
Carp liver	TSK-GEL G3000 PW_{XL} / 30 mM TRIS/HCl, pH 7.4	ICP-TOFMS	[935]
Mussel hepatopancreas	Sephadex G-75 / 10 mM TRIS/HCl, pH 7.4	ICP-MS	[919]
Mussel hepatopancreas	Sephadex G-75 / 10 mM TRIS/HCl, pH 7.4	ICP-TOFMS	[920]
Mussel hepatopancreas	Sephadex G-75 / 10 mM TRIS/HCl, pH 7.4	SE preparative for anion exchange HPLC	[936]
Plants extracts	Superdex peptide HR / 30 mM TRIS/HCl, pH 7.5	ICP-MS	[937]
	Anion exchange HPLC		
Rabbit liver	DEAE-5PN / gradient (A: 2 mM TRIS/HCl, pH 7.4 and B: 200 mM TRIS/HCl, pH 7.4)	HG-ICP-MS	[938]
Rat liver	Shodex Asahipak ES-502N7C / gradient (A: 2 mM TRIS/HCl, pH 7.2 and B: 50 mM TRIS/HCl, pH 7.2)	ICP-MS	[924]
Rat liver	Fractogel EMD DEAE-650 / gradient (A: 20 mM TRIS/HCl, pH 7.4 and B: 120 mM TRIS/HCl, pH 7.4)	AAS	[939]
Human liver	Fractogel EMD DEAE / gradient (A: 20 mM TRIS/HCl, pH 7.4 and B: 250 mM TRIS/HCl, pH 7.4)	ICP-OES	[926]
Catfish liver	Mono Q (HR 5/5) / gradient (A: 20 mM TRIS/HCl, pH 7.4 and B: 20 mM TRIS/HCl + 0.4 M ammonium acetate, pH 7.4)	ICP-MS	[932]
Pearl cichlid liver	Mono Q (HR 5/5) / gradient (A: 20 mM TRIS/HCl, pH 7.4 and B: 20 mM TRIS/HCl + 0.4M ammonium acetate, pH 7.4)	ICP-MS	[932]
Carp liver	ProteinPak DEAE-5PW / gradient (A: 2 mM TRIS/HCl, pH 7.4 and B: 200 mM TRIS/HCl, pH 7.4)	ICP-TOFMS	[933]
Carp liver	ProteinPak DEAE-5PW / gradient (A: 2 mM TRIS/HCl, pH 7.4 and B: 200 mM TRIS/HCl, pH 7.4)	ICP-TOFMS	[934]
Carp liver	ProteinPak DEAE-5PW / gradient (A: 2 mM TRIS/HCl, pH 7.4 and B: 200 mM TRIS/HCl, pH 7.4)	ICP-TOFMS	[940]
Eel liver	Mono Q (HR 5/5) / gradient (A: 4 mM TRIS/HCl, pH 7.4 and B: 10 mM TRIS/HCl + 0.25 M ammonium acetate,	ICP-MS	[941]

	pH 7.4)		
Mussel hepatopancreas	Mono Q (HR 5/5) / gradient (A: 4 mM TRIS/HCl and B: 10 mM TRIS/HCl + 250 mM ammonium acetate)	ICP-TOFMS	[920]
Mussel hepatopancreas	Mono Q (HR 5/5) / gradient (A: 4 mM TRIS/HCl and B: 10 mM TRIS/HCl + 250 mM ammonium acetate)	ICP-MS	[936]
	Reversed-phase HPLC		
Rabbit liver	Vydac C8 / gradient (A: 5 mM ammonium acetate, pH 6.0 and B: 5 mM ammonium acetate, pH 6.0 + 50% acetonitrile)	ICP-MS / ESI-MS	[904]
Rabbit liver	Microbore C8 / gradient (A: 5 mM ammonium acetate, pH 6.0 and B: 5 mM ammonium acetate, pH 6.0 + 50% methanol)	UV / ICP-MS / ESI-MS	[942]
Rabbit liver	Sperisorb C18 modified with DDAB / gradient (A: 2 mM TRIS/HCl, pH 7.4 and B: 200 mM TRIS/HCl, pH 7.4)	HG-ICP-MS	[943]
Rabbit liver	Vydac C8 / gradient (A: 10 mM TRIS/HCl in water, pH 7.4 and B: 10 mM TRIS/HCl in 50% methanol, pH 7.4)	ICP-TOFMS	[944]
Porcine liver	HyPurity C4 / gradient (A: 10 mM ammonium acetate in water, pH 7.5 and B: 10 mM ammonium acetate in methanol, pH 7.5)	ICP-MS / ESI-MS	[925]
Rat liver	Vydac C8 / gradient (A: 5 mM ammonium acetate, pH 6.0 and B: 5 mM ammonium acetate, pH 6.0 + 50% acetonitrile)	ICP-MS / ESI-MS	[905]
Carp liver	Spherisorb ODS 2 / gradient (A: 30 mM ammonium acetate in 1% methanol, pH 7.4 and B: 30 mM ammonium acetate in methanol, pH 7.4)	ICP-TOFMS	[935]
Eel liver	Sperisorb C18 modified with DDAB / gradient (A: 2 mM TRIS/HCl, pH 7.4 and B: 200 mM TRIS/HCl, pH 7.4)	HG-ICP-MS	[943]
Eel kidney	Sperisorb C18 modified with DDAB / gradient (A: 2 mM TRIS/HCl, pH 7.4 and B: 200 mM TRIS/HCl, pH 7.4)	HG-ICP-MS	[943]
Mussel hepatopancreas	Vydac C8 / gradient (A: 10 mM TRIS/HCl in water, pH 7.4 and B: 10 mM TRIS/HCl in 50% methanol, pH 7.4)	IDA ICP-MS	[936]

AAS, atomic absorption spectrometry; ESI-MS, electrospray ionization - mass spectrometry; HG-ICP-MS, hydride generation - inductively couple plasma – mass spectrometry; ICP-OES, inductively couple plasma – optical emission spectrometry; ICP-MS, inductively couple plasma – mass spectrometry; IDA-ICP-MS, isotopic dilution analysis - inductively couple plasma – mass spectrometry; ICP-TOFMS, inductively couple plasma – time of flight mass spectrometry; RP-HPLC, reverse-phase high performance liquid chromatography; SE, size exclusion, UV, ultraviolet

Table 26. Cadmium bound to MTs/MLPs separation by capillary electrophoresis

Sample type	Conditions	Detection	Reference
Rabbit liver	Capillary ID, 75 μm; length 100 cm / voltage, 20 V / buffer, 5 mM ammonium acetate, pH 6.0	ICP-MS	[899,904]
Rabbit liver	Capillary ID, 75 μm; length 100 cm / voltage, 20 V / buffer, 12 mM TRIS/HCl, pH 7.5	ICP-MS	[901]
Rabbit liver	Capillary ID, 75 μm; length 150 cm / voltage, 30.1 V / buffer, 20 mM TRIS/HCl, pH 7.1	ICP-MS	[945]
Rabbit liver	Capillary ID, 75 μm; length 120 cm / voltage, 30.1 V / buffer, 20 mM TRIS/HCl, pH 7.8	ICP-MS	[946]
Rabbit liver	Capillary ID, 75 μm; length 85 cm / voltage, 18-25 V / buffer, 15 mM TRIS/HCl, pH 6.8	ICP-MS	[947]
Rabbit liver	Capillary ID, 50 μm; length 77 cm / voltage, 25 V / buffer, 20 mM TRIS/HCl, pH 7.8	ICP-MS	[948]
Rabbit liver	Capillary ID, 50 μm; length 110 cm / voltage, 25 V / buffer, 50 mM TRIS/HCl, pH 9.1	ICP-MS	[949]
Rabbit liver	Capillary ID, 50 μm; length 10 cm / voltage, 25 V / buffer, 50 mM TRIS/HCl, pH 9.1	ICP-MS	[950]
Rabbit liver	Capillary ID, 75 μm; length 70 cm / voltage, 30 V / buffer, 20 mM TRIS/HCl, pH 7.4	ICP-MS	[951]
Rabbit liver	Capillary ID, 75 μm; length 70 cm / voltage, 30 V / buffer, 20 mM TRIS/HCl, pH 7.4	ICP-MS	[952]
Rabbit liver	Capillary ID, 75μm; length 120cm / voltage, 30V / buffer, 20mM TRIS/HCl, pH 7.4	ICP-MS	[953]
Rabbit liver	Capillary ID, 75 μm; length 68-72 cm / voltage, 20 kV / buffer, 70 mM TRIS/HCl in 5% methanol, pH 7.4	HG-ICP-MS	[954]
Rabbit liver	Capillary ID, 75 μm; length 74-79 cm / voltage, 20 kV / buffer, 70 mM TRIS/HCl in 5% methanol, pH 7.4	ICP-MS	[955]
Rabbit liver		ID-ICP-MS	[956]
Rat liver	Capillary ID, 75 μm; length 70 cm / voltage, 30 V / buffer, 20 mM TRIS/HCl, pH 7.4	ICP-MS	[902]
Horse kidney	Capillary ID, 50 μm; length 100cm / voltage, 25 V / buffer, 50 mM TRIS/HCl, pH 9.0	ICP-MS	[957]
Human tissue	Capillary ID, 75 μm; length 70 cm / voltage, 30 kV / buffer, 20 mM TRIS/HNO_3, pH 7.4	ICP-MS	[958]
Bramer liver	Capillary ID, 75 μm; length 70 cm / voltage, 30 kV / buffer, 10 mM ammonia/HNO_3, pH 7.4	ICP-MS	[959]
Bramer liver	Capillary ID, 75 μm; length 70 cm / voltage, 30 kV / buffer, 20 mM TRIS/HNO_3, pH 7.0- 7.4	ICP-MS	[960]
Eel liver	Capillary ID, 75 μm; length 74-79 cm / voltage, 20 kV / buffer, 70 mM TRIS/HCl in 5% methanol, pH 7.4	ICP-MS	[961]
Mussel hepatopancreas	Capillary ID, 75 μm; length 81 cm / voltage, 20 kV / buffer, 70 mM TRIS/HCl in 5% methanol, pH 7.4	ICP-MS	[962]

HG-ICP-MS, hydride generation – inductively couple plasma – mass spectrometry; ICP-MS, inductively couple plasma – mass spectrometry.

Chapter VII

OFFICIAL METHODS FOR CADMIUM DETERMINATION IN ENVIRONMENTAL SAMPLES

A great number of different standard and official analytical methods are available for the determination of cadmium in environmental samples. Although standard and official guidelines for sampling, preservation and handling of natural and waste water, sediment and soil, and air (PM10 and exhaust gases) have been commented in Chapters II and III, those guidelines are summarized in Table 27. Finally, Tables 28-30 show a brief description of several standard methods for environmental sample treatment and cadmium determination.

Table 27. Selected standard methods for environmental sampling, preservation and handling of environmental samples

	Details of method	Ref.
Natural and waste water		
ISO 5667-1: 2006. Water quality. Sampling Part 1: Guidance on the design of sampling programmes and sampling techniques	This norm sets out the general principles for, and provides guidance on, the design of sampling programmes and sampling techniques for all aspects of sampling of water (including waste waters, sludges, effluents and bottom deposits).	[9]
ISO 5667-4. Water quality. Sampling Part 4: Guidance on sampling from lakes, natural and man-made	This norm presents detailed principles to be applied to the design of programmes, techniques and the handling and preservation of samples of water.	[11]
ISO 5667-5. Water quality. Sampling Part 5: Guidance on sampling of drinking water from treatment works and piped distribution systems	This norm establishes principles to be applied to the techniques of sampling water intended for human consumption (all water either in its original state or after treatment, intended for drinking, cooking, food preparation, or other domestic purposes, regardless of its origin, plus all water used in any production undertaking for the manufacture, processing, preservation or marketing of products or substances intended for human consumption).	[12]

Table 27. Continued

	Details of method	Ref.
Natural and waste water		
ISO 5667-6. Water quality. Sampling Part 6: Guidance on sampling of rivers and streams	This norm sets out the principles to be applied to the design of sampling programmes, sampling techniques and the handling of water samples from rivers and streams for physical and chemical assessment. It is not applicable to the sampling of estuarine or coastal waters and has limited applicability to microbiological sampling.	[13]
ISO 5667-8. Water quality. Sampling Part 8: Guidance on the sampling of wet deposition	This norm provides guidance on the design of sampling programmes and the choice of instrumentation and techniques for the sampling of the quality of wet deposition. Does not cover measurement of the quantity of rain, dry deposition or other types of wet deposition such as mist, fog and cloud-waters.	[31]
ISO 5667-9. Water quality. Sampling Part 9: Guidance on sampling from marine waters	This norm provides guidance on the principles to be applied to the design of sampling programmes, sampling techniques and the handling and preservation of samples of sea water from tidal waters. The main objectives are quality characterization measurement, quality control measurement, measurements for specific reasons, and examination of the effects of man-made structures.	[15]
ISO 5667-10. Water quality. Sampling Part 10: Guidance on sampling of waste waters	This norm contains details on the sampling of domestic and industrial waste water, i.e. the design of sampling programmes and techniques for collection of samples including safety aspects. Covers waste water in all its forms. Sampling of accidental spillages is not included, although the methods described in certain cases may also be applicable to spillages.	[14]
ISO 5667-11. Water quality. Sampling Part 11: Guidance on sampling of groundwaters	This norm provides guidance on the sampling of groundwaters. It informs the user of the necessary considerations when planning and undertaking groundwater sampling to survey the quality of groundwater supply, to detect and assess groundwater contamination and to assist in groundwater resource management, protection and remediation.	[16]
ISO 5667-3: 2003. Water quality. Sampling Part 3: Guidance on the preservation and handling of water samples	This norm gives general guidelines on the precautions to be taken to preserve and transport all water samples.	[33]
1060. Collection and preservation of samples	This section addresses the collection and preservation of water and wastewater samples; the general principles also apply to the sampling of solid or semisolid matrices. Topics covered include collection of samples and sample storage and preservation.	[34]
Soil, sediment and sludge		
ISO 5667-12. Water quality. Sampling Part 12: Guidance on sampling of bottom sediments	Provides guidance on the sampling of sediments from rivers, streams, lakes and similar standing waters and estuarines. Sampling of industrial and sewage works sludges and ocean sediments are excluded.	[18]

ISO 5667-19. Water quality. Sampling Part 19: Guidance on sampling of marine sediments	This norm provides guidance for the sampling of sediments in marine areas for analyses of their physical and chemical properties for monitoring purposes and environmental assessments. It encompasses sampling strategy, requirements for sampling devices, observations made and information obtained during sampling, handling sediment samples, and packaging and storage of sediment samples.	[19]
ISO 10381-1. Soil quality. Sampling Part 1: Guidance on the design of sampling programmes	This norm sets out the general principles to be applied in the design of sampling programmes for the purpose of characterizing and controlling soil quality and identifying sources and effects of contamination of soil and related material, with emphasis on procedures required to locate points from which samples may be taken for examination or at which instruments may be installed for in situ measurement including statistical implications, procedures for determining how much sample to collect and whether to combine samples, methods of collecting samples, methods for containing, storing and transporting samples to prevent deterioration/contamination.	[20]
ISO 10381-2. Soil quality. Sampling Part 2: Guidance on sampling techniques	This norm gives guidance on techniques for taking and storing soil samples, information on typical equipment and selection of the equipment and the techniques to use to enable both disturbed and undisturbed samples to be correctly taken at different depths that is applicable in particular sampling situations.	[21]
ISO 5667-15. Water quality. Sampling Part 15: Guidance on preservation and handling of sludge and sediment samples	This norm gives general guidelines on the precautions to be taken to preserve and handling sludge and sediment samples.	[35]
ISO 23909. Soil quality. Preparation of laboratory samples from large samples	This norm specifies a method for the preparation of laboratory samples from large samples.	[36]
ISO 11464. Soil quality. Pretreatment of samples for physico-chemical analysis	This norm specifies the pretreatments required for soil samples that are to be subjected to physico-chemical analyses of stable and non-volatile parameters and describes the following five types of pretreatment of samples: drying, crushing, sieving, dividing and milling.	[37]
ISO 16720. Soil quality. Pretreatment of samples by freeze-drying for subsequent analysis	This norm specifies a method for pretreatment of soil, sludges and sediments samples by freeze-drying for subsequent analysis.	[38]
3030. Preliminary treatment of samples	This section describes general pretreatment for samples in which metals are to be determined. Section provides tips for preventing contamination to samples during pretreatment Methods of pretreatment described are filtration for dissolved and suspended metals, treatment for acid-extractable metals, digestion for metals, nitric acid digestion, nitric acid-hydrochloric acid digestion, nitric acid-sulfuric acid digestion, nitric-acid-perchloric acid digestion, nitric acid-perchloric acid-hydrofluoric acid digestion, and microwave-assisted digestion.	[39]

Table 27. Continued

	Details of method	Ref.
Air		
EN 12341. Air quality. Determination of the PM10 fraction of suspended particulate matter. Reference method and field test procedure to demonstrate reference equivalence of measurement methods	The norm describes a reference method for the determination of the PM10 fraction of suspended particulate matter as well as a field test procedure which helps to state the equivalence of commercial PM10 measuring devices compared to the reference method.	[29]

Table 28. Selected standard methods for natural waters treatment and cadmium determination in natural waters

	Details of method	Analytical characteristics	Ref.
Dissolved Cd and total Cd determination in natural and waste water by FAAS.	Dissolved Cd and / or total Cd is measured by FAAS aspirating the sample into the air/C_2H_2 flame and measuring the absorbance at 228.8 nm. Dissolved Cd; sample must be filtered as soon as possible after sampling through a membrane filter of 0.45 μm and then acidified with HNO_3 (pH < 2). Total Cd (after mineralization for water with high organic matter content); to 100mL of sample at pH < 2 was added 1.0 mL of conc. HNO_3 and 1.0 mL of conc. H_2O_2, the mixture is heated in a hot plate (near to dryness ~ 0.5 mL) and then dilute to 100 mL.	Applicable to Cd concentrations between 0.05 – 1.0 mg L^{-1} in natural and waste water. High concentration may be determined before sample dilution. Low concentration may determine after careful boiling previous sample acidification by nitric acid. Applicable to sediment and sludge after acid digestion. SO_4^{2-}, Cl^-, PO_4^{3-}, Na, K, Mg, Ca, Fe, Ni, Co, Pb, Si and Ti may interfere.	[963]
Dissolved Cd determination by FAAS after LLE	Cd is measured by FAAS aspirating the sample directly into the air/C_2H_2 flame and measuring the absorbance at 228.8 nm after chelation with APDC and extraction in IBMK or after chelation with HMA-HMDTC and extraction in DIBK-xylene.	Applicable to natural waters.	[964]
Dissolved Cd and total Cd determination in natural and waste water by ETAAS.	Dissolved Cd and / or total Cd is measured by ETAAS at 228.8 nm using an injection volume of 10 μL. Pd (15 and 50 μg for natural water of low and high salinity, respectively) and NH_4NO_3 (150 and 500 μg for natural water of low and high salinity, respectively) was used as chemical modifiers. Pyrolytic coated graphite tube with L'vov platform is recommended.	Applicable to Cd concentrations between 0.3 – 3.0 μg L^{-1} in natural and waste water. High concentration may be determined before sample dilution or by using injection volume <10 μL. Applicable to sediment and sludge after acid digestion. Fe, Cu, Ni, Co, Pb, Na, K, Ca, Mg, Cl^- and SO_4^{2-} may interfere.	[963]

Dissolved Cd and total Cd determination in natural and waste water by ETAAS.	Dissolved Cd and / or total Cd (ISO-15587-1 and ISO 15587-2) is measured by ETAAS at 228.8 nm using an injection volume of 20 μL. The mixtures Pd (15 μg) + $Mg(NO_3)_2$ (10 μg) or $NH_4H_2PO_4$ (200 μg) + $Mg(NO_3)_2$ (10 μg) were used as chemical modifiers. Pyrolytic coated graphite tube with L'vov platform is recommended.	Applicable to Cd concentrations between 0.4 – 4.0 μg L^{-1} in surface, ground, tap and waste water. A high Cl^- concentration interferes. To minimize the matrix effect, the chemical modification, standard addition method and background correction systems may be used.	[965]
Water digestion by aqua regia	Water digestion for total metal analysis by using aqua regia (ISO-15587-1): 6.0 mL of HCl (12.0 M) and 2.0 mL of HNO_3 (16.8 M) is added to 25 mL of sample; the mixture is heated at 103°C for 2 h to 175°C for 8 h, in a open or close digestion vessel. N-dodecane may add if foam formation occurs. After filtration (if it is necessary) sample is dilute to a fixed volume. Microwave energy can also be used for heating.	Applicable to waters with < 20 g L^{-1} of suspended solid and < 5 g L^{-1} of COT (expressed as C).	[966]
Water digestion by nitric acid	Water digestion for total metal analysis by using nitric acid (ISO-15587-2): 6.25 mL of HNO_3 (16.8 M) is added to 25 mL of sample; the mixture is heated at 103°C for 2 h to 175°C for 8 h, in a open or close digestion vessel. N-dodecane may add if foam formation occurs. After filtration (if it is necessary) sample is dilute to a fixed volume. Microwave energy can also be used for heating.	Applicable to waters with < 20 g L^{-1} of suspended solid and < 5 g L^{-1} of COT (expressed as C).	[967]
Dissolved Cd and total Cd determination in natural and waste water by ICP-OES.	Dissolved Cd and / or suspended Cd and / or total Cd is measured by ICP-OES at 214,438, 226.502 or 228.802 nm using the manufacturer's recommended conditions. Dissolved Cd; sample must be filtered as soon as possible after sampling through a membrane filter of 0.45 μm and then acidified with HNO_3 (pH < 2). Suspended Cd; 4.0 mL of conc. HNO_3 is added to the filter and the mixture heated in a hot plate near to dryness, the procedure is repeated by using 3.0 mL of conc HNO_3. Then 10 mL of HCl (0.2 M) and 15 mL of water is added and heated for 15 min. Finally, the residue is removed by filtration or centrifugation and dilute to 100 mL.	Applicable to tap and waste water. Fe, As and Co may interfere.	[968]

Table 28. Continued

	Details of method	Analytical characteristics	Ref.
	Total Cd; to 100 mL of un-filtered sample (pH < 2) was added 0.5 mL of cocn. HNO_3, the mixture is evaporated near to dryness in a hot plate, the residue is dissolved in 1.0 mL of conc HNO_3 and then the mixture is dilute 100 mL.		
Dissolved Cd and total Cd determination in natural and waste water by ICP-MS.	Dissolved Cd and / or total Cd (ISO-15587-1 and ISO 15587-2 is measured by ICP-MS using the manufacturer's recommended conditions (^{111}Cd or ^{114}Cd). Dissolved Cd determination; sample must be filtered as soon as possible after sampling through a membrane filter of 0.45 µm and then acidified with HNO_3 (pH < 2).	Applicable to 0.1 – 1.0 µg L^{-1} in tap, surface, ground and waste water. Sn^+, MoO, MoOH and ZrOH may interfere.	[969, 970]
Dissolved Cd and total Cd determination in natural waste water by FAAS	Dissolved Cd and / or suspended Cd and / or total Cd is measured by FAAS aspirating the sample into the air/C_2H_2 flame and measuring the absorbance at 328.1nm. Dissolved Cd; sample must be filtered as soon as possible after sampling through a membrane filter of 0.45 µm and then acidified with 3.0 mL of HNO_3. Suspended Cd is determined after dissolution of membrane filtration (addition of 3.0 mL of conc. HNO_3 to the filter and heat near to dryness the mixture; cool, and add 3.0 mL of conc HNO_3 and heat until digestion is complete; add 2.0 mL of HCl (1:1) and heat gently to dissolve the residue; finally dilute water. Total Cd; add of 3.0 mL of conc. HNO_3 to an un-filtered aliquot and heat and evaporate near to dryness (do not boil); cool, and add 3.0 mL of conc HNO_3 and heat until digestion is complete; add 2.0 mL of HCl (1:1) and heat gently to dissolve the residue; finally dilute water. Low Cd concentrations are measured after chelation with 2.5 mL of APDC (1.0 % w/v) to 100 mL of sample at pH 2.5 and extraction with 10 mL of IBMK after 1.0 min of shaking.	Applicable to surface, saline and waste waters. Ionization interferences are generally removed by the addition of $LaCl_3$.	[971]

Cd determination in natural and waste water by FAAS.	Cd is determined by FAAS aspirating the sample directly into the air/C_2H_2 flame.	Applicable to natural waters. At low Cd concentrations a previous extraction is required.	[972]
Cd determination in natural and waste water by ETAAS.	Cd is measured by ETAAS using the manufacturer's recommended conditions.	This method is suitable for determination of micro quantities of cadmium.	[973]
Cd determination in natural and waste water by ICP-OES.	Cd is measured by ICP-OES using the manufacturer's recommended conditions.	Applicable to tap and waste water.	[974]
Cd determination in natural and waste water by ICP-MS.	Cd is measured by ICP-MS using the manufacturer's recommended conditions.	Applicable to 0.1 – 1.0 µg L^{-1} in tap, surface, ground and waste water. It may also be suitable for wastewater, soils, sediments, sludge, and biological samples after suitable digestion followed by dilution and/or cleanup.	[975]
Cd determination in natural and waste water by ASV.	Cd is measured by ASV using the manufacturer's recommended conditions.	The method requires no sample extraction or pre-concentration. Several interferences and potential trace background contamination may occur.	[976]

APDC, ammonium pyrrolidin dithiocarbamate; ASV, anodic stripping voltammetry; COT, total organic carbon; DIBK, 2,6-dimethyl-heptan-4-one; ETAAS, electrothermal atomic absorption spectrometry; FAAS, flame atomic absorption spectrometry; HMA-HMDTC, hexamethyleneammonium hexamethylenedithiocarbamate; IBMK, 4-methyl-2-pentanone; ICP-MS, inductively coupled plasma-mass spectrometry; ICP-OES, inductively coupled plasma-optical emission spectrometry; LLE, liquid-liquid extraction.

Table 29. Selected standard methods for soil, sediment and sludge treatment and Cd determination in soil, sediment and sludge

	Details of method	Analytical characteristics	Ref.
Cd extraction form sediment and sludge by nitric acid	Nitric acid extraction: to a maximum of 0.1000 g dry sediment or equivalent of wet sample, 20 mL of HNO_3 (7.0 M) was added and the mixture heated at 120°C for 1 h in close vessel. Heating must be done using microwave energy. After cooling, the acid liquid phase was decanted, filtrated or centrifuged. The solution was made up to 100 mL. Aqua regia extraction: to a maximum of 0.1000 g dry sediment or equivalent of wet sample, 5.0 mL of HNO_3 (7.0 M) and 15 mL of HCl (7.0 M) were added and the mixture heated at 120°C for 1 h in close vessel. Heating must be done using microwave energy. After cooling, the acid liquid phase was decanted, filtrated or centrifuged. The solution was made up to 100 mL.	After extraction, Cd is determined by FAAS (ISO 5961, reference 965) or ETAAS (ISO 15586, reference 967).	[965]
Cd extraction from soil by aqua regia	Samples (3.0 g) were digested at room temperature with 21 mL HCl (12.0 M) and 7.0 mL of HNO_3 (16.8 M) mixture (28 mL per 3 g of sample) for 16 h. After this, the suspension was digested at 130°C, for 2 h under reflux conditions. The suspension was then filtered and diluted to 100 mL with HNO_3 0.5 M.	Applicable to soils and similar materials containing < 20 % (m/m) of organic carbon. For samples containing > 20 % an additional treatment with HNO_3 is required.	[977]

Table 29. Continued

	Details of method	Analytical characteristics	Ref.
Cd extraction from soil by hydrofluoric acid and perchloric acid	This procedure involves two step; first organic matter removal and then silicate matrix decomposition. Organic matter removal by calcination (0.250 g of sample placed in a Pt crisol is heated up 450°C (1 h) and then remain at 450°C for 3 h, after cooling at room temperature, few mL of water is added to fly ash obtained) or HNO_3 treatment (to 0.250 g of sample 5.0 mL of HNO_3 (15.2 M) was added and the mixture heated and evaporate at 150°C (hot plate) up ~ 1.0 mL, then the mixture is cool at room temperature. Silicate matrix decomposition by addition of 5.0 mL of HF (27.8 M) and 1.5 mL of $HClO_4$ (11.6 M) and heating (hot plate) until vapor releasing finish, add 1.0 mL of HNO_3 (15.2 M), 1.0 mL of HCl (12.1 M) and 5.0 mL of water and heat gently. Finally the mixture was made up to 50 mL.	Applicable for total dissolution of soils and similar materials by silicate matrix decomposition in an open vessel system by a mixture of HF and $HClO_4$ after organic matter removal by calcination or by nitric acid treatment.	[978]
Cd determination by FAAS in aqua regia extracts of soil	Extracted Cd is measured by FAAS aspirating the sample into the air/C_2H_2 flame and measuring the absorbance at 228.8 nm. Deuterium lamp may used as background corrector.	Applicable to Cd concentrations > 2.0 mg Kg^{-1} in dry soil. Fe may interfere.	[979]
Cd determination by ETAAS in aqua regia extracts of soil	Extracted Cd is measured by ETAAS at 228.8 nm using manufacturer's recommended conditions. Pyrolytic coated graphite tube with L'vov platform is recommended. Deuterium lamp and Zeeman effect background correctors may use.	Applicable to Cd concentrations < 2.0 mg Kg^{-1} in dry soil.	[979]
Cd determination by ICP-OES in aqua regia extracts of soil	Extracted Cd is measured by ICP-OES using the manufacturer's recommended conditions.	Applicable to soil extracts obtained with aqua regia in accordance with ISO 11466 (reference 971), with DTPA in accordance with ISO 14870 (reference 974) or other weak extractants, or soil extracts for the determination of total element contents using the acid digestion method of ISO 14869-1 (reference 972).	[980]

ETAAS, electrothermal atomic absorption spectrometry; FAAS, flame atomic absorption spectrometry; ICP-OES, inductively coupled plasma-optical emission spectrometry.

Table 30. Selected standard methods for cadmium determination in air sample (PM10 fraction) and exhaust gases

	Details of method	Analytical characteristics	Ref.
Sampling and Cd determination by AAS or AES in exhaust gases	A fixed volume of effluent from a gas pipe or chimney is isokinetic extracted for a certain period of time at a flow rate controlled (EN 13284-1, reference 961). Particles from the sampled gas are captured on a filter. Subsequently the gas is passed through a series of impingers containing a mixture of HNO_3 (3.3 % m/v) and H_2O_2 (1.5 % m/v) where metals are captured. Then, these solutions are analyzed by AAS or AES. Filter is digested in a close PTFE vessel after analysis.	This method is applicable to the determination of metals in exhaust gases from hazardous and municipal waste incinerators in the concentration range of 0.005 mg m^{-3} to 0.5 mg m^{-3}. Unless otherwise stated, concentrations are expressed at volumes under dry conditions, normalised to 273 K, 101.3 kPa, and oxygen content with a volume fraction of 11%.	[981]
Sampling and Cd determination by AAS or AES in PM10 fraction	Cd-containing particles are captured by aspiration of a measured volume of air (for 24 h) through a filter mounted on a system designed to capture the PM10 fraction of suspended particulate matter (EN 12341, reference 962). After sampling, filter is dissolved in closed vessel by using microwave energy (180°C for 20 min and 220°C for 20 min) and HNO_3 and H_2O_2.	This method is applicable to the determination of Cd in PM10 by ETAAS or ICP-MS in the concentration range of 0.1 ng m^{-3} to 50 ng m^{-3} after microwave extraction of samples.	[982]

AAS, atomic absorption spectrometry; AES, atomic emission spectrometry; ETAAS, electrothermal atomic absorption spectrometry; ICP-MS, inductively coupled plasma-mass spectrometry; PTFE, polytetrafluoroethylene.

REFERENCES

[1] Davis, A.C.; Wu, P.; Zhang, X.; Hou, X.; Jones, B.T. Determination of cadmium in biological samples. *Appl Spectrosc Rev* 2006, *41*, 35-75.

[2] Watanabe, T.; Shimbo, S.; Yasumoto, M.; lmai, Y.; Kimura, K.; Yamamoto, K.; Kawamura, S.; lkeda, M. Reduction to one half in dietary intake of cadmium and lead among Japanese populations. *Bull Environ Contam Toxicol* 1994, *52*, 196-202.

[3] World Health Organisation (WHO). Environmental Health Criteria 134 - Cadmium International Programme on Chemical Safety (IPCS) Monograph, 1992.

[4] Elinder, C.-G. In *Cadmium:Uses, Occurrence, and Intake*; Cadmium and Health: A Toxicological and Epidemiological Appraisal; CRC Press: Boca Raton, FL, 1985.

[5] American Conference of Governmental Industrial Hygienists (ACGIH). TLVs and BEls, threshold limit values and biological exposure indices for chemical substances and physical agents, 1996

[6] Jensen, A.; Bro-Rasmussen, F. Environmental contamination in Europe. *Rev Environ Cont Toxicol* 1992, *125*, 101-181.

[7] Cook, M.E.; Morrow, H. Anthropogenic sources of cadmium in Canada, National Workshop on Cadmium Transport Into Plants, Canadian Network of Toxicology Centres, Ottawa, Ontario, Canada, June 20-21, 1995.

[8] Keith, *L.H. Environmental Sampling and Analysis: A Practical Guide*; Lewis Publishers: Boca Raton, FL, 1991.

[9] International Standard Organization. 2006. Water quality. Sampling Part 1: Guidance on the design of sampling programmes and sampling techniques. ISO 5667-1. International Organization for Standardization, Case Postale 56, CH-1211, Geneva 20 Switzerland

[10] Krajča, J.M. *Water Sampling*; Ellis Horwood Limited. Chichester; 1989.

[11] International Standard Organization. 1987. Water quality. Sampling Part 1: Water quality. Sampling Part 4: Guidance on sampling from lakes, natural and man-made. ISO 5667-4. International Organization for Standardization, Case Postale 56, CH-1211, Geneva 20 Switzerland.

[12] International Standard Organization. 2006. Water quality. Sampling Part 5: Guidance on sampling of drinking water from treatment works and piped distribution systems ISO 5667-5. International Organization for Standardization, Case Postale 56, CH-1211, Geneva 20 Switzerland

[13] International Standard Organization. 2005. Water quality. Sampling Part 6: Guidance on sampling of rivers and streams. ISO 5667-6. International Organization for Standardization, Case Postale 56, CH-1211, Geneva 20 Switzerland.

[14] International Standard Organization. 1992. Water quality. Sampling Part 10: Guidance on sampling of waste waters. ISO 5667-10. International Organization for Standardization, Case Postale 56, CH-1211, Geneva 20 Switzerland.

[15] International Standard Organization. 1992. Water quality. Sampling Part 9: Guidance on sampling from marine waters. ISO 5667-9. International Organization for Standardization, Case Postale 56, CH-1211, Geneva 20 Switzerland.

[16] International Standard Organization. 2009. Water quality. Sampling Part 11: Guidance on sampling of groundwaters. ISO 5667-11. International Organization for Standardization, Case Postale 56, CH-1211, Geneva 20 Switzerland.

[17] Ruda, T.; Farrar, J. Environmental drilling for soil sampling, rock coring, borehole logging and monitoring well installation, in Nielsen, D. (ed.) *Practical Handbook of Environmental Site Characterization and Groundwater Monitoring*. CRC Press: Boca Raton, FL; 2006; 297-344.

[18] International Standard Organization. 1995. Water quality. Sampling Part 12: Guidance on sampling of bottom sediments. ISO 5667-12. International Organization for Standardization, Case Postale 56, CH-1211, Geneva 20 Switzerland.

[19] International Standard Organization. 2004. Water quality. Sampling Part 19: Guidance on sampling of marine sediments. ISO 5667-19. International Organization for Standardization, Case Postale 56, CH-1211, Geneva 20 Switzerland.

[20] International Standard Organization. 2002. Soil quality. Sampling Part 1: Guidance on the design of sampling programmes. ISO 10381-1. International Organization for Standardization, Case Postale 56, CH-1211, Geneva 20 Switzerland.

[21] International Standard Organization. 2002. Soil quality. Sampling Part 2: Guidance on sampling techniques. ISO 10381-2. International Organization for Standardization, Case Postale 56, CH-1211, Geneva 20 Switzerland.

[22] Garner, F.C.; Stapanian, M.A.; Williams, L.R. Composite sampling for environmental monitoring, in L.H. Keith (ed.) *Principles of Environmental Sampling*. ACS: Washington, D.C; 1988; 363-374.

[23] Albert, R.; Horwitz, W. Coping with sampling variability in biota. Percentiles and other strategies, in L.H. Keith (ed.) *Principles of Environmental Sampling*. ACS: Washington, D.C; 1988; 337-353.

[24] Gupta, A.; Kumar, R.; Kumari, K.M.; Srivastava, S.S. Measurement of NO_2, HNO_3, NH_3 and SO_2 and related particulate matter in rural site in Rampur, India. *Atmos Environ* 2003, *37*, 4837-4846.

[25] EPA Fourth External Review draft of Air Quality Criteria for Particulate Matter Office of Research and Development, Washington, DC; June 2003.

[26] Pope, C.A.; Burnett, R.T.; Thun, M.J.; Calle, E.E.; Krewski, D.; Ito, K.; Thurston, G.D. Lung cancer, cardiopulmonary mortality, and long-term exposure to fine particulate air pollution. *J Am Med Assen* 2002, *287*, 1132-1141.

[27] Rizzio, E.; Giaveri, G.; Gallorini, M. Some analytical problems encountered for trace elements determination in the airborne particulate matter of urban and rural areas. *Sci Total Environ* 2000, *256*, 11-22.

[28] McCormac, B.H. *Introduction to the Scientific Study of Atmospheric Pollution*; Reide: Dordrect, 1971.

[29] European Standard 1998. Air quality. Determination of the PM10 fraction of suspended particulate matter. Reference method and field test procedure to demonstrate reference equivalence of measurement methods. EN 12341. European Committee for Standardization, Brussels, Belgium.

[30] European Community, Council Decision 2004/470/EC of 29 April 2004 concerning guidance on a provisional reference method for the sampling and measurement of PM2.5. Off. J. L160 (2004) 53.

[31] International Standard Organization. 1993. Water quality. Sampling Part 8: Guidance on the sampling of wet deposition. ISO 5667-8. International Organization for Standardization, Case Postale 56, CH-1211, Geneva 20 Switzerland.

[32] ASTM D 5012-89 (1994). Guide for preparation of material used for the collection and preservation of atmospheric wet deposition.

[33] International Standard Organization. 2003. Water quality. Sampling Part 3: Guidance on the preservation and handling of water samples. ISO 5667-3. International Organization for Standardization, Case Postale 56, CH-1211, Geneva 20 Switzerland.

[34] APHA. Standard Methods for the Examination of Water and Wastewater. 1060 Collection and Preservation of Samples. American Public Health Association (APHA), American Water Works Association (AWWA), Water Environment Federation publication (WPCF). APHA, Washington, DC.

[35] International Standard Organization. 1999. Water quality. Sampling Part 15: Guidance on preservation and handling of sludge and sediment samples. ISO 5667-15. International Organization for Standardization, Case Postale 56, CH-1211, Geneva 20 Switzerland.

[36] International Standard Organization. 2008. Soil quality. Preparation of laboratory samples from large samples. ISO 23909. International Organization for Standardization, Case Postale 56, CH-1211, Geneva 20 Switzerland.

[37] International Standard Organization. 2006. Soil quality. Pretreatment of samples for physico-chemical analysis. ISO 11464. International Organization for Standardization, Case Postale 56, CH-1211, Geneva 20 Switzerland.

[38] International Standard Organization. 2005. Soil quality. Pretreatment of samples by freeze-drying for subsequent analysis. ISO 16720. International Organization for Standardization, Case Postale 56, CH-1211, Geneva 20 Switzerland.

[39] APHA. 2004. Standard Methods for the Examination of Water and Wastewater. 3030 Preliminary Treatment of Samples. American Public Health Association (APHA), American Water Works Association (AWWA), Water Environment Federation publication (WPCF). APHA, Washington, DC.

[40] Heinz-Helmut, P. *UV-VIS Spectroscopy and its Applications;* Springer-Verlag: Berlin; 1992.

[41] Harris, D.A.; Bashford, C.L. *Spectrophotometry and Spectrofluorimetry: a Practical Approach*; Oxford University Press: Oxford, 1988.

[42] Sandell, E.B.; Onishi, H. *Photometric Determination of Trace Metals. General Aspects*; John Wiley and Sons: New York, 1978.

[43] Singh, A.K.; Ratnam, B.K. Spectrophotometric determination of cadmium with dithizone in water analysis-improvement in sensitivity by surfactant-induced sensitization. *Microchem J* 1989, *39*, 241-243.

[44] Otomo, M.; Singh, R.B. 2,2'-Diquinolyl ketone 2-quinolylhydrazone as a highly sensitive reagent for the extractive spectrophotometric determination of cadmium(II). *Anal Sci* 1985, *1*, 165-168.

[45] López García, I.; Navarro, P.; Hernández Córdoba, M. Manual and FIA methods for the determination of cadmium with Malachite Green and iodide. *Talanta* 1988, *35*, 885-889.

[46] Gomes Neto, J.A.; Oliveira, A.P.; Freshi, G.P.G.; Dakuzaku, C.S.; de Moraes, M. Minimization of lead and copper interferences on spectrophotometric determination of cadmium using electrolytic deposition and ion-exchange in multi-commutation flow system. *Talanta* 2000, *53*, 497-503.

[47] Agrawal, Y.K.; Desai, T.A. Extraction-spectrophotometric determination of cadmium. *Analyst* 1986, *111*, 305-307.

[48] Grudpan, K.; Taylor, C.G. Some azo-dye reagents for the spectrophotometric determination of cadmium. *Talanta* 1989, *36*, 1005-1009.

[49] Salim, R.; Shraydeh, B. Spectrophotometric determination of cadmium(II) using 2-(5-bromo-2-pyridylazo)-5-diethylaminophenol. *Microchem J* 1986, *34*, 251-253.

[50] Raman, B.; Shinde, V.M. Extraction, separation and spectrophotometric determination of cadmium and mercury using triphenylphosphine oxide and its application to environmental samples. *Analyst* 1990, *115*, 93-98.

[51] Chakravarty, S.; Deb, M.K.; Mishra, R.K. Hydroxyamidines as new extracting reagents for spectrophotometric determination of cadmium with 4-(2-pyridylazo)naphthol in industrial effluents, coal, and fly ash. *J AOAC Int* 1993, *76*, 604-608.

[52] Mathew, L. Rao, T.P.; Iyer, C.S.P.; Damodaran, A.D. Liquid-liquid extraction and spectrophotometric determination of cadmium with 1,10-phenanthroline and thymol blue. *Mikrochim Acta* 1993, *111*, 231-237.

[53] Ishizuki, T.; Matsumoto, K.; Yuchi, A.; Ozawa, T.; Yamada, H.; Wada, H. Flow-injection spectrophotometric determination of cadmium with quinolylazo compound after online separation using a silica gel column. *Talanta* 1994, *41*, 799-803.

[54] Hashem, E.Y. Spectrophotometric studies on the simultaneous determination of cadmium and mercury with 4-(2-pyridylazo)-resorcinol. *Spectrochim Acta A* 2002, *58*, 1401-1410.

[55] Akl, M.A.; Khalifa, M.E.; Ghazy, S.E.; Hassanien, M.M. Selective flotation-separation and spectrophotometric determination of cadmium using phenanthraquinone monophenylthiosemicarbazone. *Anal Sci* 2002, *18*, 1235-1240.

[56] Bulgariu, L.; Bulgariu, D.; Sarghie, I. Spectrophotometric determination of cadmium(II) using p,p'-dinitro-SYM-diphenylcarbazid in aqueous solutions. *Anal Lett* 2005, *38*, 2365-2375.

[57] Vaidya, B.; Porter, M.D.; Utterback, M.D.; Bartsch, R.A. Selective determination of cadmium in water using a chromogenic crown ether in a mixed micellar solution. *Anal Chem* 1997, *69*, 2688-2693.

[58] Hsu, C.; Wang, W.; Yang, L.; Pan, J.; Wang, Y. Spectrophotometric and first-derivative spectrophotometric determination of cadmium with o-hydroxybenzenediazo amino azobenzene (HDAA). *Mikrochim Acta* 1989, *1*, 313-320.

[59] Zhu, Y.-r.; Wang, C.-c.; Chen, J.; Jiang, W.-q.; Jin, G. Spectrophotometric determination of cadmium by the chromogenic reagent 2-pyridinediazoaminoazo benzene in the presence of Triton X-100. *Analyst* 1995, *120*, 2853-2856.

[60] Amin, A.S. Spectrophotometric determination of cadmium using thiazolylazo chromogenic reagents in the presence of Triton X-100: Application in environmental samples. *Anal Lett* 2001, *34*, 163-176.

[61] Welz, B.; Sperling, M. *Atomic Absorption Spectrometry*; Wiley–VCH Verlag: Weinheim, 1999.

[62] Ebdon, L.; Evans, E.H.; Fisher, A.; Hill, S.J. *An Introduction to Analytical Atomic Spectrometry*; Wiley and Sons: West Sussex, 1998.

[63] Welz, B.; Becker-Ross, H.; Florek, S.; Heitmann, U. *High Resolution Continuum Source AAS;* Wiley-VCH Verlag: Weinheim, 2005.

[64] Ivanova, E.; Schaldach, G.; Berndt, H. Hydraulic high-pressure nebulization sample introduction for direct analysis or on-line matrix separation and trace preconcentration in flame AAS. *Fresenius J Anal Chem* 1992, *342*, 47-50.

[65] Todolí, J.L.; Canals, A.; Hernandis, V. Characterization of a new single-bore high-pressure pneumatic nebulizer for atomic spectrometry. I. Drop size distribution, transport variables and analytical signal in flame atomic absorption spectrometry. *Spectrochim Acta B* 1993, *48*, 373-386.

[66] Todolí, J.L.; Canals, A.; Hernandis, V. Characterization of a new single-bore high-pressure pneumatic nebulizer for atomic spectrometry-II. Discrete sample introduction in flame atomic absorption spectrometry. *Spectrochim Acta B* 1993, *48*, 1461-1470.

[67] Matusiewicz, H. Atom trapping and in situ pre-concentration techniques for flame atomic absorption spectrometry. *Spectrochim Acta B* 1997, *52*, 1711-1736.

[68] Delves, H.T. A micro-sampling method for the rapid determination of lead in blood by atomic-absorption spectrophotometry. *Analyst* 1970, *95*, 431-438.

[69] Watling, R.J. The use of a slotted quartz tube for the determination of arsenic, antimony, selenium and mercury. *Anal Chim Acta* 1977, *94*, 181-186.

[70] Watling, R.J. The use of a slotted quartz tube for the determination of lead, zinc, cadmium, bismuth, cobalt, manganese and silver by atomic absorption spectrometry. *Anal Chim Acta* 1978, *97*, 395-398.

[71] Yaman, M. The improvement of sensitivity in lead and cadmium determinations using flame atomic absorption spectrometry. *Anal Biochem* 2005, *339*, 1-8.

[72] Santiago-Rivas, S.; Moreda-Piñeiro, A.; Bermejo-Barrera, A.; Bermejo-Barrera, P. Sensitive STAT-FAAS determination of cadmium in seafood products. *At Spectrosc* 2005, *26*, 165-172.

[73] Korkmaz, D.; Mahmut, M.; Helles, R.; Ertas, N.; Ataman, O.Y. Interference studies in slotted silica tube trap technique. *J Anal At Spectrom* 2003, *18*, 99-104.

[74] Yaman, M. Determination of cadmium and lead in human urine by STAT-FAAS after enrichment on activated carbon. *J Anal At Spectrom* 1999, *14*, 275-278.

[75] Gáspár, A.; Berndt, H. Beam injection flame furnace atomic absorption spectrometry: A new flame method. *Anal Chem* 2000, *72*, 240-246

[76] Küllmer, G.; Milner, B.A. Enhancement procedure for flame-AAS. *Labor Praxis* 1984, *8*, 50-55.

[77] Brown, A.A.; Milner, B.A.; Taylor, A. Use of a slotted quartz tube to enhance the sensitivity of conventional flame atomic-absorption spectrometry. *Analyst* 1985, *110*, 501-505.

[78] Yaman, M. Simultaneous enrichment of Cd, Pb, Ni, and Al and their determination in water by STAT-FAAS. *Spectrosc Lett* 2001, *34*, 763-773.

[79] Senkal, B.F.; Ince, M.; Yavuz, E.; Yaman, M. The synthesis of new polymeric sorbent and its application in preconcentration of cadmium and lead in water samples. *Talanta* 2007, *72*, 962-967.

[80] Ertas, N.; Korkmaz, D.K.; Kumser, S.; Ataman, O.Y. Novel traps and atomization techniques for flame AAS. *J Anal At Spectrom* 2002, *17*, 1415-1420.

[81] Yaman, M.; Bakirdere, S. Identification of chemical forms of lead, cadmium and nickel in sewage sludge of waste water treatment facilities. *Mikrochim Acta* 2003, *141*, 47-54.

[82] Yaman, M.; Dilgin, Y. AAS determination of cadmium in fruits and soils. *At Spectrosc* 2002, *23*, 59-64.

[83] Bakirdere, S.; Yaman, M. Determination of lead, cadmium and copper in roadside soil and plants in Elazig, Turkey. *Environ Monit Assess* 2008, *136*, 401-410.

[84] Kaya, G.; Yaman, M. Online preconcentration for the determination of lead, cadmium and copper by slotted tube atom trap (STAT)-flame atomic absorption spectrometry. *Talanta* 2008, *75*, 1127-1133.

[85] Khalighie, J.; Ure, A.M.; West, T.S. Atom-trapping absorption spectrometry with water-cooled metal collector tubes. *Anal Chim Acta* 1982, *134*, 271-281.

[86] Lau, C.M.; Ure, A.M.; West, T.S. The determination of selenium by atom-trapping atomic absorption spectrometry. *Anal Chim Acta* 1982, *141*, 213-224.

[87] Korkmaz, D.; Kumser, S.; Ertas, N.; Mahmut, M.; Ataman, O.Y. Investigations on nature of re-volatilization from atom trap surfaces in flame AAS. *J Anal At Spectrom* 2002, *17*, 1610-1614.

[88] Khalighie, J.; Ure, A.M.; West, T.S. Some observations on the mechanisms of atomization in atomic absorption spectrometry with atom-trapping and electrothermal techniques. *Anal Chim Acta* 1980, *117*, 257-266.

[89] Khalighie, J.; Ure, A.M.; West, T.S. Atom-trapping atomic absorption spectrometry of arsenic, cadmium, lead, selenium and zinc in air-acetylene and air-propane flames. *Anal Chim Acta* 1981, *131*, 27-36.

[90] Brown, A.A.; Roberts, D.J.; Kahokola, K.V. Methods for improving the sensitivity in flame atomic absorption spectrometry. *J Anal At Spectrom* 1987, *2*, 201-204.

[91] Roberts, D.J.; Kahokola, K.V. Improving sensitivity and precision by automated dual silica tube atom trapping. *J Anal At Spectrom* 1989, *4*, 185-189.

[92] Turner, A.D.; Roberts, D.J.; Le Cor, Y. Improving the design of a water-cooled atom trap to increase sensitivity and precision. *J Anal At Spectrom* 1995, *10*, 721-725.

[93] Ellis, L.A.; Roberts, D.J. Further development of the "bent" tube water cooled atom trap for arsenic, antimony copper and manganese. *J Anal At Spectrom* 1996, *11*, 259-263.

[94] Hallam, C.; Thompson, K.C. Determination of lead and cadmium in potable waters by atom-trapping atomic-absorption spectrophotometry. *Analyst* 1985, *110*, 497-500.

[95] Lau, C.M.; Ure, A.M.; West, T.S. Atom trapping atomic absorption spectrometric determination of some trace elements in soils, natural waters, seawater, and bovine liver. *Bull Chem Soc Jpn* 1988, *61*, 79-85.

[96] Lau, C.M.; Ure, A.M.; West, T.S. The determination of lead and cadmium in soils by atom-trapping atomic absorption spectrometry. *Anal Chim Acta* 1983, *146*, 171-179.

[97] Fraser, S.M.; Ure, A.M.; Mitchell, M.C.; West, T.S. Determination of cadmium in calcium chloride extracts of soils by atom-trapping atomic absorption spectrometry. *J Anal At Spectrom* 1986, *1*, 19-21.

[98] Turner, A.D.; Roberts, D.J. Metal determinations with a novel slotted-tube water-cooled atom trap. *J Anal At Spectrom* 1996, *11*, 231-234.

[99] Matusiewicz, H; Kopras, M. Methods for improving the sensitivity in atom trapping flame atomic absorption spectrometry: analytical scheme for the direct determination of trace elements in beer. *J Anal At Spectrom* 1997, *12*, 1287-1291.

[100] Matusiewicz, H.; Krawczyk, M. On-line hyphenation of hydride generation with in situ trapping flame atomic absorption spectrometry for arsenic and selenium determination. *Anal Sci* 2006, *22*, 249-253.

[101] Matusiewicz, H.; Krawczyk, M. Determination of total antimony and inorganic antimony species by hydride generation in situ trapping flame atomic absorption spectrometry: a new way to (ultra)trace speciation analysis. *J Anal At Spectrom* 2008, *23*, 43-53.

[102] Gáspár, A.; Berndt, H. Thermospray flame furnace atomic absorption spectrometry (TS-FF-AAS) - a simple method for trace element determination with microsamples in the µg/l concentration range. *Spectrochim Acta B* 2000, *55*, 587-597.

[103] Petrucelli, G.A.; Stocco, P.K.; Bueno, M.I.M.S.; Pereira-Filho, E. Tube atomizers in thermospray flame furnace atomic absorption spectrometry: characterization using X-ray fluorescence, scanning electron microscopy and chemometrics. *J Anal At Spectrom* 2006, *21*, 1298-1304.

[104] Doherty, M.P.; Hieftje, G.M. Jet-impact nebulization for sample introduction in inductively coupled plasma spectrometry. *Appl Spectrosc* 1984, *38*, 405-412.

[105] Gáspár, A.; Berndt, H. Beam-injection flame-furnace atomic-absorption spectrometry (BIFF-AAS) with low-pressure sample-jet generation. *Anal Bioanal Chem* 2002, *372*, 695-699.

[106] Ratka, A.; Berndt, H. Beam-injection flame furnace AAS: comparison of different nozzle types for beam generation and application of sub-critical liquid carbon dioxide as carrier and gas pressure pump. *Anal Bioanal Chem* 2004, *378*, 416-422.

[107] Gáspár, A.; Szeles, E.; Berndt, H. Analysis of submicroliter samples using micro thermospray flame furnace atomic absorption spectrometry. *Anal Bioanal Chem* 2002, *372*, 136-140.

[108] Brancalion, M.L.; Sabadini, E.; Arruda, M.A.Z. Description of the thermospray formed at low flow rate in thermospray flame furnace atomic absorption spectrometry based on high-speed images. *Anal Chem* 2007, *79*, 6527-6533.

[109] Davies, J.; Berndt, H. Improvements in thermospray flame furnace atomic absorption spectrometry. *Anal Chim Acta* 2003, *479*, 215-223.

[110] Brancalion, M.L.; Sabadini, E.; Arruda, M.A.Z. Thermospray nebulization for flame furnace atomic absorption spectrometry - Correlations between spray formation and cadmium analytical sensitivity. *Spectrochim Acta B* 2009, *64*, 89-94.

[111] Aleixo, P.C.; Santos Júnior, D.; Tomazelli, A.C.; Rufini, I.A.; Berndt, H.; Krug, F.J. Cadmium and lead determination in foods by beam injection flame furnace atomic

absorption spectrometry after ultrasound-assisted sample preparation. *Anal Chim Acta* 2004, *512*, 329-337.

[112] Teixeira Tarley, C.R.; Zezzi Arruda, M.A. A sensitive method for cadmium determination using an on-line polyurethane foam preconcentration system and thermospray flame furnace atomic absorption spectrometry. *Anal Sci* 2004, *20*, 961-966.

[113] Ivanova, E.; Berndt, H.; Pulvermacher, E. Air driven on-line separation and preconcentration on a C18 column coupled with thermospray flame furnace AAS for the determination of cadmium and lead at μg l^{-1} levels. *J Anal At Spectrom* 2004, *19*, 1507-1509.

[114] González, E.; Ahumada, R.; Medina, V.; Neira, J.; González, U. Thermospray flame furnace atomic absorption spectrometry: Application for total determination of Cd, Pb and Zn in fresh waters, seawater and marine sediments. *Quimica Nova* 2004, *27*, 873-877.

[115] Wu, P.; Zhang, Y.; Lv, Y.; Hou, X. Cloud point extraction-thermospray flame quartz furnace atomic absorption spectrometry for determination of ultratrace cadmium in water and urine. *Spectrochim Acta B* 2006, *61*, 1310-1314.

[116] Coelho, L.M.; Bezerra, M.A.; Arruda, M.A.Z.; Bruns, R.E.; Ferreira, S.L.C. Determination of Cd, Cu, and Pb after cloud point extraction using multielemental sequential determination by thermospray flame furnace atomic absorption spectrometry (TS-FF-AAS). *Separation Sci Technol* 2008, *43*, 815-827.

[117] Lemos, V.A.; Bezerra, M.A.; Amorim, F.A.C. On-line preconcentration using a resin functionalized with 3,4-dihydroxybenzoic acid for the determination of trace elements in biological samples by thermospray flame furnace atomic absorption spectrometry. *J Hazar Mat* 2008, *157*, 613-619.

[118] Pereira-Filho, E.R.; Berndt, H.; Zezzi Arruda, M.A. Simultaneous sample digestion and determination of Cd, Cu and Pb in biological samples using thermospray flame furnace atomic absorption spectrometry (TS-FF-AAS) with slurry sample introduction. *J Anal At Spectrom* 2002, *17*, 1308-1315.

[119] Pereira, M.G.; Pereira-Filho, E.R.; Berndt, H.; Arruda, M.A.Z. Determination of cadmium and lead at low levels by using preconcentration at fullerene coupled to thermospray flame furnace atomic absorption spectrometry. *Spectrochim Acta B* 2004, *59*, 515-521.

[120] Brancalion, M.L.; Arruda, M.A.Z. Evaluation of medicinal plant decomposition efficiency using microwave ovens and mini-vials for Cd determination by TS-FF-AAS. *Microchim Acta* 2005, *150*, 283-290.

[121] Amorim, F.A.C.; Bezerra, M.A. Online preconcentration system for determining ultratrace amounts of Cd in vegetal samples using thermospray flame furnace atomic absorption spectrometry. *Microchim Acta* 2007, *159*, 183-189.

[122] Petrucelli, G.A.; Poppi, R.J.; Mincato, R.L.; Pereira-Filho, E.R. TS-FF-AAS and multivariate calibration: A proposition for sewage sludge slurry sample analyses. *Talanta* 2007, *71*, 620-626.

[123] L′vov, B.V. Electrothermal atomization – way toward absolute methods of atomic absorption analysis. *Spectrochim Acta B* 1978, *33*, 153-193.

[124] Ediger, R.D. Atomic absorption analysis with the graphite furnace using matrix modification. *At Absorpt Newsl* 1975, *14*, 127-130.

[125] Ebdon, L.; Lechotycki, A. The determination of cadmium in environmental samples by slurry atomization graphite furnace atomic absorption spectrometry using platforms and matrix modification. *Microchem J* 1987, *36*, 207-215.

[126] Viñas, P.; Campillo, N.; López-García, I.; Hernández-Córdoba, M. Slurry-electrothermal atomic absorption spectrometric determination of aluminum and chromium in vegetables using hydrogen peroxide as a matrix modifier. *Talanta* 1995, *42*, 527-533.

[127] Hassell, D.C.; Majidi, V.; Holcombe, J.A. Temperature programmed static secondary ion mass spectrometric study of phosphate chemical modifiers in electrothermal atomizers. *J Anal At Spectrom* 1991, *6*, 105-108.

[128] Schlemmer, G.; Welz, B. Palladium and magnesium nitrates, a more universal modifier for graphite furnace atomic absorption spectrometry. *Spectrochim Acta B* 1986, *41*, 1157-1165.

[129] Hoenig, M.; Puskaric, E.; Choisy, P.; Wartel, M. Direct determination of high- and mid-volatile metals (cadmium, lead, manganese) in seawater by electrothermal atomic absorption spectrometry: existing approaches and critical parameters. *Analusis* 1991, *19*, 285-291.

[130] Bermejo-Barrera, P.; Moreda-Piñeiro, J.; Moreda-Piñeiro, A.; Bermejo-Barrera, A. Usefulness of the chemical modification and the multi-injection technique approaches in the electrothermal atomic absorption spectrometric determination of silver, arsenic, cadmium, chromium, mercury, nickel and lead in sea-water. *J Anal At Spectrom* 1998, *13*, 777-786.

[131] Bermejo-Barrera, P.; Moreda-Piñeiro, A.; Bermejo-Barrera, A. Study of ammonium molybdate to minimize the phosphate interference in the selenium determination by electrothermal atomic absorption spectrometry with deuterium background correction. *Spectrochim Acta B* 2002, *57*, 327-337.

[132] Voth-Beach, L.M.; Shrader, D.E. Investigation of a reduced palladium chemical modifier for graphite furnace atomic absorption. *J Anal At Spectrom* 1987, *2*, 45-50.

[133] Cimadevilla Álvarez-Cabal, E.; Wróbel, K.; Sanz-Medel, A. Capabilities and limitations of different techniques in electrothermal atomic absorption spectrometry for direct monitoring of arsenic, cadmium and lead contamination of sea-water. *J Anal At Spectrom* 1995, *10*, 149-54.

[134] Arpadjan, S.; Krivan, V. Behavior of chromium in the graphite furnace during the performance of the flameless absorption spectrometry. *Fresenius'J Anal Chem* 1988, *329*, 745-749.

[135] Slavin, W.; Manning, D.C.; Carnrick, G.R. The stabilized temperature platform furnace. *At Spectrosc* 1981, *2*, 137-145.

[136] Ohlsson, K.E.A.; Frech, W. Photographic observation of molecular spectra in inverse Zeeman-effect graphite furnace atomic absorption spectrometry. *J Anal At Spectrom* 1989, *4*, 379-385.

[137] Yin, X.-f.; Schlemmer, G.; Welz, B. Cadmium determination in biological materials using graphite furnace atomic absorption spectrometry with palladium nitrate-ammonium nitrate modifier. *Anal Chem* 1987, *59*, 1462-1466.

[138] L'vov, B.V.; Yatsenko, L.F. Carbothermal reduction of zinc, cadmium, lead and bismuth oxides in graphite furnaces for atomic absorption analysis in the presence of organic substances. *Zh Anal Khim* 1984, *39*, 1773-1780.

[139] Hulanicki, A.; Bulska, E.; Wróbel, K. Effect of inorganic matrices on the determination of cadmium by atomic-absorption spectrometry with electrothermal atomization. *Analyst* 1985, *110*, 1141-1145.
[140] Slavin, W.; Manning, D.C. The L'vov platform for furnace atomic absorption analysis. *Spectrochim Acta A* 1980, *35*, 701-714.
[141] Hernández Caraballo, E.A.; Domínguez, J.R.; Alvarado, D.J. Determination of Cd in river sediment reference material by GF-ETAAS using lithium tetraborate as chemical modifier. *At Spectrosc* 2001, *22*, 295-298.
[142] Feo, J.C.; Castro, M.A.; Lumbreras, J.M.; de Celis, B.; Aller, A.J. Nickel as a chemical modifier for sensitivity enhancement and fast atomization processes in electrothermal atomic absorption spectrometric determination of cadmium in biological and environmental samples. *Anal Sci* 2003, *19*, 1631-1636.
[143] Acar, O. Determination of cadmium and lead in biological samples by Zeeman ETAAS using various chemical modifiers. *Talanta* 2001, *55*, 613-622.
[144] Arpadjan, S.; Karadjova, I. Behavior of volatile elements in the graphite furnace in the presence of silver as matrix modifier. *Spectrosc Lett* 2003, *36*, 441-447.
[145] Fuyi, W.; Zucheng, J.; Bin, H.; Tianyou, P. Comparative studies on chemical modification of polytetrafluoroethylene slurry in ETV-ICP-AES and ETAAS. *J Anal At Spectrom* 1999, *14*, 1619-1624.
[146] Acar, O. Determination of cadmium, copper and lead in soils, sediments and sea water samples by ETAAS using a Sc + Pd + NH_4NO_3 chemical modifier. *Talanta* 2005, *65*, 672-677.
[147] Imai, S.; Kanematsu, Y.; Satoh, A.; Yonetani, A. Pyrolytic and non-pyrolytic boron nitride platforms in electrothermal AAS for the determination of cadmium in sea and estuarine water without chemical modification. *Anal Sci* 2004, *20*, 1755-1758.
[148] Capelo, J.L.; Lavilla, I.; Bendicho, C. Ultrasound-assisted extraction of cadmium from slurried biological samples for electrothermal atomic absorption spectrometry. *J Anal At Spectrom* 1998, *13*, 1285-1290.
[149] Bermejo-Barrera, P.; Moreda-Piñeiro, A.; Muñiz-Naveiro, O.; Gómez-Fernández, A.M.J.; Bermejo-Barrera, A. Optimization of a microwave-pseudo-digestion procedure by experimental designs for the determination of trace elements in seafood by atomic absorption spectrometry. *Spectrochim Acta B* 2000, *55*, 1351-1371.
[150] Bermejo-Barrera, P.; Muñiz-Naveiro, O.; Moreda-Piñeiro, A.; Bermejo-Barrera, A. The multivariate optimization of ultrasonic bath-induced acid leaching for the determination of trace elements in seafood by atomic absorption spectrometry. *Anal Chim Acta* 2001, *439*, 211-227.
[151] Smichowski, P.; Valiente, L.; Piccinna, M. ETAAS determination of essential and potentially toxic trace elements in Antarctic Krill certified reference material: Evaluation of two microwave digestion procedures. *At Spectrosc* 2001, *22*, 350-355.
[152] Santamaría-Fernández, R.; Santiago-Rivas, S.; Moreda-Piñeiro, A.; Bermejo-Barrera, A. Bermejo-Barrera, P. Blending procedure for the cytosolic preparation of mussel samples for AAS determination of Cd, Cr, Cu, Pb and Zn bound to low molecular weight compounds. *At Spectrosc* 2004, *25*, 37-43.
[153] Aranda, P.R.; Gil, R.A.; Moyano, S.; de Vito, I.; Martínez, L.D. Cloud point extraction for ultra-trace Cd determination in microwave-digested biological samples by ETAAS. *Talanta* 2008, *77*, 663-666.

[154] Bikkes, M.; Polyak, K.; Hlavay, J. Fractionation of elements by particle size and chemical bonding from aerosols followed by ETAAS determination. *J Anal At Spectrom* 2001, *16*, 74-81.

[155] Arain, M.B.; Kazi, T.G.; Jamali, M.K.; Jalbani, N.; Afridi, H.I.; Shah, A. Total dissolved and bioavailable elements in water and sediment samples and their accumulation in *Oreochromis mossambicus* of polluted Manchar Lake. *Chemosphere* 2008, *70*, 1845-1856.

[156] Pancras, J.P.; Ondov, J.M.; Zeisler, R. Multi-element electrothermal AAS determination of 11 marker elements in fine ambient aerosol slurry samples collected with SEAS-II. *Anal Chim Acta* 2005, *538*, 303-312.

[157] Zemberyova, M.; Bartekova, J.; Hagarova, I. The utilization of modified BCR three-step sequential extraction procedure for the fractionation of Cd, Cr, Cu, Ni, Pb and Zn in soil reference materials of different origins. *Talanta* 2006, *70*, 973-978.

[158] Bermejo-Barrera, P.; Fernández-Nocelo, S.; Moreda-Piñeiro, A.; Bermejo-Barrera, A. Usefulness of enzymatic hydrolysis procedures based on the use of pronase E as sample pre-treatment for multi-element determination in biological samples. *J Anal At Spectrom* 1999, *14*, 1893-1900.

[159] Bermejo-Barrera, P.; Moreda-Pineiro, A.; Moreda-Pineiro, J.; Kauppila, T.; Bermejo-Barrera, A. Slurry sampling for electrothermal AAS determination of cadmium in seafood products. *At Spectrosc* 2000, *21*, 5-9.

[160] Vaciaenen, A.; Suontamo, R. Comparison of ultrasound-assisted extraction, microwave-assisted acid leaching and reflux for the determination of arsenic, cadmium and copper in contaminated soil samples by electrothermal atomic absorption spectrometry. *J Anal At Spectrom* 2002, *17*, 739-742

[161] Imai, S.; Ishikawa, S.; Kikuchi, Y.; Yonetani, A. Large-volume injection electrothermal AAS combined with tungsten-treated pyrolytic graphite furnace: Determination of cadmium in tap, snow and river water samples with phosphate modifier. *Anal Sci* 2004, *20*, 575-577

[162] Barciela-Alonso, M.C.; Pazos-Capeáns, P.; Regueira-Míguens, M.E.; Bermejo-Barrera, A.; Bermejo-Barrera, P. Study of cadmium, lead and tin distribution in surface marine sediment samples from Ría de Arousa (NW of Spain). *Anal Chim Acta* 2004, *524*, 115-120

[163] Silva, R.G.L.; Santelli, R.E.; Willie, S.N.; Sturgeon, R.E.; Sella, S.M. Alkaline solubilization of biological materials for trace element analysis by electrothermal atomic absorption spectrometry. *Analyst* 1999, *124*, 1843-1846.

[164] Scriver, C.; Kan, M.; Willie, S.; Soo, C.; Birnboim, H. Formic acid solubilization of marine biological tissues for multi-element determination by ETAAS and ICP-AES. *Anal Bioanal Chem* 2005, *381*, 1460-1466.

[165] Žemberyová, M.; Barteková, J.; Závadská, M.; Šišoláková, M. Determination of bioavailable fractions of Zn, Cu, Ni, Pb and Cd in soils and sludges by atomic absorption spectrometry. *Talanta* 2007, *71*, 1661-1668.

[166] Vassileva, E.; Baeten, H.; Hoenig, M. Advantages of the iridium permanent modifier in fast programs applied to trace-element analysis of plant samples by electrothermal atomic absorption spectrometry. *Fresenius' J Anal Chem* 2001, *369*, 491-495.

[167] Acar, O. Molybdenum, Mo-Ir and Mo-Ru coatings as permanent chemical modifiers for the determination of cadmium and lead in sediments and soil samples by electrothermal atomic absorption spectrometry. *Anal Chim Acta* 2005, *542*, 280-286.

[168] Acar, O. Evaluation of V, Ir, Ru, V-Ir, V-Ru, and W-V as permanent chemical modifiers for the determination of cadmium, lead, and zinc in botanic and biological slurries by electrothermal atomic absorption spectrometry. *Anal Chim.Acta* 2005, *545*, 244-251.

[169] Lima, E.C.; Barbosa, R.V.; Vaghetti, J.C.P.; Ferreira, L.S. Evaluation of different permanent modifiers for the determination of cadmium in environmental samples by electrothermal AAS. *At Spectrosc* 2002, *23*, 135-142.

[170] Lima, E.C.; Barbosa, R.V.; Brasil, J.L.; Santos, A.H.D.P. Evaluation of different permanent modifiers for the determination of arsenic, cadmium and lead in environmental samples by electrothermal atomic absorption spectrometry. *J Anal At Spectrom* 2002, *17*, 1523-1529.

[171] Lima, E.C., Krug, F.J.; Jackson, K.W. Evaluation of tungsten-rhodium coating on an integrated platform as a permanent chemical modifier for cadmium, lead, and selenium determination by electrothermal atomic absorption spectrometry. *Spectrochim Acta B* 1998, *53*, 1791-1804.

[172] Lima, E.C.; Barbosa, F.; Drug, F.J. The use of tungsten-rhodium permanent chemical modifier for cadmium determination in decomposed samples of biological materials and sediments by electrothermal atomic absorption spectrometry. *Anal Chim Acta* 2000, *409*, 267-274.

[173] Lima, E.C.; Barbosa, F. Jr.; Krug, F.J.; Silva, M.M.; Vale, M.G.R. Comparison of ultrasound-assisted extraction, slurry sampling and microwave-assisted digestion for cadmium, copper and lead determination in biological and sediment samples by electrothermal atomic absorption spectrometry. *J Anal At Spectrom* 2000, *15*, 995-1000.

[174] Borba da Silva, J.B.; Borges, D.L.G.; Silva da Veiga, M.A.M.; Curtius, A.J.; Welz, B. Determination of cadmium in biological samples solubilized with tetramethylammonium hydroxide by electrothermal atomic absorption spectrometry, using ruthenium as permanent modifier. *Talanta* 2003, *60*, 977-982.

[175] Brasil, J.L.; Lima, E.C.; Veses, R.C.; Tisott, M.M. Ultrasound-assisted extraction of cadmium in environmental samples for ETAAS determination. *At Spectrosc* 2004, *25*, 94-101.

[176] Acar, O. Electrothermal atomic absorption spectrometric determination of cadmium and lead in environmental, botanic and biological samples by different permanent modifiers. *J Anal At Spectrom* 2004, *19*, 709-711.

[177] de Almeida Pereira, L.; Gonçalves de Amorim, I.; Borba da Silva, J.B. Development of methodologies to determine aluminum, cadmium, chromium and lead in drinking water by ET-AAS using permanent modifiers. *Talanta* 2004, *64*, 395-400.

[178] Piascik, M.; Bulska, E. Performance of electrodeposited noble metals as permanent modifiers for the determination of cadmium in the presence of mineral acids by electrothermal atomic absorption spectrometry. *Spectrochim Acta B* 2001, *56*, 1615-1623.

[179] Nölte, J. *ICP Emission Spectrometry. A Practical Guide;* Wiley-VCH Verlag: Weinheim, 2003.

[180] Montaser, A.; Ohls K. D.; Golightly, D.W. Inductively coupled plasmas in gases other than argon; Montaser, A.; Golightly D.W.; Ed.; *Inductively Coupled Plasmas in Analytical Atomic Spectrometry*; Wiley-VCH: New York, 1998.

[181] Dean, J.R. *Practical Inductively Coupled Plasma Spectroscopy;* Wiley: Hoboken, NJ, 2005.

[182] Todolí, J.L.; Mermet, J.M. Elemental analysis of liquid microsamples through inductively coupled plasma spectrochemistry. *Trends Anal Chem* 2005, *24*, 107-116.

[183] Maestre, S.E.; Todolí, J.L.; Mermet, J.M. Evaluation of several pneumatic micronebulizers with different designs for use in ICP-AES and ICP-MS. Future directions for further improvements. *Anal Bioanal Chem* 2004, *379*, 888-899.

[184] Almagro, B.; Ganan-Calvo, A.M.; Hidalgo, M.; Canals, A. Flow focusing pneumatic nebulizers in comparison with several micronebulizers in inductively coupled plasma atomic emission spectrometry. *J Anal At Spectrom* 2006, *21*, 770-777.

[185] McLaughlin, R.L.J.; Brindle, I.D. Multimode sample introduction system. United States Patent US 6,891,605 B2, May, 10, 2005.

[186] Ding, L.; Liang, F.; Huan, Y.; Cao, Y.; Zhang, H.; Jin, Q. A low-powered microwave thermospray nebulizer for inductively coupled plasma atomic emission spectrometry. *J Anal At Spectrom* 2000, *15*,293-296.

[187] Mora, J.; Todolí, J.L.; Rico, I.; Canals, A. Aerosol desolvation studies with a thermospray nebulizer coupled to inductively coupled plasma atomic emission spectrometry. *Analyst* 1999, *123*, 1229-1234.

[188] Beres, S.A.; Bruckner, P.H.; Denoyer, E.R. Performance evaluation of a cyclonic spray chamber for ICP-MS. *At Spectrosc* 1994, *15*, 96–99.

[189] Tanner, S.D.; Baranov, B.I.; Bandura, D.R. Reaction cells and collision cells for ICP-MS: A tutorial review. *Spectrochim Acta B* 2002, *57*, 1361-1452.

[190] Koppenaal, D.W.; Eden, G.C.; Barinaga, C.J. Collision and reaction cells in atomic mass spectrometry: Development, status, and applications. *J Anal At Spectrom* 2004, *19*, 561-570.

[191] Broekaert, J.A.C. *Analytical Atomic Spectrometry with Flames and Plasmas;* Wiley-VCH: Weinheim, 2005.

[192] Sabe, R.; Rauret, G. Challenges for achieving trazability of analytical measurements of heavy metals in environmental samples by isotopic dilution mass spectrometry. *Trends Anal Chem* 2004, *23*, 273-280

[193] Xu, S.; Sturgeon, R.E.; Guo, Y.; Zang, W.; Zhao, H. Chemical vapor generation of copper: Optimization of generation media. *Ann Chim* 2005, *95*, 491-499.

[194] Peña-Vázquez, E.; Villanueva-Alonso, J.; Bermejo-Barrera, P. Optimization of a vapor generation method for metal determination using ICP-OES. *J Anal At Spectrom* 2007, *22*, 642-649

[195] Feng, J.L.; Lam, J.W.; Sturgeon, R.E. Expanding the scope of chemical vapor generation for noble and transition metals. *Analyst* 2001, *126*, 1833-1837.

[196] Feng, J.L.; Sturgeon, R.E.; Lam, J.W.; D'Ulivo, A. Insights into the mechanisms of chemical vapor generation of transition and noble metals. *J Anal At Spectrom* 2005, *20*, 255-265.

[197] Pohl, P.; Zyrnicki, W. Study of chemical vapor generation of Au, Pd and Pt by inductively coupled plasma atomic emission spectrometry. *J Anal At Spectrom* 2001, *16*, 1442-1445.

[198] Pohl, P.; Zyrnicki, W. On the transport of some metals into the inductively coupled plasma during hydride generation process. *Anal Chim Acta* 2001, *429*, 135-143

[199] Pohl, P.; Zyrnicki, W. Study of reaction of Ir, Os, Rh and Ru ions with $NaBH_4$ in the acid medium by inductively coupled plasma atomic emission spectrometry. *J Anal At Spectrom* 2002, *17*, 746-749.

[200] Pohl, P.; Zyrnicki, W. Analytical features of Au, Pd and Pt chemical vapor generation inductively coupled plasma atomic emission spectrometry. *J Anal At Spectrom* 2003, *18*, 798-801.

[201] Asfaw, A.; Wibetoe, G. Simultaneous determination of hydride (Se) and non-hydride forming (Ca, Mg, K, P, S and Zn) elements in various beverages (beer, coffee, and milk), with minimum sample preparation, by ICP-AES and use of a dual-mode sample introduction system. *Anal Bioanal Chem* 2005, *382*, 173-179.

[202] Ritschdorff, E.T.; Fitzgerald, N.; Mclaughlin, R.L.J.; Brindle, I.D. The use of a modified Multimode Sample Introduction System for the simple and rapid determination of cadmium by chemical vapor generation atomic absorption spectrometry. *Spectrochim Acta B* 2005, *60*, 139-143.

[203] Grabi, G.; Cavalli, P.; Achilli, M.; Rossi, G.; Omenetto, N. Use of the HGA-500 graphite furnace as a sampling unit for ICP emission spectrometry. *At Spectrosc* 1982, *3*, 81-88.

[204] Kántor, T.; Záray, G. Improved design and optimization of an electrothermal vaporization system for inductively coupled plasma atomic emission spectrometry. *Microchem J* 1995, *51*, 266-277.

[205] Peng Tianyou, Jiang Zucheng. Fluorination assisted slurry electrothermal vaporization in ICP-AES for the direct analysis of silicon dioxide powder. *Fresenius'J Anal Chem* 1998, *360*, 43-46.

[206] Resano, M.; Vanhaecke, F.; de Loos-Vollebregt, M.T.C. Electrothermal vaporization for sample introduction in atomic absorption, atomic emission and plasma mass spectrometry - A critical review with focus on solid sampling and slurry analysis. *J Anal At Spectrom* 2008, *23*, 1450-1475.

[207] Martin-Esteban, A.; Slowikowski, B. Electrothermal vaporization - inductively coupled plasma-mass spectrometry (ETV-ICP-MS): A valuable tool for direct multielement determination in solid samples. *Critical Rev Anal Chem* 2003, *33*, 43-55.

[208] Belarra, M.A.; Resano, M.; Vanhaecke, F.; Moens, L. Direct solid sampling with electrothermal vaporization/atomization: What for and how?. *Trends Anal Chem* 2002, *21*, 828-839.

[209] Todolí, J.L.; Mermet, J.M. Elemental analysis of liquid microsamples through inductively coupled plasma spectrochemistry. T*rends Anal Chem* 2005, *24*, 107-116.

[210] Vanhaecke, F.; Resano, M.; Moens, L. Electrothermal vaporization ICP-mass spectrometry (ETV-ICP-MS) for the determination and speciation of trace elements in solid samples - A review of real-life applications from the author's lab. *Anal Bioanal Chem* 2002, *374*, 188-195.

[211] Hu, B.; Li, S.; Xiang, G.; He, M.; Jiang, Z. Recent progress in electrothermal vaporization-inductively coupled plasma atomic emission spectrometry and inductively coupled plasma mass spectrometry. *Appl Spectrosc Rev* 2007, *42*, 203-234.

[212] Vanderpool, R.A.; Buckley, W.T. Liquid–liquid extraction of cadmium by sodium diethyldithiocarbamate from biological matrices for isotope dilution inductively coupled plasma mass spectrometry. *Anal Chem* 1999, *71*. 652–659.

[213] Jarrett, J.M.; Xiao, G.; Caldwell, K.L.; Henahan, Shakirova, G. Jones, R.L. Eliminating molybdenum oxide interference in urine cadmium biomonitoring using ICP-DRC-MS. *J Anal At Spectrom* 2008, *23*, 962-967.

[214] Rosland, E., Lund, W. Direct determination of trace metals in sea-water by inductively coupled plasma mass spectrometry. *J Anal At Spectrom* 1999, *13*, 1239-1244.

[215] Pozebon, D.; Dressler, V.L.; Curtius, A.J. Determination of copper, cadmium, lead, bismuth and selenium(IV) in sea-water by electrothermal vaporization inductively coupled plasma mass spectrometry after online separation. *J Anal At Spectrom* 1998, *13*, 363-369.

[216] Liu, H.-W.; Jiang, S.-J.; Liu, S.-H. Determination of cadmium, mercury and lead in seawater by electrothermal vaporization isotope dilution inductively coupled plasma mass spectrometry. *Spectrochim Acta B* 1999, *54*, 1367-1375.

[217] Xia, L.; Wu, Y.; Hu, B. Hollow-fiber liquid-phase microextraction prior to low-temperature electrothermal vaporization ICP-MS for trace element analysis in environmental and biological samples. *J Mass Spectrom* 2007, *42*, 803-810.

[218] Li, L.; Hu, B.; Xia, L.; Jiang, Z. Determination of trace Cd and Pb in environmental and biological samples by ETV-ICP-MS after single-drop microextraction. *Talanta* 2006, *70*, 468-473.

[219] Pu, X.; Jiang, Z.; Hu, B. Zirconia-coated graphite adsorption bar micro-extraction combined with ETV-ICP-MS for the determination of trace amounts of Cd, Hg and Pb in environmental and biological samples. *J Mass Spectrom* 2006, *41*, 887-893.

[220] Li, Y.-C.; Jiang, S.-J. Determination of Cu, Zn, Cd and Pb in fish samples by slurry sampling electrothermal vaporization inductively coupled plasma mass spectrometry. *Anal Chim Acta* 1998, *359*, 205-212.

[221] Liao, H.-C.; Jiang, S.-J. Determination of cadmium, mercury and lead in coal fly ash by slurry sampling electrothermal vaporization inductively coupled plasma mass spectrometry. *Spectrochim Acta B* 1999, *54*, 1233-1242.

[222] Liao, H.-C.; Jiang, S.-J. EDTA as the modifier for the determination of Cd, Hg and Pb in fish by slurry sampling electrothermal vaporization inductively coupled plasma mass spectrometry. *J Anal At Spectrom* 1999, *14*, 1583-1588.

[223] Lu, H.-H.; Jiang, S.-J. Organic acids as the modifier to determine Zn, Cd, Tl and Pb in soil by slurry sampling electrothermal vaporization inductively-coupled plasma mass spectrometry. *Anal Chim Acta* 2001, *429*, 247-255.

[224] Rybak, M.E.; Salin, E.D. Direct determination of metals in soils and sediments by induction heating-electrothermal vaporization (IH-ETV) inductively coupled plasma-optical emission spectrometry (ICP-OES). *Appl Spectrosc* 2001, *55*, 816-821.

[225] Maia, S.M.; Pozebon, D.; Curtius, A.J. Determination of Cd, Hg, Pb and Tl in coal and coal fly ash slurries using electrothermal vaporization inductively coupled plasma mass spectrometry and isotopic dilution. *J Anal At Spectrom* 2003, *18*, 330-337.

[226] Dias, L.F.; Miranda, G.R.; Saint'Pierre, T.D.; Maia, S.M.; Frescura, V.L.A.; Curtius, A.J. Method development for the determination of cadmium, copper, lead, selenium and thallium in sediments by slurry sampling electrothermal vaporization inductively

coupled plasma mass spectrometry and isotopic dilution calibration. *Spectrochim Acta B* 2005, *60*, 117-124.
[227] Vieira, M.A.; Ribeiro, A.S.; Dias, L.F.; Curtius, A.J. Determination of Cd, Hg, Pb and Se in sediments slurries by isotopic dilution calibration ICP-MS after chemical vapor generation using an on-line system or retention in an electrothermal vaporizer treated with iridium. *Spectrochim Acta B* 2005, *60*, 643-652.
[228] Chen, S.-F.; Jiang, S.-J. Determination of cadmium, mercury and lead in soil samples by slurry sampling electrothermal vaporization inductively coupled plasma mass spectrometry. *J Anal At Spectrom* 1998, *13*, 1113-1117.
[229] Galbacs, G.; Vanhaecke, F.; Moens, L.; Dams, R. Determination of cadmium in certified reference materials using solid sampling electrothermal vaporization inductively coupled plasma mass spectrometry supplemented with thermogravimetric studies. *Microchem J* 1996, *54*, 272-286.
[230] Bitterli, B.A.; Cousin, H.; Magyar, B. Determination of metals in airborne particles by electrothermal vaporization inductively coupled plasma mass spectrometry after accumulation by electrostatic precipitation. *J Anal At Spectrom* 1997, *12*, 957-961.
[231] Ludke, C.; Hoffmann, E.; Skole, J.; Kriews, M. Determination of trace metals in size fractionated particles from arctic air by electrothermal vaporization inductively coupled plasma mass spectrometry. *J Anal At Spectrom* 1999, *14*, 1685-1690.
[232] Masson, P. Determination of trace elements in plants by electrothermal vaporization-high frequency plasma emission spectrometry coupling. *Canadian J Anal Sci Spectrosc* 2005, *50*, 268-279.
[233] Masson, P.; Dauthieu, M.; Trolard, F.; Denaix, L. Application of direct solid analysis of plant samples by electrothermal vaporization-inductively coupled plasma atomic emission spectrometry: Determination of Cd and Si for environmental purposes. *Spectrochim Acta B* 2007, *62*, 224-230.
[234] Luedke, C.; Hoffmann, E.; Skole, J. Comparative studies on metal determination in airborne particulates by LA-ICP-MS and furnace atomization non-thermal excitation spectrometry. *Fresenius' J Anal Chem* 1994, *350*, 272-276.
[235] Tanaka, S.; Yasushi, N.; Sato, N.; Fukasawa, T.; Santosa, S.J.; Yamanaka, K.; Ootoshi, T. Rapid and simultaneous multi-element analysis of atmospheric particulate matter using inductively coupled plasma mass spectrometry with laser ablation sample introduction. *J Anal At Spectrom* 1998, *13*, 135-140.
[236] Chin, C.-J.; Wang, C.-F.; Jeng, S.-L. Multi-element analysis of airborne particulate matter collected on PTFE-membrane filters by laser ablation inductively coupled plasma mass spectrometry. *J Anal At Spectrom* 1999, *14*, 663-668.
[237] Okuda, T.; Kato, J.; Mori, J.; Tenmoku, M.; Suda, Y.; Tanaka, S., He, K.; Ma, Y.; Yang, F.; Yu, X.; Duan, F.; Lei, Y. Daily concentrations of trace metals in aerosols in Beijing, China, determined by using inductively coupled plasma mass spectrometry equipped with laser ablation analysis, and source identification of aerosols. *Sci Total Environ* 2004, *330*, 145-158.
[238] Hu, Z.; Liu, Y.; Gao, S.; Hu, S.; Dietiker, R.; Guenther, D. A local aerosol extraction strategy for the determination of the aerosol composition in laser ablation inductively coupled plasma mass spectrometry. *J Anal At Spectrom* 2008, *23*, 1192-1203.

[239] Coedo, A.G.; Padilla, I.; Dorado, M.T. Determination of minor elements in steelmaking flue dusts using laser ablation inductively coupled plasma mass spectrometry. *Talanta* 2005, *67*, 136-143.

[240] Guo, X.; Lichte, F.E. Analysis of rocks, soils and sediments for the chalcophile elements by laser ablation-inductively coupled plasma mass spectrometry. *Analyst* 1995, *120*, 2707-2711.

[241] Lee, Y.-L.; Chang, C.-C.; Jiang, S.-J. Laser ablation inductively coupled plasma mass spectrometry for the determination of trace elements in soil. *Spectrochim Acta B* 2003, *58*, 523-530.

[242] Bellotto, V.R.; Miekeley, N. Improvements in calibration procedures for the quantitative determination of trace elements in carbonate material (mussel shells) by laser ablation ICP-MS. *Fresenius' J Anal Chem* 2000, *367*, 635-640.

[243] Barats, A.; Pecheyran, C.; Amouroux, D.; Dubascoux, S.; Chauvaud, L.; Donard, O.F.X. Matrix-matched quantitative analysis of trace-elements in calcium carbonate shells by laser-ablation ICP-MS: Application to the determination of daily scale profiles in scallop shell (*Pecten maximus*). *Anal Bioanal Chem* 2007, *387*, 1131-1140.

[244] Bellotto, V.R.; Miekeley, N. Trace metals in mussel shells and corresponding soft tissue samples: A validation experiment for the use of *Perna perna* shells in pollution monitoring. *Anal Bioanal Chem* 2007, *389*, 769-776.

[245] Morrison, G.; Fatoki, O.S.; Linder, S.; Lundehn, C. Determination of heavy metal concentrations and metal fingerprints of sewage sludge from Eastern Cape Province, South Africa by inductively coupled plasma - mass spectrometry (ICP-MS) and laser ablation-inductively coupled plasma-mass spectrometry (LA-ICP-MS). *Water Air Soil Pollut* 2004, *152*, 111-127.

[246] Reinhardt, H.; Kriews, M.; Miller, H.; Schrems, O.; Ludke, C.; Hoffmann, E.; Skole, J. Laser ablation inductively coupled plasma mass spectrometry: a new tool for trace element analysis in ice cores. *Fresenius' J Anal Chem* 2001, *370*, 629-636.

[247] Jiménez, M.S.; Gómez, M.T., Castillo, J.R. Multi-element analysis of compost by laser ablation-inductively coupled plasma mass spectrometry. *Talanta* 2007, *72*, 1141-1148.

[248] Watmough, S.A.; Hutchinson, T.C.; Evans, R.D. Development of solid calibration standards for trace elemental analyses of tree rings by laser ablation inductively coupled plasma-mass spectrometry. *Environ Sci Technol* 1998, *32*, 2185-2190.

[249] Prohaska, T.; Stadlbauer, C.; Wimmer, R.; Stingeder, G.; Latkoczy, Ch.; Hoffmann, E.; Stephanowitz, H. Investigation of element variability in tree rings of young Norway spruce by laser-ablation-ICPMS. *Sci Total Environ* 1998, *219*, 29-39.

[250] Monticelli, D., Iorio, A.; Ciceri, E.; Castelletti, A.; Dossi, C. Tree ring microanalysis by LA-ICP-MS for environmental monitoring: validation or refutation? Two case histories. *Microchim Acta* 2009, *164*, 139-148.

[251] Binet, M.R.B.; Ma, R.; McLeod, C.W.; Poole, R.K. Detection and characterization of zinc- and cadmium-binding proteins in *Escherichia coli* by gel electrophoresis and laser ablation-inductively coupled plasma-mass spectrometry. *Anal Biochem* 2003, *318*, 30-38.

[252] Polatajko, A.; Azzolini, M.; Feldmann, I.; Stuezel, T.; Jakubowski, N. Laser ablation-ICP-MS assay development for detecting Cd- and Zn-binding proteins in Cd-exposed *Spinacia oleracea L. J Anal At Spectrom* 2007, *22*, 878-887.

[253] Yang, L.; Sturgeon, R.E.; Mester, Z. Quantitation of trace metals in liquid samples by dried-droplet laser ablation inductively coupled plasma mass spectrometry. *Anal Chem* 2005, *77*, 2971-2977.

[254] Bol'shov, M.A.; Boutron, C.F.; Ducroz, F.M.; Gorlach, U.; Kompanets, O.N.; Rudnev, S.N.; Hutch, B. Direct ultratrace determination of cadmium in Antarctic and Greenland snow and ice by laser atomic fluorescence spectrometry. *Anal Chim Acta* 1991, *251*, 169-75.

[255] Bol'shov, M.A.; Rudnev, S.N.; Huetsch, B. Determination of trace amounts of cadmium by laser excited atomic fluorescence spectrometry. *J Anal A. Spectrom* 1992, *7*, 1-6.

[256] Bol'shov, M.A.; Koloshnikov, V.G.; Rudnev, S.N.; Boutron, C.F.; Gorlach, U.; Patterson, C.C. Detection of trace amounts of toxic metals in environmental samples by laser-excited atomic fluorescence spectrometry. *J Anal At Spectrom* 1992, *7*, 99-104.

[257] Bol'shov, M.A.; Boutron, C.F. Determination of heavy metals in polar snow and ice by laser-excited atomic fluorescence spectrometry. *Analusis* 1994, *22*, M44-M46.

[258] Cheam, V.; Lawson, G.; Lechner, J.; Desrosiers, R.; Nriagu, J. Thallium and cadmium in recent snow and firn layers in the Canadian Arctic by atomic fluorescence and absorption spectrometries. *Fresenius' J Anal Chem* 1996, *355*, 332-335.

[259] Zhou, J.X.; Hou, X.; Yang, K.X.; Michel, R.G. Laser-excited atomic fluorescence spectrometry in a graphite furnace with an optical parametric oscillator laser for sequential multi-element determination of cadmium, cobalt, lead, manganese and thallium in Buffalo river sediment. *J Anal At Spectrom* 1998, *13*, 41-47.

[260] Duan, Y.; Du, X.; Li, Y.; Jin, Q. Characterization of a modified, low-power argon microwave plasma torch (MPT) as an atomization cell for atomic fluorescence spectrometry. *Appl Spectrosc* 1995, *49*, 1079-1085.

[261] Frentiu, T.; Darvasi, E.; Senila, M.; Ponta, M.; Cordos, E. Preliminary investigation of a medium power argon radiofrequency capacitively coupled plasma as atomization cell in atomic fluorescence spectrometry of cadmium. *Talanta* 2008, *76*, 1170-1176.

[262] Dědina J.; Tsalev, D. L. *Hydride Generation Atomic Absorption Spectrometry*. Wiley and Sons: Surrey, U.K., 1995.

[263] Sturgeon, R.E.; Guo, X.; Mester, Z. Chemical vapor generation: Are further advances yet possible?. *Anal Bioanal Chem* 2005, *382*, 881-883.

[264] Luna, A.S.; Borges Pereira, H.; Takase, I.; Araujo Goncalves, R.; Sturgeon, R.E.; Calixto de Campos, R. Chemical vapor generation-electrothermal atomic absorption spectrometry: New perspectives. *Spectrochim Acta B* 2002, *57*, 2047-2056.

[265] Guo, X.; Sturgeon, R.E.; Mester, Z.; Gardner, G.J. Vapor generation by UV irradiation for sample introduction with atomic spectrometry. *Anal Chem* 2004, *76*, 2401-2405.

[266] Pohl, P. Recent advances in chemical vapor generation via reaction with sodium tetrahydroborate. *Trends Anal Chem* 2004, *23*, 21-27.

[267] Pohl, P. Hydride generation – recent advances in atomic emission spectrometry. *Trends Anal Chem* 2004, *23*, 87-101.

[268] Narsito, Agterdenbos, J. Santosa, S.J. Study of processes in the hydride generation atomic absorption spectrometry of antimony, arsenic and selenium. *Anal. Chim. Acta* 1990, *237*, 189-199.

[269] D'Ulivo, A. Chemical vapor generation by tetrahydroborate (III) and other borane complexes in aqueous media – A critical discussion of fundamental processes and

mechanisms involved in reagent decomposition and hydride formation. *Spectrochim Acta B* 2004, *59*, 793-825

[270] Willie, S.N. First order speciation of As using flow injection hydride generation atomic absorption spectrometry with in-situ trapping of the arsine in a graphite furnace. *Spectrochim. Acta B* 1996, *51*, 1781-1790.

[271] Laborda, F.; Bolea, E.; Castillo, J.R. Electrochemical hydride generation as a sample introduction technique in atomic spectrometry: Fundamentals, interferences and applications. *Anal Bioanal Chem* 2007, *388*, 743-775.

[272] Chaudhry, M.M.; Ure, A.M.; Cooksey, B.G.; Littlejohn, D.; Halls, D.J. Investigation of *in situ* concentration of hydride forming elements in a graphite furnace atomizer. *Anal. Proc.* 1991, *28*, 44-46.

[273] Sturgeon, R.E.; Willie, S.N.; Sproule, G.I.; Berman, S.S. Sorption and atomization of metallic hydrides in a graphite furnace. *J Anal At Spectrom* 1987, *2*, 719-722.

[274] Dědina, J. Evaluation of hydride generation and atomization for atomic absorption spectrometry. *Prog Analyt Spectrosc* 1988, *11*, 251-360.

[275] D'Ulivo, A.; Chen, Y.-W. Determination of cadmium in aqueous samples by vapor generation with sodium tetraethylborate(III) reagent. *J Anal At Spectrom* 1989, *4*, 319-322.

[276] Cacho, J.; Beltrán, I.; Nerín, C. Generation of a volatile cadmium species in an organic medium. *J Anal At Spectrom* 1989, *4*, 661-663.

[277] Lampugnani, L.; Salvetti, C.; Tsalev, D. L. Hydride generation atomic absorption spectrometry with different flow systems and in-atomizer trapping for determination of cadmium in water and urine - Overview of existing data on cadmium vapour generation and evaluation of critical parameters. *Talanta* 2003, *61*, 683-698.

[278] Valdés-Hevia y Temprano, M.C.; Fernández de la Campa, M.R.; Sanz-Medel, A. Sensitive inductively coupled plasma atomic emission spectrometric determination of cadmium by continuous alkylation with sodium tetraethylborate. *J Anal At Spectrom* 1994, *9*, 231-236.

[279] Sanz-Medel, A.; Valdés-Hevia y Temprano, M.C.; Bordel García, N.; Fernández de la Campa, M.R. The use of surfactants to obtain cadmium atoms at room temperature and its application for the cold vapor AAS determination of the metal. *Anal Proc* 1995, *32*, 49-52.

[280] Sanz-Medel, A.; Valdés-Hevia y Temprano, M.C.; Bordel García, N.; Fernández de la Campa, M.R. Generation of cadmium atoms at room temperature using vesicles and its application to cadmium determination by cold vapor atomic spectrometry. *Anal Chem* 1995, *67*, 2216-2223.

[281] Guo, X.-W.; Guo, X.-M. Studies on the reaction between cadmium and potassium tetrahydroborate in aqueous solution and its application in atomic fluorescence spectrometry. *Anal Chim Acta* 1995, *310*, 377-385.

[282] Guo, X.-W.; Guo, X.-M. Determination of cadmium at ultratrace levels by cold vapour atomic absorption spectrometry. *J Anal At Spectrom* 1995, *10*, 987-991.

[283] Goenaga Infante, H.; Fernández Sánchez, M.L.; Sanz-Medel, A. Ultratrace determination of cadmium by atomic absorption spectrometry using hydride generation with in situ preconcentration in a palladium-coated graphite atomizer. *J Anal At Spectrom* 1996, *11*, 571-575.

[284] Goenaga Infante, H.; Fernández Sánchez, M.L.; Sanz-Medel, A. Vesicular hydride generation-in situ preconcentration-electrothermal atomic absorption spectrometry determination of sub-parts-per-billion levels of cadmium. *J Anal At Spectrom* 1997, *12*, 1333-1336.

[285] Bermejo-Barrera, P.; Moreda-Piñeiro, J.; Moreda-Piñeiro, A.; Bermejo-Barrera, A. Use of flow injection cold vapor generation and preconcentration on coated graphite tubes for the determination of cadmium in sea-water by electrothermal atomic absorption spectrometry. *J Anal At Spectrom* 1996, *11*, 1081-1086.

[286] Bermejo-Barrera, P.; Moreda-Piñeiro, J.; Moreda-Piñeiro, A.; Bermejo-Barrera, A Iridium-coated graphite tubes for the direct determination of As, Cd, Hg and Pb in seawater by vapor generation ETAAS. *At Spectrosc* 1998, *19*, 100-106.

[287] Hwang, T.-J.; Jiang, H.-J. Determination of cadmium by flow injection isotope dilution inductively coupled plasma mass spectrometry with vapor generation sample introduction. *J Anal At Spectrom* 1997, *12*, 579-584.

[288] Liu, M.; Xu, S. The determination of trace cadmium by flow injection cold vapor generation AAS. *At Spectrosc* 1997, *18*, 195-201.

[289] Matusiewicz, H.; Kopras, M.; Sturgeon, R.E. Determination of cadmium in environmental samples by hydride generation with in situ concentration and atomic absorption detection. *Analyst* 1997, *122*, 331-336.

[290] Liva Garrido, M.; Muñoz-Olivas, R.; Cámara, C. Determination of cadmium in aqueous media by flow injection cold vapor atomic absorption spectrometry. Application to natural water samples. *J Anal At Spectrom* 1998, *13*, 295-300.

[291] Liva Garrido, M.; Muñoz-Olivas, R.; Cámara, C. Interference removal for cadmium determination in waste water and sewage sludge by flow injection cold vapor generation atomic absorption spectrometry. *J Anal At Spectrom* 1998, *13*, 1145-1149.

[292] Tyson, J.F. High-performance, flow-based, sample pre-treatment and introduction procedures for analytical atomic spectrometry. *J Anal At Spectrom* 1999, *14*, 169-178.

[293] Vargas-Razo, C.; Tyson, J.F. Determination of cadmium by flow injection-chemical vapor generation-atomic absorption spectrometry. *Fresenius J Anal Chem* 2000, *366*, 182-190.

[294] Liva, M.; Muñoz-Olivas, R.; Cámara, C. Determination of Cd in sonicated slurries and leachates of biological and environmental materials by FI-CV-AAS. *Talanta* 2000, *51*, 381-387.

[295] Liang, J.; Wang, Q.; Huang, B. Concentrations of hazardous heavy metals in environmental samples collected in Xiamen, China, as determined by vapor generation non-dispersive atomic fluorescence spectrometry. *Anal Sci* 2004, *20*, 85-88.

[296] Sun, H.-W.; Ran, S. Enhancement reagents for simultaneous vapor generation of zinc and cadmium with intermittent flow system coupled to atomic fluorescence spectrometry. *Anal Chim Acta* 2004, *509*, 71-76.

[297] Li, G.; Wu, L.; Xin, J.; Hou, X. Chemical vapor generation by reaction of cadmium with potassium tetrahydroborate and sodium iodate in acidic aqueous solution for atomic fluorescence spectrometric application. *J Anal At Spectrom* 2004, *19*, 1010-1013.

[298] Korkmaz, D.; Demir, C.; Aydin, F.; Ataman, O.Y. Cold vapor generation and on-line trapping of cadmium species on quartz surface prior to detection by atomic absorption spectrometry. *J Anal At Spectrom* 2005, *20*, 46-52.

[299] Chuachuad, W.; Tyson, J.F. Determination of cadmium by flow injection atomic absorption spectrometry with cold vapor generation by a tetrahydroborate-form anion-exchanger. *J Anal At Spectrom* 2005, *20*, 273-281.

[300] Vieira, M.A.; Ribeiro, A.S.; Dias, L.F.; Curtius, A.J. Determination of Cd, Hg, Pb and Se in sediments slurries by isotopic dilution calibration ICP-MS after chemical vapor generation using an on-line system or retention in an electrothermal vaporizer treated with iridium. *Spectrochim Acta B* 2005, *60*, 643-652.

[301] Chuachuad, W.; Tyson, J.F. Determination of cadmium by electrothermal atomic absorption spectrometry with flow injection chemical vapor generation from a tetrahydroborate-form anion-exchanger and in-atomizer trapping. *Canadian J Anal Sci Spectrosc* 2004, *49*, 362-373.

[302] Ke, Y.; Sun, Q.; Yang, Z.; Xin, J.; Chen, L.; Hou, X. Determination of trace cadmium and zinc in corn kernels and related soil samples by atomic absorption and chemical vapor generation atomic fluorescence after microwave-assisted digestion. *Spectrosc Lett* 2006, *39*, 29-43.

[303] Manzoori, J.L.; Abdolmohammad-Zadeh, H.; Amjadi, M. Ultratrace determination of cadmium by cold vapor atomic absorption spectrometry after preconcentration with a simplified cloud point extraction methodology. *Talanta* 2007, *71*, 582-587.

[304] Fu, Q.; Yang, L.; Wang, Q. Determination of cadmium in seawater by vapor generation atomic fluorescence spectrometry after online preconcentration with a novel alkyl phosphinic acid resin. *Spectrosc Lett* 2007, *40*, 547-557.

[305] Cankur, O.; Yavuz Ataman, O. Chemical vapor generation of Cd and on-line preconcentration on a resistively heated W-coil prior to determination by atomic absorption spectrometry using an unheated quartz absorption cell. *J Anal At Spectrom* 2007, *22*, 791-799.

[306] Zhang, S.-J.; Gan, W.-E. Determination of cadmium using thermo-nebulous phase chemical vapor generation with atomic fluorescence spectrometry. *At Spectrosc* 2007, *28*, 108-112.

[307] Chen, M.-L.; Tian, Y.; Wang, J.-H. Integrating preconcentration, tetrahydroborate immobilization, elution and chemical vapor generation onto a cellulose surface for the determination of cadmium with atomic fluorescence spectrometry. *J Anal At Spectrom* 2008, *23*, 876-880.

[308] Ebdon, L.; Goodall, P.; Hill, S.J.; Stockwell, P.B.; Thompson, K.C. Ultra-trace determination of cadmium by vapor generation atomic fluorescence spectrometry. *J Anal At Spectrom* 1993, *8*, 723-729.

[309] Tao, S.; Kumamaru, T. Inductively coupled plasma atomic emission spectrometric determination of cadmium in biological and environmental materials using electrothermal vaporization after in situ alkylation. *Anal Chim Acta* 1995, *310*, 369-375.

[310] Valles Mota, J.P.; Fernández de la Campa, M.R.; García Alonso, J.I.; Sanz-Medel, A. Determination of cadmium in biological and environmental materials by isotope dilution inductively coupled plasma mass spectrometry: Effect of flow sample introduction methods. *J Anal At Spectrom* 1999, *14*, 113-120.

[311] Arbab-Zavar, M.H.; Chamsaz, M.; Youssefi, A.; Aliakbari, M. Mechanistic aspects of electrochemical hydride generation for cadmium. *Anal Chim Acta* 2006, *576*, 215-220.

[312] Klockenkamper, R. *Total Reflection X-ray Fluorescence Analysis;* John Wiley: New York, 1997.

[313] Jenkins, R.; Gould, R.W.; Gedcke, D. *Quantitative X-ray Spectrometry;* Marcel Dekker: New York, 1995.

[314] Van Grieken, R.E.; Marckowicz, A.A. *Handbook of X-ray Spectrometry: Methods and Techniques;* Marcel Dekker: New York, 1993.

[315] Knoth, J.; Prange, A.; Schneider, H.; Schwenke, H. Variable X-ray excitation for total reflection X-ray fluorescence spectrometry using an Mo/W alloy anode and a tunable double multilayer monochromator. *Spectrochimica Acta B* 1997, *52*, 907-913.

[316] Van Meel, K.; Fontàs, C.; Van Grieken, R.; Queralt, I.; Hidalgo, M.; Marguí, E. Application of high-energy polarised beam energy dispersive X-ray fluorescence spectrometry to cadmium determination in saline solutions. *J Anal At Spectrom* 2008, *23*, 1034-1037.

[317] Marguí, E.; Fontàs, C.; Van Meel, K.; Van Grieken, R.; Queralt, I.; Hidalgo, M. High-energy polarized-beam energy-dispersive X-ray fluorescence analysis combined with activated thin layers for cadmium determination at trace levels in complex environmental liquid samples. *Anal Chem* 2008, *80*, 2357-2364.

[318] Potts, P.J.; Ellis, A.T.; Kregsamer, P.; Streli, C.; Vanhoof, C.; West, M.; Wobrauschek, P. Atomic spectrometry update. X-Ray fluorescence spectrometry. *J Anal At Spectrom* 2005, *20*, 1124-1154.

[319] Krivan, V.; Franek, M.; Baumann, H.; Pavel, J. Sequential multielement analysis of a single aerosol filter by different instrumental and wet-chemical methods. *Fresenius' J Anal Chem* 1990, *338*, 583-587.

[320] Steinhoff, G.; Haupt, O.; Dannecker, W. Fast determination of trace elements on aerosol-loaded filters by X-ray fluorescence analysis considering the inhomogeneous elemental distribution. *Fresenius' J Anal Chem* 2000, *366*, 174-177.

[321] Kyotani, T.; Iwatsuki, M. Multi-element analysis of environmental samples by X-ray fluorescence spectrometry using a simple thin-layer sample preparation technique. *Analyst* 1998, *123*, 1813-1816.

[322] Sinem Atgin, R.; El-Agha, O.; Zararsiz, A.; Kocatas, A.; Parlak, H.; Tuncel, G. Investigation of the sediment pollution in Izmir Bay: Trace elements. *Spectrochim Acta B* 2000, *55*, 1151-1164.

[323] Krishna, A.K.; Murthy, N.N.; Govil, P.K. Multielement analysis of soils by wavelength-dispersive X-ray fluorescence spectrometry. *At Spectrosc* 2007, *28*, 202-214.

[324] Stephens, W.E.; Calder, A. Analysis of non-organic elements in plant foliage using polarized X-ray fluorescence spectrometry. *Anal Chim Acta* 2004, *527*, 89-96.

[325] Hou, X.; Peters, H.L.; Yang, Z.; Wagner, K.A.; Batchelor, J.D.; Daniel, M.M.; Jones, B.T. Determination of trace metals in drinking water using solid-phase extraction disks and X-ray fluorescence spectrometry. *Appl Spectrosc* 2003, *57*, 338-342.

[326] Marguí, E.; Fontàs, C.; Hidalgo, M.; Queralt, I. Improved instrumental sensitivity for Cd determination in aqueous solutions using wavelength dispersive X-ray fluorescence spectrometry, Rh-target tube instrumentation. *Spectrochim Acta B* 2008, *63*, 1329-1332.

[327] Fontàs, C.; Marguí, E.; Hidalgo, M.; Queralt, I. Improvement approaches for the determination of Cr(VI), Cd(II), Pd(II) and Pt(IV) contained in aqueous samples by conventional XRF instrumentation. *X-Ray Spectrom* 2009, *38*, 9-17.

[328] Eksperiandova, L.P.; Blank, A.B.; Makarovskaya, Y.N. Analysis of waste water by X-ray fluorescence spectrometry. *X-Ray Spectrom* 2002, *31*, 259-263.

[329] Heckel, J.; Brumme, M.; Weinert, A.; Irmer, K. Multi-element trace analysis of rocks and soils by EDXRF using polarized radiation. *X-Ray Spectrom* 1991, *20*, 287-292.

[330] Hettipathirana, T.D. Simultaneous determination of parts-per-million level Cr, As, Cd and Pb, and major elements in low level contaminated soils using borate fusion and energy dispersive X-ray fluorescence spectrometry with polarized excitation. *Spectrochim Acta B* 2004, *59*, 223-229.

[331] Akyuez, T.; Bassari, A.; Saltoglu, T.; Kurtcebe, T. Determination of the concentration of some elements in the Black Sea sediment samples using energy dispersive X-ray fluorescence analyzer. *Toxicol Environ Chem* 1995, *48*, 125-133.

[332] Akyuz, T.; Akyuz, S.; Bassari, A. Radioisotope excited EDXRF analysis of sediment core samples from the southern part of the Black Sea. *J Radioanal Nucl Ch* 2001, *250*, 129-137.

[333] Manso, M.; Carvalho, M.L.; Nunes, M.L. Characterization of essential and toxic elements in cephalopod tissues by EDXRF and AAS. *X-Ray Spectrom* 2007, *36*, 413-418.

[334] Russell, P.A.; James, R. Determination of toxic elements in liquid hazardous waste using high-resolution energy-dispersive X-ray fluorescence spectrometry. *J Anal At Spectrom* 1997, *12*, 25-32.

[335] Nkono, N.A.; Asubiojo, O.I. Elemental composition of drinking water supplies in three states in Southeastern Nigeria. *J Radioanal Nucl Ch* 1998, *227*, 117-119.

[336] Pepelnik, R.; Prange, A.; Niedergesaess, R. Comparative study of multi-element determination using inductively coupled plasma mass spectrometry, total reflection X-ray fluorescence spectrometry and neutron activation analysis. *J Anal At Spectrom* 1994, *9*, 1071-1074.

[337] Gasparics, T.; Csato, I.; Gy Zaray. Analysis of Antarctic marine sediment by inductively coupled plasma atomic emission and total reflection X-ray fluorescence spectrometry. *Microchem J* 1997, *55*, 56-63.

[338] Pepelnik, R.; Erbsloeh, B.; Michaelis, W.; Prange, A. Determination of trace element deposition into a forest ecosystem using total-reflection X-ray fluorescence. *Spectrochim Acta B* 1993, *48*, 223-229.

[339] Theisen, M.; Niessner, R. Elemental analysis of airborne dust samples with TXRF. Comparison of oxygen-plasma ashing on sapphire carriers and acid digestion for sample preparation. *Fresenius' J Anal Chem* 1999, *365*, 332-337.

[340] Markert, B.; Reus, U.; Herpin, U. The application of TXRF in instrumental multielement analysis of plants, demonstrated with species of moss. *Sci Total Environ* 1994, *152*, 213-220.

[341] Mages, M.; Bandow, N.; Kuester, E.; Brack, W.; von Tuempling, W. Zinc and cadmium accumulation in single zebrafish (*Danio rerio*) embryos - A total reflection X-ray fluorescence spectrometry application. *Spectrochim Acta B* 2008, *63*, 1443-1449.

[342] Prange, A.; Boeddeker, H.; Kramer, K. Determination of trace elements in river-water using total-reflection X-ray fluorescence. *Spectrochim Acta B* 1993, *48*, 207-215.

[343] Reus, U.; Markert, B.; Hoffmeister, C.; Spott, D.; Guhr, H. Determination of trace metals in river water and suspended solids by TXRF spectrometry. A methodical study on analytical performance and sample homogeneity. *Fresenius' J Anal Chem* 1993, *347*, 430-435.

[344] Vértes, A.; Nagy, S.; Süvegh, K. *Nuclear Methods in Mineralogy and Geology*; Plenum Press: New York, 1998.

[345] Alfassi, Z. *Activation Analysis*; CRC Press: Boca Raton, FL, 1990.

[346] Simonits, A.; De Corte, F.; Hoste, J. Single-comparator methods in reactor neutron activation analysis. *J Radioanal Chem* 1975, *24*, 31–46.

[347] Acharya, R.N.; Nair, A.G.C.; Reddy, A.V.R.; Manohar, S.B. Validation of a neutron activation analysis method using k_0-standardization. *Appl Radiat Isot* 2002, *57*, 391-398.

[348] Dinescu, L.C.; Steinnes, E.; Duliu, O.G.; Ciortea, C.; Sjobakk, T.E.; Dumitriu, D.E.; Gugiu, M.M.; Haralambie, M. Distribution of some major and trace elements in Danube Delta lacustrine sediments and soil. *J Radioanal Nucl Ch* 2004, *262*, 345-354.

[349] Coskun, M.; Steinnes, E.; Frontasyeva, M.V.; Sjobakk, T. E.; Demkina, S. Heavy Metal Pollution of Surface Soil in the Thrace Region, Turkey. *Environ Monit Assess* 2006, *119*, 545-556.

[350] Grosheva, E.; Zaichick, D.; Zaichick, V. Application of INAA in the assessment of chemical elements in soils of the Khamar-Daban mountain range. *J Radioanal Nucl Ch* 2007, *271*, 565-572.

[351] Akyuz, S.; Akyuz, T.; Algan, A.O.; Mukhamedshina, N.M.; Mirsagatova, A.A. Energy dispersive X-ray fluorescence and neutron activation analysis of surficial sediments of the Sea of Marmara and the Black Sea around Istanbul. *J Radioanal Nucl Ch* 2002, *254*, 569-575.

[352] Tanaka, Y.; Kuno, A.; Matsumo, M. A study of environmental analysis of urban river sediments using activation analysis. *J Radioanal Nucl Ch* 2003, *255*, 239-243.

[353] Kataoka, M.; Kuno, A.; Matsumo, M. A study on vertical distribution of elements and their chemical states in Yatsu tideland sediments. *J Radioanal Nucl Ch* 2003, *255*, 283-286.

[354] Yusof, A.M.; Thanapalasingham, V.; Wood, A.K.H. Assessment of the health of a river ecosystem due to the impact of pollution from industrial discharge using instrumental neutron activation analysis and inductively coupled plasma-mass spectrometry. *J Radioanal Nucl Ch* 2007, *273*, 525-531.

[355] Bin Saion, E.; Wood, A.K.H.; Sulaiman, A.Z.A.; Alzahrany, A.A.; Elias, Md S.; Siong, W.B. Determination of heavy metal pollution in depth profile of marine sediment samples from the Strait of Malacca. *Fresenius Environ Bull* 2007, *16*, 1279-1287.

[356] Adomako, D.; Nyarko, B.J.B.; Dampare, S.B.; Serfor-Armah, Y.; Osae, S.; Fianko, J.R.; Akaho, E.H.K. Determination of toxic elements in waters and sediments from River Subin in the Ashanti Region of Ghana. *Environ Monit Assess* 2008, *141*, 165-175.

[357] Bem, H.; Gallorini, M.; Rizzio, E.; Krzeminska, M. Comparative studies on the concentrations of some elements in the urban air particulate matter in Lodz City of Poland and in Milan, Italy. *Environ Int* 2003, *29*, 423-428.

[358] Karakas, D.; Oelmez, I.; Tosun, S.; Tuncel, G. Trace and major element compositions of Black Sea aerosol. *J Radioanal Nucl Ch* 2004, *259*, 187-192.

[359] Athari, M.; Sohrabpour, M.; Shahriari, M.; Rostami, S. Elemental characterization of TSP and two size fractions of airborne particulate matter from Tehran by INAA and AAS. *J Radioanal Nucl Ch* 2004, *260*, 351-356.

[360] Bergamaschi, L.; Rizzio, E.; Profumo, A.; Gallorini, M. Determination of trace elements by INAA in urban air particulate matter and transplanted lichens. *J Radioanal Nucl Ch* 2005, *263*, 745-750.

[361] Giaveri, G.; Bergamaschi, L.; Rizzio, E.; Verza, G.; Zambelli, G.; Brandone, A.; Profumo, A.; Baudo, R.; Tartari, G.; Gallorini, M. INAA at the top of the world: Elemental characterization and analysis of airborne particulate matter collected in the Himalayas at 5100 m high. *J Radioanal Nucl Ch* 2005, *263,* 725-732.

[362] Waheed, S.; Rahman, A.; Daud, M.; Rahman, S.; Islam, Z.; Ahmad, S. Air quality evaluation of some industrial cities of Pakistan using INAA and AAS. *Radiochim Acta* 2005, *93*, 487-495.

[363] Freitas, M.C; Pacheco, A.M.G. Elemental concentrations of aerosols near Portuguese power plants by INAA and PIXE. *J Radioanal Nucl Ch* 2007, *271*, 185-189.

[364] Avino, P.; Capannesi, G.; Rosada, A. Heavy metal determination in atmospheric particulate matter by instrumental neutron activation analysis. *Microchem J* 2008, *88*, 97-106.

[365] Landsberger, S.; Massicotte, A.; Braisted, J.; Gong, S. Determination of cadmium in Arctic air filters by epithermal neutron activation analysis and Compton suppression. *J Radioanal Nucl Ch* 2008, *276*, 193-197.

[366] Ni, B.; Tian, W.; Wang, P.; Zhang, L.; Zhang, G.; Liu, C.; Huang, D.; Li, D. Study on air pollution in Beijing's major industrial areas using multielements in biomonitors and NAA techniques. *Int J Environ Pollut* 2008, *32*, 456-466.

[367] Marrero, J.; Polla, G.; Jiménez Rebagliati, R.; Pla, R.; Gómez, D.; Smichowski, P. Characterization and determination of 28 elements in fly ashes collected in a thermal power plant in Argentina using different instrumental techniques. *Spectrochim Acta B* 2007, *62*, 101-108.

[368] Papaefthymiou, H.; Symeopoulos, B.D.; Soupioni, M. Neutron activation analysis and natural radioactivity measurements of lignite and ashes from Megalopolis basin, Greece. *J Radioanal Nucl Ch* 2007, *274*, 123-130.

[369] Yusof, A.M.; Yanta, N.F.; Wood, A.K.H. The use of bivalves as bio-indicators in the assessment of marine pollution along a coastal area. *J Radioanal Nucl Ch* 2004, *259*, 119-127.

[370] Serfor-Armah, Y.; Carboo, D.; Akuamoah, R.K.; Chatt, A. Determination of selected elements in red, brown and green seaweed species for monitoring pollution in the coastal environment of Ghana. *J Radioanal Nucl Ch* 2006, *269*, 711-718.

[371] Bergamaschi, L.; Rizzio, E.; Giaveri, G.; Giordani, L.; Profumo, A.; Gallorini, M. INAA for the determination of trace elements and evaluation of their enrichment factors in lichens of high altitude areas. *J Radioanal Nucl Ch* 2005, *263*, 721-724.

[372] Smodis, B.; Bleise, A. IAEA quality control study on determining trace elements in biological matrices for air pollution research. *J Radioanal Nucl Ch* 2007, *271*, 269-274.

[373] Bergamaschi, L.; Rizzio, E.; Giaveri, G.; Loppi, S.; Gallorini, M. Comparison between the accumulation capacity of four lichen species transplanted to a urban site. *Environ Pollut* 2007, *148*, 468-476.

[374] Saiki, M.; Alves, E.R.; Marcelli, M.P. Analysis of lichen species for atmospheric pollution biomonitoring in the Santo Andre municipality, Sao Paulo, Brazil. *J Radioanal Nucl Ch* 2007, *273*, 543-547.

[375] Ayrault, S.; Galsomies, L.; Amblard, G.; Sciarretta, M.D.; Bonhomme, P.; Gaudry, A. Instrumental neutron activation analysis (INAA) and inductively coupled plasma/mass spectrometry (ICP-MS) for trace element biomonitoring using mosses. *Int J Environ Anal Chem* 2002, *82*, 463-473.
[376] Frontasyeva, M.V.; Galinskaya, T.Y.; Krmar, M.; Matavuly, M.; Pavlov, S.S.; Povtoreyko, E.A.; Radnovic, D.; Steinnes, E. Atmospheric deposition of heavy metals in northern Serbia and Bosnia-Herzegovina studied by the moss biomonitoring, neutron activation analysis and GIS technology. *J Radioanal Nucl Ch* 2004, *259,* 141-144.
[377] Cucu-Man, S.; Mocanu, R.; Culicov, O.; Steinnes, E.; Frontasyeva, M. Atmospheric deposition of metals in Romania studied by biomonitoring using the epiphytic moss *Hypnum cupressiforme*. *Int J Environ Anal Chem* 2004, *84*, 845-854.
[378] Ermakova, E.V.; Frontasyeva, M.V.; Pavlov, S.S.; Povtoreiko, E.A.; Steinnes, E.; Cheremisina, Y.N. Air pollution studies in Central Russia (Tver and Yaroslavl regions) using the moss biomonitoring technique and neutron activation analysis. *J Atmos Chem* 2004, *49*, 549-561.
[379] Culicov, O.A.; Mocanu, R.; Frontasyeva, M.V.; Yurukova, L.; Steinnes, E. Active moss biomonitoring applied to an industrial site in Romania: relative accumulation of 36 elements in moss-bags. 2005. *Environ Monit Assess* 2005, *108*, 229-240.
[380] Barandovski, L.; Cekova, M.; Frontasyeva, M.V.; Pavlov, S.S.; Stafilov, T.; Steinnes, E.; Urumov, V. Atmospheric deposition of trace element pollutants in Macedonia studied by the moss biomonitoring technique. *Environ Monit Assess* 2008, *138*, 107-118.
[381] Bard, A.J.; Faulkner, L.R. *Electrochemical Methods. Fundamentals and Applications*; John Wiley & Sons: New York, 2001.
[382] Wang, J. *Stripping Analysis. Principles, Instrumentation and Applications;* VCH: Weinheim, 1985.
[383] Brainina, K.H.; Neyman, E. *Electroanalytical Stripping Method*; John Wiley & Sons: New York, 1993.
[384] Economou, A. Bismuth-film electrodes: Recent developments and potentialities for electroanalysis. *Trends Anal Chem* 2005, *24*, 334-340.
[385] Prego, R.; Cobelo-García, A. Cadmium, copper and lead contamination of the seawater column on the Prestige shipwreck (NE Atlantic Ocean). *Anal Chim Acta* 2004, *524*, 23-26.
[386] Nedeltcheva, T.; Atanassova, M.; Dimitrov, J.; Stanislavova, L. Determination of mobile form contents of Zn, Cd, Pb and Cu in soil extracts by combined stripping voltammetry. *Anal Chim Acta* 2005, *528*, 143-146.
[387] Vargas Mamani, M.C.; Aleixo, L.M.; Ferreira de Abreu, M.; Rath, S. Simultaneous determination of cadmium and lead in medicinal plants by anodic stripping voltammetry. *J Pharmaceutical Biomed Anal* 2005, *37*, 709-713.
[388] Afzali, D.; Mostafavi, A.; Taher, M.A.; Rezaeipour, E.; Khayatzadeh Mahani, M. Natural analcime zeolite modified with 5-Br-PADAP for the preconcentration and anodic stripping voltammetric determination of trace amount of cadmium. *Anal Sci* 2005, *21*, 383-386.
[389] Buzica, D.; Gerboles, M.; Borowiak, A.; Trincherini, P.; Passarella, R.; Pedroni, V. Comparison of voltammetry and inductively coupled plasma-mass spectrometry for the

determination of heavy metals in PM10 airborne particulate matter. *Atmos Environ* 2006, *40*, 4703-4710.

[390] Vieira dos Santos, A.C.; Masini, J.C. Development of a sequential injection anodic stripping voltammetry (SI-ASV) method for determination of Cd(II), Pb(II) and Cu(II) in wastewater samples from coatings industry. *Anal Bioanal Chem* 2006, *385*, 1538-1544.

[391] Carvalho, L.M.; do Nascimento, P.C.; Koschinsky, A.; Bau, M.; Stefanello, R.F.; Spengler, C.; Bohrer, D.; Jost, C. Simultaneous determination of cadmium, lead, copper, and thallium in highly saline samples by anodic stripping voltammetry (ASV) using mercury-film and bismuth-film electrodes. *Electroanal* 2007, *19*, 1719-1726.

[392] Honary, S.; Ebrahimi, P.; Naghibi, F.; Mosaddegh, M.; Shahhoseini, S. Study on the simultaneous determination of Pb and Cd in some commercial medicinal plants by both atomic absorption and voltametry methods. *Anal Lett* 2007, *40*, 2405-2414.

[393] Monterroso, S.C.C.; Carapuca, H.M.; Duarte, A.C. Mixed polyelectrolyte coatings on glassy carbon electrodes: Ion-exchange, permselectivity properties and analytical application of poly-L-lysine-poly(sodium 4-styrenesulfonate)-coated mercury film electrodes for the detection of trace metals. *Talanta* 2006, *68*, 1655-1662.

[394] McGaw, E.A.; Swain, G.M. A comparison of boron-doped diamond thin-film and Hg-coated glassy carbon electrodes for anodic stripping voltammetric determination of heavy metal ions in aqueous media. *Anal Chim Acta* 2006, *575*, 180-189.

[395] Truzzi, C.; Annibaldi, A.; Illuminati, S.; Bassotti, E.; Scarponi, G. Square-wave anodic-stripping voltammetric determination of Cd, Pb, and Cu in a hydrofluoric acid solution of siliceous spicules of marine sponges (from the Ligurian Sea, Italy, and the Ross Sea, Antarctica). *Anal Bioanal Chem* 2008, *392*, 247-262.

[396] Hutton, E.A.; Van Elteren, J.T.; Ogorevc, B.; Smyth, M.R. Validation of bismuth film electrode for determination of cobalt and cadmium in soil extracts using ICP-MS. *Talanta* 2004, *63*, 849-855.

[397] Kefala, G.; Economou, A.; Voulgaropoulos, A. A study of Nafion-coated bismuth-film electrodes for the determination of trace metals by anodic stripping voltammetry. *Analyst* 2004, *129*, 1082-1090.

[398] Kirgoez, U.A.; Marín, S.; Pumera, M.; Merkoci, A.; Alegret, S. Stripping voltammetry with bismuth modified graphite-epoxy composite electrodes. *Electroanal.* 2005, *17*, 881-886.

[399] Legeai, S.; Soropogui, K.; Cretinon, M.; Vittori, O.; Heeren de Oliveira, A.; Barbier, F.; Grenier-Loustalot, M.-L. Economic bismuth-film microsensor for anodic stripping analysis of trace heavy metals using differential pulse voltammetry. *Anal Bioanal Chem* 2005, *383*, 839-847.

[400] Svancara, I.; Baldrianova, L.; Tesarova, E.; Hocevar, S.B.; Elsuccary, S.A.A.; Economou, A.; Sotiropoulos, S.; Ogorevc, B.; Vytras, K. Recent advances in anodic stripping voltammetry with bismuth-modified carbon paste electrodes. *Electroanal* 2006, *18*, 177-185.

[401] Legeai, S.; Vittori, O. A Cu/Nafion/Bi electrode for on-site monitoring of trace heavy metals in natural waters using anodic stripping voltammetry: An alternative to mercury-based electrodes. *Anal Chim Acta* 2006, *560*, 184-190.

[402] Kefala, G.; Economou, A. Polymer-coated bismuth film electrodes for the determination of trace metals by sequential-injection analysis/anodic stripping voltammetry. *Anal Chim Acta* 2006, *576*, 283-289.

[403] Zhu, W.W.; Li, N.B.; Luo, H.Q. Anodic stripping voltammetry determination of Pb(II) and Cd(II) at a bismuth/poly(aniline) film electrode. *Anal Lett* 2006, *39*, 2273-2284.

[404] Baldrianova, L.; Svancara, I.; Economou, A.; Sotiropoulos, S. Anodic stripping voltammetry at in situ bismuth-plated carbon and gold microdisc electrodes in variable electrolyte content unstirred solutions. *Anal Chim Acta* 2006, *580*, 24-31.

[405] Economou, A; Voulgaropoulos, A. On-line stripping voltammetry of trace metals at a flow-through bismuth-film electrode by means of a hybrid flow-injection/sequential-injection system. *Talanta* 2007, *71*, 758-765.

[406] Kachoosangi, R.T.; Banks, C.E.; Ji, X.; Compton, R.G. Electroanalytical determination of cadmium(II) and lead(II) using an in-situ bismuth film modified edge plane pyrolytic graphite electrode. *Anal Sci* 2007, *23*, 283-289.

[407] Baldrianova, L.; Svancara, I.; Sotiropoulos, S. Anodic stripping voltammetry at a new type of disposable bismuth-plated carbon paste mini-electrodes. *Anal Chim Acta* 2007, *599*, 249-255.

[408] Adraoui, I.; Rhazi, M.E.; Amine, A. Fibrinogen-coated bismuth film electrodes for voltammetric analysis of lead and cadmium using the batch injection analysis. *Anal Lett* 2007, *40*, 349-368.

[409] Jia, J.; Cao, L.; Wang, Z. Nafion/poly(sodium 4-styrenesulfonate) mixed coating modified bismuth film electrode for the determination of trace metals by anodic stripping voltammetry. *Electroanal* 2007, *19*, 1845-1849.

[410] Hwang, G.H.; Han, W.K.; Park, J.S.; Kang, S.G. Determination of trace metals by anodic stripping voltammetry using a bismuth-modified carbon nanotube electrode. *Talanta* 2008, *76*, 301-308.

[411] Wang, N.; Dong, X. Stripping voltammetric determination of Pb(II) and Cd(II) based on the multiwalled carbon nanotubes-nafion-bismuth modified glassy carbon electrodes. *Anal Lett* 2008, *41*, 1267-1278.

[412] Hočevar, S.B.; Švancara, I.; Vytřas, K.; Ogorevc, B. Novel electrode for electrochemical stripping analysis based on carbon paste modified with bismuth powder. *Electrochim Acta* 2005, *51*, 706-710.

[413] Hoyer, B.; Jensen, N. Use of sodium dodecyl sulfate as an antifouling and homogenizing agent in the direct determination of heavy metals by anodic stripping voltammetry. *Analyst* 2004, *129*, 751-754.

[414] Farghaly, O.A.; Ghandour, M.A. Square-wave stripping voltammetry for direct determination of eight heavy metals in soil and indoor-airborne particulate matter. *Environ Res* 2005, *97*, 229-235.

[415] Monterroso, S.C.C.; Carapuca, H.M.; Duarte, A.C. Ion-exchange and permselectivity properties of poly(sodium 4-styrenesulfonate) coatings on glassy carbon: Application in the modification of mercury film electrodes for the direct voltammetric analysis of trace metals in estuarine waters. *Talanta* 2005, *65*, 644-653.

[416] Heitzmann, M.; Basaez, L.; Brovelli, F.; Bucher, C.; Limosin, D.; Pereira, E.; Rivas, B.L.; Royal, G.; Saint-Aman, E.; Moutet, J.-C. Voltammetric sensing of trace metals at a poly(pyrrole-malonic acid) film modified carbon electrode. *Electroanal* 2005, *17*, 1970-1976.

[417] Shams, E.; Torabi, R. Determination of nanomolar concentrations of cadmium by anodic-stripping voltammetry at a carbon paste electrode modified with zirconium phosphated amorphous silica. *Sensors Actuators B* 2006, *117*, 86-92.

[418] Carregalo, S.; Merkoci, A.; Alegret, S.. Application of graphite-epoxy composite electrodes in differential pulse anodic stripping voltammetry of heavy metals. *Microchim Acta* 2004, *147*, 245-251.

[419] Garnier, C.; Lesven, L.; Billon, G.; Magnier, A.; Mikkelsen, O.; Pizeta, I. Voltammetric procedure for trace metal analysis in polluted natural waters using homemade bare gold-disk microelectrodes. *Anal Bioanal Chem* 2006, *386,* 313-323.

[420] Zhu, W.W.; Li, N.B.; Luo, H.Q. Simultaneous determination of chromium(III) and cadmium(II) by differential pulse anodic stripping voltammetry on a stannum film electrode. *Talanta* 2007, *72*, 1733-1737.

[421] Toghill, K.E.; Wildgoose, G.G.; Moshar, A.; Mulcahy, C.; Compton, R.G. The fabrication and characterization of a bismuth nanoparticle modified boron doped diamond electrode and its application to the simultaneous determination of cadmium(II) and lead(II). *Electroanal* 2008, *20*, 1731-1737.

[422] Parat, C.; Betelu, S.; Authier, L.; Potin-Gautier, M. Determination of labile trace metals with screen-printed electrode modified by a crown-ether based membrane. *Anal Chim Acta* 2006, *573+574*, 14-19.

[423] Cooper, J.; Bolbot, J.A.; Saini, S.; Setford, S.J. Electrochemical method for the rapid on site screening of cadmium and lead in soil and water samples. *Soil Water Air Soil Pollut* 2007, *179*, 183-195.

[424] Guell, R.; Aragay, G.; Fontas, C.; Antico, E.; Merkoci, A. Sensitive and stable monitoring of lead and cadmium in seawater using screen-printed electrode and electrochemical stripping analysis. *Anal Chim Acta* 2008, *627*, 219-224.

[425] Meucci, V.; Laschi, S.; Minunni, M.; Pretti, C.; Intorre, L.; Soldani, G.; Mascini, M. An optimized digestion method coupled to electrochemical sensor for the determination of Cd, Cu, Pb and Hg in fish by square wave anodic stripping voltammetry. *Talanta* 2009, *77*, 1143-1148.

[426] Evans, A. *Potentiometry and Ion Selective Electrodes*; John Wiley & Sons: New York, 1987.

[427] Bakker, E.; Pretsch, E. Potentiometric sensors for trace-level analysis. *Trend Anal Chem* 2005, *24*, 199-207.

[428] Kloock, J.P.; Moreno, L.; Bratov, A.; Huachupoma, S.; Xu, J.; Wagner, T.; Yoshinobu, T.; Ermolenko, Y.; Vlasov, Y.G.; Schoening, M.J. PLD-prepared cadmium sensors based on chalcogenide glasses-ISFET, LAPS and mISE semiconductor structures. *Sensors and Actuators B Chemical* 2006, *118*, 149-155.

[429] Bizerea-Spiridon, O.; Bizerea, M. Cd-ion selective membrane electrode based microprobe for trace elements detection. *J Trace Microprobe Tech* 2002, *20*, 473-479.

[430] Javanbakht, M.; Shabani-Kia, A.; Darvich, M.R.; Ganjali, M. R.; Shamsipur, M. Cadmium(II)-selective membrane electrode based on a synthesized tetrol compound. *Anal Chim Acta* 2000, *408*, 75-81.

[431] Gupta, V.K.; Jain, A.K.; Ludwig, R.; Maheshwari, G. Electroanalytical studies on cadmium(II) selective potentiometric sensors based on t-butyl thiacalix[4]arene and thiacalix[4]arene in poly(vinyl chloride). *Electrochim Acta* 2008, *53*, 2362-2368.

[432] Panwar, A.; Baniwal, S.; Sharma, C.L.; Singh, A.K. A polystyrene based membrane electrode for cadmium(II) ions. *Fresenius' J Anal Chem* 2002, *368*, 768-772.

[433] Abbas, M.N.; Zahran, E. Novel solid-state cadmium ion-selective electrodes based on its tetraiodo- and tetrabromo-ion pairs with cetylpyridinium. *J Electroanal Chem* 2005, *576*, 205-213.

[434] Khan, A.A.; Alam, M.M. Synthesis, characterization and analytical applications of a new and novel organic-inorganic composite material as a cation exchanger and Cd(II) ion-selective membrane electrode: Polyaniline Sn(IV) tungstoarsenate. *Reactive Functional Polymers* 2003, *55*, 277-290.

[435] Nabi, S.A.; Alam, Z. Inamuddin. A cadmium ion-selective membrane electrode based on strong acidic organic-inorganic composite cation-exchanger: Polyaniline Ce(IV) molybdate. *Sensors Transducers J* 2008, *92*, 87-98.

[436] Mashhadizadeh, M.H.; Sheikhshoaie, I.; Saeid-Nia S. Asymmetrical Schiff bases as carriers in PVC membrane electrodes for cadmium(II) ions. *Electroanalysis* 2005, *17*, 648-654.

[437] Gupta, V.K.; Singh, A. K.; Gupta, B. Schiff bases as cadmium(II) selective ionophores in polymeric membrane electrodes. *Anal Chim Acta* 2007, *583*, 340-348.

[438] Zamani, H.A.; Ganjali, M.R.; Adib, M. Cd(II) PVC-based membrane sensor based on N'-[1-(2-furyl)methylidene]-2-furohydrazide. *Sensor Letters* 2006, *4*, 345-350.

[439] Rezaei, B.; Meghdadi, S.; Sarandi, R.F. A fast response cadmium-selective polymeric membrane electrode based on N,N'-(4-methyl-1,2-phenylene)diquinoline-2-carboxamide as a new neutral carrier. *J Hazard Mat* 2008, *153*, 179-186.

[440] Ion, A.C.; Bakker, E.; Pretsch, E. Potentiometric Cd^{2+}-selective electrode with a detection limit in the low ppt range. *Anal Chim Acta* 2001, *440*, 71-79.

[441] Gupta, K.C.; D'Arc, M.J. Effect of concentration of ion exchanger, plasticizer and molecular weight of cyanocopolymers on selectivity and sensitivity of Cd(II) ion selective electrode. *Talanta* 2000, *52*, 1087-1103.

[442] Gupta, K.C.; D'Arc, M.J. Cadmium ion-selective electrode based on cyanocopolymer. *Electroanalysis* 2000, *12*, 1408-1413.

[443] Shamsipur, M.; Mashhadizadeh, M.H. Cadmium ion-selective electrode based on tetrathia-12-crown-4. *Talanta* 2001, *53*, 1065-1071.

[444] Gupta, V.K.; Chandra, S.; Mangla, R. Dicyclohexano-18-crown-6 as active material in PVC matrix membrane for the fabrication of cadmium selective potentiometric sensor. *Electrochim Acta* 2002, *47*, 1579-1586.

[445] Gupta, V.K.; Jain, A.K.; Kumar, P. PVC-based membranes of dicyclohexano-24-crown-8 as Cd(II) selective sensor. *Electrochim Acta* 2006, *52*, 736-741.

[446] Singh, A.K;. Mehtab, S.; Singh, U.P.; Aggarwal, V. Comparative studies of tridentate sulfur and nitrogen-containing ligands as ionophores for construction of cadmium ionselective membrane sensors. *Electroanalysis* 2007, *19*, 1213-1221.

[447] Valcárcel M.; Luque de Castro M.D. *Automatic Methods of Analysis*; Elsevier: New York, 1988.

[448] Namieśnik, J. Green analytical chemistry - some remarks. *J Sep Sci* 2001, *24*, 151–153.

[449] Anastas, P.T. Green Chemistry and the role of analytical methodology development. *Crit Rev Anal Chem* 1999, *29*, 167-175.

[450] Thomson G. Ocean chemical processes. www.water encyclopedia.com/Mi-Oc/Ocean-Chemical-Processes.html.

[451] Boutron, C.; Martin, S. Preconcentration of dilute solutions at the 10^{-12} g/g level by nonboiling evaporation with variable variance calibration curves. *Anal Chem* 1979, *51*, 140-145.

[452] Nishikawa, M., Ambe, Y.; Mizoguchi, T. Evaporation preconcentration of trace elements in rainwater for inductively coupled plasma emission spectrometry. *Bunseki Kagaku* 1985, *34*, 659-664.

[453] Gorlach, U.; Boutron, C.F. Preconcentration of lead, cadmium, copper and zinc in water at the pg g^{-1} level by non-boiling evaporation. *Anal Chim Acta* 1990, *236*, 391-398.

[454] Qian, X.; Li, X. Comparison of two preconcentration techniques for neutron activation analysis of natural water. *Huanjing Huaxue* 1986, *5*, 60-64.

[455] Dekersabiec, A. M.; Blanc, G.; Pinta, M. Water analysis by Zeeman atomic-absorption spectrometry. *Fresenius'Z Anal Chem* 1985, *322*, 731-735.

[456] Harrison, S.H.; LaFleur, P.D.; Zoller, W.H. Evaluation of lyophilization for the preconcentration of natural water samples prior to neutron activation analysis. *Anal Chem* 1975, *47*, 1685-1688.

[457] Burba, P.; Lieser, K.H.; Neitzert, V.; Roeber, H.M. Preconcentration and determination of trace elements in fresh water and sea water. Comparison of results obtained by different methods (X-ray fluorescence, neutron activation and atomic absorption). *Fresenius' Z Anal Chem* 1978, *291*, 273-277.

[458] Kusaka, Y.; Tsuji, H.; Imai, S.; Ohmori, S. Neutron activation analysis of trace elements in sea water samples. *Radioisotopes* 1979, *28*, 139-144.

[459] Suzuki, S.; Okada, Y.; Hirai, S. Determination of trace elements in certified river water reference materials by instrumental NAA using a freeze-dried preconcentration. *Bunseki Kagaku* 1997, *46*, 223-227.

[460] Hiraide, M.; Ito, T.; Baba, M.; Kawaguchi, H.; Mizuike, A. Multielement preconcentration of trace heavy metals in water by coprecipitation and flotation with indium hydroxide for inductively coupled plasma-atomic emission spectrometry. *Anal Chem* 1980, *52*, 804-807.

[461] Mizuike, A.; Hiraide, M.; Mizuno, K. Preconcentration of trace heavy metals in large aqueous samples by coprecipitation-flotation in a flow system. *Anal Chim Acta* 1983, *148*, 305-309.

[462] Hiraide, M.; Chen, Z.S.; Kawaguchi, H. Co-precipitation of traces of heavy metals with indium hydroxide for graphite-furnace atomic-absorption spectrometry. *Anal Sci* 1991, *7*, 65-68.

[463] Hiraide, M.; Usami, T.; Kawaguchi, H. Minimization of the amount of indium carrier in co-precipitation for the determination of cadmium by graphite-furnace atomic-absorption spectrometry. *Anal Sci* 1992, *8*, 31-34.

[464] Kyung-Hee, P.; Yong-Nam, P. A new online coprecipitation preconcentration technique for trace metal analysis by ICP-AES. *Bull Korean Chem Soc* 1995, *16*, 422-427.

[465] Ueda, J.; Yamazaki, N. Flameless atomic-absorption spectrometric and differential pulse polarographic determination of cadmium after co-precipitation with hafnium hydroxide. *Bull Chem Soc Jpn* 1986, *59*, 1845-1848.

[466] Quigley, M.N.; Vernon, F. Comparison of coprecipitation and chelating ion exchange for the preconcentration of selected heavy metals from seawater. *Anal Proc* 1991, *28*, 175-176.

[467] Hiraide, M.; Chen, Z.S.; Sugimoto, K.; Kawaguchi, H. Coprecipitation with tin(IV) hydroxide followed by removal of tin carrier for the determination of trace heavy metals by graphite-furnace atomic-absorption spectrometry. *Anal Chim Acta* 1995, *302*, 103-107.

[468] Kim, Y.-S.; Kim, K.-C. Simultaneous preconcentration and determination of trace elements in water samples by coprecipitation-flotation with lanthanum hydroxide [$La(OH)_3$]. *Bull Korean Chem Soc* 1995, *16*, 582-587.

[469] Rahmi, D.; Zhu, Y.; Umemura, T.; Haraguchi, H.; Itoh, A.; Chiba, K. Determination of 56 elements in Lake Baikal water by high-resolution ICP-MS with the aid of a tandem preconcentration method. *Anal Sci* 2008, *24*, 1513-1517.

[470] Nakamura, T.; Oka, H.; Ishii, M.; Sato, J. Direct atomization atomic-absorption-spectrometric determination of beryllium, chromium, iron, cobalt, nickel, copper, cadmium and lead in water with zirconium hydroxide co-precipitation. *Analyst* 1994, *119*, 1397-1401.

[471] Wu, J.; Boyle, E.A. Low blank preconcentration technique for the determination of lead, copper and cadmium in small-volume sea water samples by isotope-dilution ICP MS. *Anal Chem* 1997, *69*, 2464-2470.

[472] Peker, D.S.K.; Turkoglu, O.; Soylak, M. Dysprosium(III) hydroxide coprecipitation system for the separation and preconcentration of heavy metal contents of table salts and natural waters. *J Hazard Mater* 2007, *143*, 555-560.

[473] Kagaya, S.; Miwa, S.; Mizuno, T.; Tohda, K. Rapid coprecipitation technique using yttrium hydroxide for the preconcentration and separation of trace elements in saline water prior to their ICP-AES determination. *Anal Sci* 2007, *23*, 1021-1024.

[474] Soylak, M.; Kars, A.; Narin, I. Coprecipitation of Ni^{2+}, Cd^{2+} and Pb^{2+} for preconcentration in environmental samples prior to flame atomic absorption spectrometric determinations. *J Hazard Mater* 2008, *159*, 435-439.

[475] Umashankar, V.; Radhamani, R.; Ramadoss, K.; Murty, D. S. R. Simultaneous separation and preconcentration of trace elements in water samples by coprecipitation on manganese dioxide using D-glucose as reductant for $KMnO_4$. *Talanta* 2002, *57*, 1029-1038.

[476] Nakashima, S.; Sturgeon, R.E.; Willie, S.N.; Berman, S.S. Determination of trace elements in sea-water by graphite-furnace atomic-absorption spectrometry after pre-concentration by tetrahydroborate reductive precipitation. *Anal Chim Acta* 1988, *207*, 291-299.

[477] Sella, S.; Sturgeon, R.E.; Willie, S.N.; Campos, R.C. Flow injection on-line reductive precipitation preconcentration with magnetic collection for electrothermal atomic absorption spectrometry. *J Anal At Spectrom* 1997, *12*, 1281-1285.

[478] Zhuang, Z.X.; Yang, C.L.; Wang, X.R.; Yang, P.Y.; Huang, B.L. Preconcentration of trace elements from natural water with palladium precipitation. *Fresenius' J Anal Chem* 1996, *355*, 277-280.

[479] Shan, X.; Tie, J.; Xie, G. Determination of trace elements in water, sea-water and biological samples by inductively coupled plasma atomic-emission spectrometry after pre-concentration with ammonium pyrrolidine-1-carbodithioate precipitation. *J Anal At Spectom* 1988, *3*, 259-263.

[480] Vircavs, M.; Pelne, A.; Rone, V.; Vircava, D. Oxidation product of ammonium pyrrolidin-1-yldithioformate as a coprecipitator for the preconcentration of vanadium,

cobalt, zinc, arsenic, iron, cadmium, selenium and mercury from aqueous solution. *Analyst* 1992, *117*, 1013-1017.

[481] Rao, R.R.; Chatt, A. Preconcentration neutron activation analysis for trace elements in seawater [and biological material] by coprecipitation with 1-(2-thiazolylazo)-2-naphthol, pyrrolidinedithiocarbamate, and N-nitrosophenylhydroxylamine. *J Radioanal Nucl Ch* 1993, *168*, 439-448.

[482] Souza, A.S.; dos Santos, W.N.L.; Ferreira, S.L.C. Application of Box–Behnken design in the optimisation of an on-line pre-concentration system using knotted reactor for cadmium determination by flame atomic absorption spectrometry. *Spectrochim Acta B* 2005, *60*, 737 -742.

[483] Delayette-Mills, M.; Karm, L.; Janauer, G.E.; Chan, P.K.; Bernier, W.E. Selective separations by reactive ion exchange. IV. Pre-concentration of cadmium and zinc by in situ precipitation as hexacyanoferrate(II) salts on gel and macro-porous ion-exchange resins. *Anal Chim Acta* 1981, *124*, 365-372.

[484] Moore, R.V. Dibenzylammonium and sodium dibenzyldithiocarbamates as precipitants for pre-concentration of trace elements in water for analysis by energy-dispersive X-ray fluorescence. *Anal Chem* 1982, *54*, 895-897.

[485] Zmijewska, W.; Polkowska-Motrenko, H.; Stokowska, H. Preconcentration of trace elements from water by coprecipitation and ion exchange. *J Radioanal Nucl Ch* 1984, *84*, 319-328.

[486] Akatsuka, K.; Atsuya, I. Direct analysis of solid samples by atomic-absorption spectrometry, following pre-concentration of trace elements from sea-water with quinolin-8-ol. *Fresenius' J Anal Chem* 1987, *329*, 453-456.

[487] Akatsuka, K.; Nobuyema, N.; Atsuya, I. Atomic-absorption spectrometry of nanogram amounts of cadmium, lead and zinc after precipitation with quinolin-8-ol. *Anal Sci* 1988, *4*, 281-285.

[488] Beinrohr, E.; Hofbauerova, H. Preconcentration of trace metals from acidic solutions by coprecipitation with dithizone. *Mikrochim Acta* 1989, *98*, 119-128.

[489] Garrido, I.; Soto, R.M.; Carlosena, A.; López-Mahía, P.; Muniategui, S.; Padra, D. Flame atomic absorption spectrometry with flow-injection on-line adsorption preconcentration using a knotted reactor for cadmium determination in aqueous samples. *Anal Lett* 2001, *34*, 1763-1779.

[490] Krishnamurty, K.V.; Reddy, M.M. Tris(pyrrolidine dithiocarbamato)cobalt(III) chelate matrix for trace metal preconcentration from aqueous solution by coprecipitation. *Anal Chem* 1977, *49*, 222-226.

[491] Zhuang, Z.; Wang, X.; Yang, P.; Yang, C.; Huang, B. Online flow injection cobalt-ammonium pyrrolidin-1-yldithioformate coprecipitation for preconcentration of trace amounts of metals in waters with simultaneous determination by inductively coupled plasma atomic emission spectrometry. *J Anal At Spectrom* 1994, *9*, 779-784.

[492] Laumond, F.; Copin-Montegut, G.; Courav, P. ; Nicolas, G. Cadmium, copper and lead in the western Mediterranean Sea. *Mar Chem* 1984, *15*, 251-261.

[493] Beazley, P.I.; Rao, R.R.; Chatt, A. Preconcentration neutron activation analysis of trace elements in surface waters by coprecipitation with pyrrolidinedithiocarbamate in the presence of Bi(III). *J Radioanal Nucl Ch* 1994, *179*, 267-276.

[494] Min, R.W.; Hansen, E.H. Flow injection - flame atomic absorption spectrometric determination of trace levels of cadmium with online preconcentration by means of coprecipitation. *Chem Anal (Warsaw)* 1995, *40*, 361-371.

[495] Stafilov, T.; Pavlovska, G.; Cundeva, K.; Zendelovska, D.; Paneva, V. Separation, preconcentration, and determination of cadmium in drinking waters. *J Environ Sci Heal A* 2001, *36*, 735-746.

[496] Pavlovska, G.; Stafilov, T.; Cundeva, K. Preconcentration and separation of cadmium by use of cobalt(III) hexamethylenedithiocarbamate as a collector prior to its determination by atomic-absorption spectrometry. *Fresenius'J Anal Chem* 2001, *369*, 670-673.

[497] Chen, Z.S.; Hiraide, M.; Kawaguchi, H. Preconcentration of trace heavy metals in water by coprecipitation with magnesium oxinate for electrothermal atomic absorption spectrometry. *Mikrochim Acta* 1996, *124*, 27-34.

[498] Chen, H.; Jin, J.; Wang, Y. Flow injection on-line coprecipitation-preconcentration system using copper(II) diethyldithiocarbamate as carrier for flame atomic absorption spectrometric determination of cadmium, lead and nickel in environmental samples. *Anal Chim Acta* 1997, *353*, 181-188.

[499] Komjarova, I.; Blust, R. Comparison of liquid-liquid extraction, solid-phase extraction and co-precipitation preconcentration methods for the determination of cadmium, copper, nickel, lead and zinc in seawater. *Anal Chim Acta* 2006, *576*, 221-228.

[500] Elçi, L.; Sahin, U.; Oztas, S. Determination of trace amounts of some metals in samples with high salt content by atomic-absorption spectrometry after cobalt / diethyldithiocarbamate coprecipitation. *Talanta* 2007, *44*, 1017-1023.

[501] Liu, J.; Chen, H.; Mao, X.; Jin, X. Determination of trace copper, lead, cadmium, and iron in environmental and biological samples by flame atomic absorption spectrometry coupled to flow injection on-line coprecipitation preconcentration using DDTC-nickel as coprecipitate carrier. *Int J Environ An Ch* 2000, *76*, 267-282.

[502] Sato, H.; Ueda, J. Electrothermal atomic-absorption-spectrometric determination of cadmium after coprecipitation with nickel diethyldithiocarbamate. *Anal Sci* 2000, *16*, 99-301.

[503] Kim, Y.S.; Kim, K.C.; Lee, C.W. Preconcentration and determination of trace Cd(II) and Pb(II) in a water sample by organic precipitate flotation with 8-hydroxyquinoline. *Bull Korean Chem Soc* 1999, *20*, 431-435.

[504] Cundeva, K.; Stafilov, T.; Pavlovska, G.; Karadjova, I.; Arpadjan, S. Preconcentration procedures for trace cadmium determination in natural aqueous systems prior to Zeeman ETAAS. *Int J Environ An Ch* 2003, *83*, 1009-1019.

[505] Soylak, M.; Erdogan, N.D. Copper(II)-rubeanic acid coprecipitation system for separation-preconcentration of trace metal ions in environmental samples for their flame atomic absorption spectrometric determinations. *J Hazard Mater* 2006, *137*, 1035-1041.

[506] Fang, Z., Sperling, M.; Welz, B. Flame atomic-absorption spectrometric determination of lead in biological samples using a flow-injection system with online pre-concentration by co-precipitation without filtration. *J Anal At Spectrom* 1991, *6*, 301-306.

[507] Matsubara, C.; Izumi, S.; Takamura, K.; Yoshioka, H.; Mori, Y. Determination of trace amounts of phosphate in water after pre-concentration using a thermally reversible polymer. *Analyst* 1993, *118*, 553-556.
[508] Ohyama, T.; Arai, K.; Nakagawa, T.; Matsubara, C.; Takamura, K. Collection of chemical substances in water with use of a thermally reversible polymer. *Bunseki Kagaku* 1997, *46*, 59-62.
[509] Saitoh, T.; Yoshida, Y.; Matsudo, T.; Fujiwara, S.; Dobashi, A.; Iwaki, K.; Suzuki, Y.; Matsubara, C. Concentration of hydrophobic organic compounds by polymer-mediated extraction. *Anal Chem* 1999, *71*, 4506-4512.
[510] Saitoh, T.; Yoshida, Y.; Matsubara, C. Concentration of nonylphenol and its polyethoxylated derivatives by polymer-mediate extraction. *J Chromatogr A* 2000, *891*, 69-74.
[511] Matsubara, C.; Kikuchi, N.; Takamura, K. Preconcentration on poly(methyl vinyl ether) for the determination of polycyclic aromatic hydrocarbons in air. *Bunseki Kagaku* 1995, *44*, 311-312.
[512] Hiraide, M.; Morishima, A. Collection of traces of heavy metals on a thermally reversible polymer for the analysis of river water and sea water. *Anal Sci* 1997, *13*, 829-831.
[513] Saitoh, T.; Satoh, F.; Hiraide, M. Concentration of heavy metal ions in water using thermoresponsive chelating polymer. *Talanta* 2003, *61*, 811-817.
[514] Bakircioğlu, Y.; Şeren, G.; Akman, S. Concentration of cadmium, copper and zinc using water soluble polyacrylic acid polymer. *Spectrochim Acta B* 2000, *55*, 1129-1133.
[515] Lund, W.; Larsen, B.V. The application of electrodeposition techniques to flameless atomic absorption spectrometry: Part I. The determination of cadmium with a tungsten filament. *Anal Chim Acta* 1974, *70*, 299-310.
[516] Lund, W.; Larsen, B.V. The application of electrodeposition techniques to flameless atomic absorption spectrometry: Part II. Determination of cadmium in sea water. *Anal Chim Acta* 1974, *72*, 57-62.
[517] Jensen, F.O.; Dolezal, J.; Langmyhr, F.J. Atomic-absorption spectrometric determination of cadmium, lead and zinc in salts or salt solutions by hanging mercury drop electrodeposition and atomization in a graphite furnace. *Anal Chim Acta* 1974, *72*, 245-250.
[518] Lund, W.; Thomassen, Y.; Doevle, P. Flame-atomic absorption analysis for trace metals after electrochemical preconcentration on a wire filament. *Anal Chim Acta* 1977, *93*, 53-60.
[519] Long, S.E.; Snook, R.D. Electrochemical preconcentration technique for use with inductively coupled plasma atomic-emission spectrometry. Part I. *Analyst* 1983, *108*, 1331-1338.
[520] Veber, M.; Gomišček, S.; Streško, V. Electrothermal atomic absorption spectrometry of elements after electrochemical deposition on graphite electrodes. *Anal Chim Acta* 1987, *193*, 157-167.
[521] Vrana, A.; Komarek, J. Determination of cadmium and copper with ET AAS after electrochemical deposition on a graphite electrode. *Fresenius' J Anal Chem* 1996, *355*, 321-323.

[522] Zhang, G.; Li, J.; Fu, D.; Hao, D.; Xiang, P. Atomic absorption determination of traces of cadmium in urine after electrodeposition onto a tungsten wire. *Talanta* 1993, *40*, 409-413.

[523] Komárek, J.; Holý, J. Determination of heavy metals by electrothermal atomic absorption spectrometry after electrodeposition on a graphite probe. *Spectrochim Acta B*. 1999, *54*, 733-738.

[524] Abdullin, I.F.; Turova, E.N.; Budnikov, G. K. Determination of copper and cadmium by atomic absorption spectrometry with electrochemical and sorption preconcentration. *J Anal Chem* 2000, *55*, 567-569.

[525] Bulska, E.; Wałcerz, M.; Jędral, W.; Hulanicki, A. On-line preconcentration of lead and cadmium for flame atomic absorption spectrometry using a flow-through electrochemical microcell. *Anal Chim Acta* 1997, *357*, 133-140.

[526] Knápek, J.; Komárek, J.; Krásenský, P. Determination of cadmium by electrothermal atomic absorption spectrometry using electrochemical separation in a microcell. *Spectrochim Acta B* 2005, *60*, 393-398.

[527] Riddick, J.A.; Bunger, W.B.; Saikano, T.K. *Organic Solvents. Physical Properties and Methods of Purification, vol. II*; Willey: New York, 1986.

[528] Brooks, R.R.; Presley, B. J.; Kaplan, I. R. APDC-MIBK extraction system for the determination of trace metals in saline waters by atomic-absorption spectrometry. *Talanta* 1967, *14*, 809-816.

[529] Danielsson, L.-G.; Magnusson, B.; Westerlund, S.; Zhang, K. Trace-metal determinations in estuarine waters by electrothermal atomic-absorption spectrometry after extraction of dithiocarbamate complexes into Freon [1,1,2-trichlorotrifluoroethane]. *Anal Chim Acta* 1982, *144*, 183-188.

[530] Statham, P.J. The determination of dissolved manganese and cadmium in sea water at low nmol l^{-1} concentrations by chelation and extraction followed by electrothermal atomic absorption spectrometry. *Anal Chim Acta* 1985, *169*, 149-159.

[531] Chakraborti, D.; Adams, F.; Van Mol, W.; Irgolic, K.J. Determination of trace metals in natural waters at nanogram per litre levels by electrothermal atomic-absorption spectrometry after extraction with sodium diethyldithiocarbamate. *Anal Chim Acta* 1987, *196*, 23-31.

[532] Jung, W.-T.; Kim, S.-K.; Lee, K.-J.; Sohn, D.-H. Multi-element determination of heavy metals in the Han River of South Korea using liquid-liquid extraction-flame atomic absorption spectrometry. *Jpn. J Toxicol Environ Health* 1997, *43*, 243-250.

[533] Jan, T.K.; Young, D.R. Determination of microgram amounts of some transition metals in sea water by methyl isobuthyl ketone-nitric acid successive extraction and flameless atomic absorption spectrometry. *Anal Chem* 1978, *50*, 1250-1253.

[534] Apte, S.C.; Gunn, A.M. Rapid determination of copper, nickel, lead and cadmium in small samples of estuarine and coastal waters by liquid - liquid extraction and electrothermal atomic-absorption spectrometry. *Anal Chim Acta* 1987, *193*, 147-156.

[535] Kinrade, J.D.; Van Loon, J.C. Solvent extraction for use with flame atomic absorption spectrometry. *Anal Chem* 1974, *46*, 1894-1898.

[536] Danielsson, L.G.; Magnusson, B.; Westerlund, S. An improved metal extraction procedure for the determination of trace metals in sea water by atomic absorption spectrometry with electrothermal atomization. *Anal Chim Acta* 1978, *98*, 47-57.

[537] Bone, K.M.; Hibbert, W.D. Solvent extraction with ammonium pyrrolidinedithiocarbamate and 2,6-dimethyl-4-heptanone for the determination of trace metals in effluents and natural waters. *Anal Chim Acta* 1979, *107*, 219-229.

[538] Sturgeon, R.E.; Berman, S.S.; Desaulniers, A.; Russell, D.S. Pre-concentration of trace metals from sea-water for determination by graphite-furnace atomic-absorption spectrometry. *Talanta* 1980, *27*, 85-94.

[539] Magnusson, B.; Westerlund, S. Solvent extraction procedures combined with back-extraction for trace metal determinations by atomic absorption spectrometry. *Anal Chim Acta* 1981, *131*, 63-72.

[540] Lo, M.J.; Yu, J.C.; Hutchison, F.I.; Wai, C.M. Solvent extraction of dithiocarbamates complexes and back-extraction with mercury (II) for determination of trace metals in sea water by atomic absorption spectrometry. *Anal Chem* 1982, *54*, 2536-2539.

[541] Batterham, G.J.; Parry, D.L. Improved dithiocarbamate-oxine solvent-extraction method for the preconcentration of trace metals from sea water using metal-exchange back-extraction. *Mar Chem* 1996, *55*, 381-388.

[542] Batterham, G.J.; Munksgaard, N.C.; Parry, D.L. Determination of trace metals in sea-water by inductively coupled plasma mass spectrometry after off-line dithiocarbamate solvent extraction. *J Anal At Spectrom* 1997, *12*, 1277-1280.

[543] Ndung'u, K.; Franks, R.P.; Bruland, K.W.; Flegal, A.R. Organic complexation and total dissolved trace metal analysis in estuarine waters: Comparison of solvent-extraction graphite furnace atomic absorption spectrometric and chelating resin flow injection inductively coupled plasma-mass spectrometric analysis. *Anal Chim Acta* 2003, *481*, 127-138.

[544] Itoh, Y.; Kawamoto, H.; Akaiwa, H. Pre-concentration of cadmium(II) by synergic extraction with dithizone and tributylphosphine oxide. *Bunseki Kagaku* 1987, *36*, T119-T122.

[545] Facchin, I.; Pasquín, C. Two-phase liquid-liquid extraction in monosegmented flow analysis. Determination of cadmium with 1-(2'-pyridylazo) naphthol. *Anal Chim Acta* 1995, *308*, 231-237.

[546] Suzuki, T.; Kobayashi, H.; Sawada, K. The determination of cadmium in sea-water by solvent-extraction using capriquat and carbon furnace atomic-absorption spectrometry. *Nippon Kagaku Kaishi* 1982, *7*, 1167-1170.

[547] Ueda, K.; Kitahara, S.; Kubo, K.; Yamamoto, Y. Pre-concentration and graphite-furnace AAS determination of ultra-trace elements in natural waters by utilizing synergic extraction. *Bunseki Kagaku* 1987, *36*, 728-734.

[548] Ueda, K.; Kubo, K.; Yoshimura, O.; Yamamoto, Y. Determination of trace metals in sea-water by graphite-furnace atomic-absorption spectrometry after pre-concentration using synergic extraction. *Bull Chem Soc Jpn* 1988, *61*, 2791-2795.

[549] Granzhan, A.V.; Kuchuk, G.M.; Charykov, A.K. Solvent extraction - atomic-absorption spectrometry determination of beryllium and cadmium in some natural samples. *Zh Anal Khim*1990, *45*, 2015-2018.

[550] Nord, L., Karlberg, B. Automated extraction system for flame atomic-absorption spectrometry. *Anal Chim Acta* 1981, *125*, 199-202.

[551] Nord, L.; Karlberg, B. Sample pre-concentration by continuous-flow extraction with a flow-injection atomic-absorption detection system. *Anal Chim Acta* 1983, *145*, 151-158.

[552] Fang, Z. *Flow Injection Atomic Absorption Spectrometry*; Wiley: West Sussex, 1995.
[553] Backstrom, K.; Danielsson, L.-G. Design of a continuous-flow two-step extraction sample work-up system for graphite-furnace atomic-absorption spectrometry. *Anal Chim Acta* 1990, *232*, 301-315.
[554] Anthemidis, A.N.; Zachariadis, G.A.; Farastelis, C.G.; Stratis, J.A. On-line liquid-liquid extraction system using a new phase separator form flame atomic absorption spectrometric determination of ultra-trace cadmium in natural waters. *Talanta* 2004, *62*, 437-443.
[555] Miro, M.; Estela, J. M.; Cerda, V. Recent advances in on-line solvent extraction exploiting flow injection/sequential injection analysis. *Curr Anal Chem* 2005, *1*, 329-343.
[556] Anthemidis, A.N.; Miro, M. Recent developments in flow injection/sequential injection liquid-liquid extraction for atomic spectrometric determination of metals and metalloids. *Appl Spectros Rev* 2009, *44*, 140-167.
[557] Wang, J.; Hansen, E.H. Development of an automated sequential injection on-line solvent extraction-back extraction procedure as demonstrated for the determination of cadmium with detection by electrothermal atomic absorption spectrometry. *Anal Chim Acta* 2002, *456*, 283-292.
[558] Majors, R.E. Liquid extraction techniques for sample preparation. *LC-GC* 1996, *14*, 936-938.
[559] Murray, D.A.J. Rapid micro extraction procedure for analyses of trace amounts of organic compounds in water by gas choromatography and comparisons with macro extraction methods. *J Chromatogr A* 1979, *177*, 135-140.
[560] Zapf, A.; Heyer, R.; Stan, H. Rapid micro liquid-liquid extraction method for trace analysis of organic contaminants in drinking water. *J Chromatogr A* 1995, *694*, 453-461.
[561] Barrio, C.S.; Melgosa, E.R.; Asensio, J.S.; Bernal, J.G. Extraction of pesticides from aqueous samples: A comparative study. *Mikrochim Acta* 1996, *122*, 267-277.
[562] Carasek, E.A low-cost flame atomic absorption spectrometry method for determination of trace metals in aqueous samples. *Talanta* 2000, *51*, 173-178.
[563] Luque-Pérez, E.; Rios, A.; Valcárcel, M. Development of a liquid-liquid (micro) extraction method for online monitoring of lead. *Quim Anal* 1997, *16*, 107-112.
[564] Sachsenberg, S.; Klenke, T.; Krumbein, W.E.; Zeeck, E. Back-extraction procedure for the dithiocarbamate solvent extraction method. Rapid determination of metals in sea-water matrices. *Fresenius' J Anal Chem* 1992, *342*, 163-166.
[565] Carasek, E.; Tonjes, J.W.; Scharf, M. A new method of microvolume back-extraction procedure for enrichment of Pb and Cd and determination by flame atomic absorption spectrometry. *Talanta* 2002, *56*, 185-191.
[566] Carasek, E.; Tonjes, J.W.; Scharf, M. A liquid-liquid microextraction system for Pb and Cd enrichment and determination by flame atomic absorption spectrometry. *Quim Nova* 2002, *25*, 748-752.
[567] Miura, J.; Ishii, H.; Watanabe, H. Extraction and separation of nickel chelate of 1-(2-thiazolylazo)-2-naphthol in nonionic surfactant solution. *Bunseki Kagaku* 1976, *25*, 808-809.
[568] Watanabe, H. *Solution Behavior of Surfactants*; Mittal, K.L., Fendler E.J.; Ed.; Plenum Press: New York, 1982.

[569] Pelizzetti, E.; Pramauro, E. Analytical applications of organized molecular assemblies. *Anal Chim Acta* 1985, *169*, 1-29.
[570] Hinze, W.L.; Pramauro, E. A critical-review of surfactant-mediated phase separations (cloud-point extractions) - theory and applications. *Critical Rev Anal Chem* 1993, *24*, 133-177.
[571] Stalikas, C.D. Micelle-mediated extraction as a tool for separation and preconcentration in metal analysis. *Trends Anal Chem* 2002, *21*, 343-355.
[572] Burguera, J.L.; Burguera, M. Analytical applications of organized assemblies for on-line spectrometric determinations: present and future. *Talanta* 2004, *64*, 1099-1108.
[573] Bezerra, M.D.; Arruda, M.A.Z.; Ferreira, S.L.C. Cloud point extraction as a procedure of separation and pre-concentration for metal determination using spectroanalytical techniques: A review. *Appl Spectros Rev* 2005, *40*, 269-299.
[574] Silva, M.F.; Cerutti, E.S.; Martinez, L.D. Coupling cloud point extraction to instrumental detection systems for metal analysis. *Microchim Acta* 2006, *155*, 349-364.
[575] Carabias-Martínez, R.; Rodríguez-Gonzalo, E.; Moreno-Cordero, B.; Pérez-Pavón, J.L.; García-Pinto, C.; Fernández Laespada, E. Surfactant cloud point extraction and preconcentration of organic compounds prior to chromatography and capillary electrophoresis. *J Chromatogra A* 2000, *902*, 251-265.
[576] Sosa-Ferrera, Z.; Padrón-Sanz, C.; Mahugo-Santana, C.; Santana-Rodríguez, J.J. The use of micellar systems in the extraction and pre-concentration of organic pollutants in environmental samples. *Trends Anal Chem* 2004, *23*, 469-479.
[577] Paleólogos, E.K.; Giokas, D.L.; Karayannis, M.I. Micelle-mediated separation and cloud-point extraction. *Trends Anal Chem* 2005, *24*, 426-436.
[578] Baghdadi, M.; Shemirani, F. Cold-induced aggregation microextraction: A novel sample preparation technique based on ionic liquids. *Anal Chim Acta* 2008, *613*, 56-63.
[579] Attwood, D.; Florence, A.T. *Surfactants Systems-Their Chemistry Pharmacy and Biology*; Chapman and Hall: London, 1985.
[580] Gullickson, N.D.; Scamehom, J.F.; Harwell, J.H. *Surfactant-Based Separation Processes*; Schamehom, J.F., Harwell, J.H.; Ed.; Marcel Dekker: New York, 1989.
[581] García Pinto, C.; Pérez Pavón, J.L.; Moreno Cordero, B.; Romero Beato, E.; García Sánchez, S. Cloud point preconcentration and flame atomic absorption spectrometry: application to the determination of cadmium. *J Anal At Spectrom* 1996, *11*, 37-41.
[582] Zhu, X.; Zhu, X.; Wang, B. Determination of trace cadmium in water samples by graphite furnace atomic absorption spectrometry after cloud point extraction. *Microchim Acta* 2006, *154*, 95-100.
[583] Borkowska-Burnecka, J.; Jakubiel, M.; Zyrnicki, W. Cloud point extraction as sample preparation procedure prior to multielemental analysis by inductively coupled plasma-optical emission spectrometry. *Chem Anal (Warsaw)* 2008, *53*, 335-346.
[584] da Silva, M.A.M.; Frescura, V.L.A.; Curtius, A.J. Determination of trace elements in water samples by ultrasonic nebulization inductively coupled plasma mass spectrometry after cloud point extraction. *Spectrochim Acta B* 2000, *55*, 803-813.
[585] Coelho, L.M.; Arruda, M.A.Z. Preconcentration procedure using cloud point extraction in the presence of electrolyte for cadmium determination by flame atomic absorption spectrometry. *Spetrochim Acta B* 2005, *60*, 743-748.

[586] Chen, J.; Teo, K.C. Determination of cadmium, copper, lead and zinc in water samples by flame atomic absorption spectrometry after cloud point extraction. *Anal Chim Acta* 2001, *450*, 215-222

[587] Bezerra, M.A.; Maêda, S.M.N.; Oliveira, E.P.; Batista de Carvalho, M.F.; Santelli, R.E. Internal standardization for the determination of cadmium, cobalt, chromium and manganese in saline produced water from petroleum industry by inductively coupled plasma optical emission spectrometry after cloud point extraction. *Spectrochim Acta B* 2007, *62*, 985-991.

[588] Manzoori, J.L.; Karim-Nezhad, G. Development of a cloud point extraction and preconcentration method for Cd and Ni prior to flame atomic absorption spectrometric determination. *Anal Chim Acta* 2004, *521*, 173-177.

[589] Yuan, C.G.; Jiang, G.B.; Cai, Y.Q.; He, B.; Liu, J.F. Determination of cadmium at the nanogram per liter level in seawater by graphite furnace AAS using cloud point extraction, *Atom Spectrom* 2004, *25*, 170-176.

[590] Giokas, D.L.; Eksperiandova, L.P.; Blank, A.B.; Karayannis, M.I. Comparison and evaluation of cloud point extraction and low-temperature directed crystallization as preconcentration tools for the determination of trace elements in environmental samples. *Anal Chim Acta* 2004, *505*, 51-58.

[591] Liang, P.; Li, J.; Yang, X. Cloud point extraction preconcentration of trace cadmium as 1-phenyl-3-methyl-4-benzoyl-5-pyrazolone complex and determination by flame atomic absorption spectrometry. *Microchim Acta* 2005, *152*, 47-51.

[592] Shemirani, F.; Abkenar, S.D.; Jamali, M.R. Determination of cadmium(II), copper(II) and zinc(II) in water samples by flame atomic absorption spectrometry after cloud point extraction *Indian J Chem Sect A* 2005, *44*, 1211-1214.

[593] Farajzadeh, M.A.; Fallahi, M.R. Simultaneous cloud-point extraction of nine cations from water samples and their determination by flame atomic absorption spectrometry. *Anal Sci* 2006, *22*, 635-639.

[594] Bezerra, M.A.; Bruns, R.E.; Ferreira, S.L.C. Statistical design-principal component analysis optimization of a multiple response procedure using cloud point extraction and simultaneous determination of metals by ICP OES. *Anal Chim Acta* 2006, *580*, 251-257.

[595] Yamini, Y.; Faraji, M.; Shariati, S.; Hassani, R.; Ghambarian, M. On-line metals preconcentration and simultaneous determination using cloud point extraction and inductively coupled plasma optical emission spectrometry in water samples. *Anal Chim Acta* 2008, *612*, 144-151.

[596] Kilinc, E.; Cetin, A.; Togrul, M.; Hosgoren, H. Synthesis of bis(amino alcohol)oxalamides and their usage for the preconcentration of trace metals by cloud point extraction. *Anal Sci* 2008, *24*, 763-768.

[597] Filik, H.; Dondurmacioglu, F.; Apak, R. Micelle mediated extraction of cadmium from water and tobacco samples with glyoxal-bis(2-hydroxyanil) and determination by electrothermal atomic absorption spectrometry. *Int J Environ Anal Chem* 2008, *88*, 637-648.

[598] Afkhami, A.; Madrakian, T.; Siampour, H. Flame atomic absorption spectrometric determination of trace quantities of cadmium in water samples after cloud point extraction in Triton X-114 without added chelating agents. *J Hazard Mater* 2006, *138*, 269-272.

[599] Manzoori, J.L.; Abdolmohammad-Zadeh, H.; Amjadi, M. Ultratrace determination of cadmium by cold vapor atomic absorption spectrometry after preconcentration with a simplified cloud point extraction methodology. *Talanta* 2007, *71*, 582-587.

[600] Candir, S.; Narin, I.; Soylak, M. Ligandless cloud point extraction of Cr(III), Pb(II), Cu(II), Ni(II), Bi(III), and Cd(II) ions in environmental samples with Tween 80 and flame atomic absorption spectrometric determination. *Talanta* 2008, *77*, 289-293.

[601] Gu, T.; Galera-Gómez, P.A. Clouding of Triton X-114: The effect of added electrolytes on the cloud point of Triton X-114 in the presence of ionic surfactants. *Colloid Surf A* 1995, *104*, 307-312.

[602] Komaromy-Hiller, G.; Calkins, N.; Wandruszka, R. Changes in polarity and aggregation number upon clouding of a nonionic detergent: Effect of ionic surfactant and sodium chloride. *Langmuir* 1996, **12**, 919-920.

[603] Armstrong, J.K.; Chowdhry, B.Z.; Snowden, M.J.; Leharne, S.A. Effect of sodium chloride upon micellization and phase separation transitions in aqueous solutions of triblock copolymers: A high sensitivity differential scanning calorimetry study. *Langmuir* 1998, *14*, 2004-2010.

[604] Fernández-Laespada, M.E.; Pérez-Pavón, J.L.; Moreno-Cordero, B. Micelle-mediated methodology for the pre-concentration of uranium prior to its determination by flow injection. *Analyst* 1993, *118*, 209-212.

[605] Fang, Q.; Du, M.; Huie, C.W. Online incorporation of cloud point extraction to flow-injection analysis. *Anal Chem* 2001, *73*, 3502-3505.

[606] Pena-Pereira, F.; Lavilla, I.; Bendicho, C. Miniaturized preconcentration methods based on liquid–liquid extraction and their application in inorganic ultratrace analysis and speciation: A review. *Spectrochim Acta B* 2009, *64*, 1-15.

[607] Liu, S.; Dasgupta, P.K. Liquid droplet. A renewable gas sampling interface. *Anal Chem* 1995, *67*, 2042-2049.

[608] Liu, H.; Dasgupta, P.K. Analytical chemistry in a drop. Solvent extraction in a microdrop. *Anal Chem* 1996, *68*, 1817-1821.

[609] Jeannot, M.A.; Cantwell, F.F. Solvent microextraction into a single drop. *Anal Chem* 1996, *68*, 2236-2240.

[610] Jeannot, M.A.; Cantwell, F.F. Mass transfer characteristics of solvent extraction into a single drop at the tip of a syringe needle. *Anal Chem* 1997, *69*, 235-239.

[611] Dietz, C.; Sanz, J.; Cámara, C. Recent developments in solid-phase microextraction coatings and related techniques. *J Chromatog A* 2006, *1103*, 183-192.

[612] Chamsaz, M.; Arbab-Zavar, M.H.; Nazari, S. Determination of arsenic by electrothermal atomic absorption spectrometry using headspace liquid phase microextraction after in situ hydride generation. *J Anal At Spectrom* 2003, *18*, 1279-1282.

[613] Xu, L.; Basheer, C.; Lee, H.K. Developments in single-drop microextraction. *J Chromatog A* 2007, *1152*, 184–192.

[614] Psillakis, E.; Kalogerakis, N. Developments in single-drop microextraction. *Trends Anal Chem* 2002, *21*, 53-63

[615] Fan, Z.; Zhou, W. Dithizone-chloroform single drop microextraction system combined with electrothermal atomic absorption spectrometry using Ir as permanent modifier for the determination of Cd in water and biological samples. *Spectrochim Acta B* 2006, *61*, 870-874.

[616] Jiang, H.; Hu, B. Determination of trace Cd and Pb in natural waters by direct single drop microextraction combined with electrothermal atomic absorption spectrometry. *Microchim Acta* 2008, *161*, 101-107.

[617] Nazari, S. Determination of trace amounts of cadmium by modified graphite furnace atomic absorption spectrometry after liquid phase microextraction. *Microchem J* 2008, *90*, 107-112.

[618] Anthemidis Aristidis, N.; Adam Ibrahim, S.I. Development of on-line single-drop micro-extraction sequential injection system for electrothermal atomic absorption spectrometric determination of trace metals. *Anal Chim Acta* 2009, *632*, 216-220.

[619] Cantwell, F.F.; Losier, M. *Liquid–liquid Extraction*; Pawliszyn J.; Ed.; Sampling and Sample Preparation for Field and Laboratory; Elsevier: Amsterdam, 2002.

[620] Xia, L.; Hu, B.; Jiang, Z.; Wu, Y.; Liang, Y. Single-drop microextraction combined with low-temperature electrothermal vaporization ICPMS for the determination of trace Be, Co, Pd, and Cd in biological samples. *Anal Chem* 2004, *76*, 2910-2915.

[621] Rezaee, M.; Assadi, Y.; Milani Hosseini, M.R.; Aghaee, E.; Ahmadi, F.; Berijani, S. Determination of organic compounds in water using dispersive liquid–liquid microextraction. *J Chromatogr A* 2006, *1116*, 1-9.

[622] Schramm, L. L. *Emulsions, Foams, and Suspensions. Fundamentals and Applications*; Willey-VCH: Weinheim, 2005.

[623] Takahiko, B.; Fumio, K.; Susumu, N.; Katsuroku, T. Study of drop coalescence behavior for liquid–liquid extraction operation. *Chem Eng Sci* 2000, *55*, 5385-5391.

[624] Elham, Z.J.; Araz, B.; Yaghoub, A.; Mohammad-Reza, M.H.; Mohammad-Reza, J. Dispersive liquid-liquid microextraction combined with graphite furnace atomic absorption spectrometry: Ultra trace determination of cadmium in water samples. *Anal Chim Acta* 2007, *585*, 305-311.

[625] Moghimi, A. Preconcentration ultra trace of Cd(II) in water samples using dispersive liquid-liquid microextraction with salen(N,N'-bis(salicylidene)-ethylenediamine) and determination graphite furnace atomic absorption spectrometry. *J Chin Chem Soc (Taipei)* 2008, *55*, 369-376.

[626] Pedersen-Bjergaard, S.; Rasmussen, K.E. Liquid-liquid-liquid microextraction for sample preparation of biological fluids prior to capillary electrophoresis. *Anal Chem* 1999, *71*, 2650-2656.

[627] Psillakis, E.; Kalogerakis, N. Developments in liquid-phase microextraction. *Trends Anal Chem* 2003, *22*, 565-574.

[628] Rasmussen, K.E.; Pedersen-Bjergaard, S. Developments in hollow fibre-based, liquid-phase microextraction, *Trends Anal Chem* 2004, *23*, 1-10.

[629] Peng, F.; Liu, R.; Liu, J.F.; He, B.; Hu, X.L.; Jiang, G.B. Ultrasensitive determination of cadmium in seawater by hollow fiber supported liquid membrane extraction coupled with graphite furnace atomic absorption spectrometry. *Spectrochim Acta B* 2007, 62, 499-503.

[630] Danielsson, L.G.; Magnusson, B.; Zhang, K. Matrix interference in determination of trace metals by graphite-furnace AAS after Chelex-100 pre-concentration. *At Spectrosc* 1982, *2*, 39-40.

[631] Baffi, F.; Frache, R.; Dadone, A. Comparison between pre-concentration methods for determination of the chemical forms of cadmium in inshore sea-water by atomic-

absorption spectrophotometry with electrothermal atomization. *Ann Chim* 1984, *74*, 385-397.

[632] Isozaki, A.; Veki, K.; Sasaki, H.; Utsumi, S. Simultaneous determination of cadmium and lead by a graphite furnace-dual channel AAS using direct heating of metal-adsorbed chelating resin. *Bunseki Kagaku* 1987, *36*, 672-677.

[633] Pai, S.C. Pre-concentration efficiency of Chelex-100 resin for heavy metals in seawater: Part 2. Distribution of heavy metals on a Chelex-100 column and optimization of the column efficiency by a plate simulation method. *Anal Chim Acta* 1988, *211*, 271-280.

[634] Pai, S.C.; Fang, T.H.; Chem, C.T.; Jeng, K.L. A low contamination Chelex-100 technique for shipboard pre-concentration of heavy metals in seawater. *Mar. Chem.* 1990, *29*, 295-306.

[635] Pyrzynska, K.; Calatayud, J.M.; Mateo, J.V.G. Preconcentration of cadmium on different sorbents in flow-injection system. *Chem Anal (Warsaw)* 2001, 46, 539-546.

[636] Soylak, M.; Narin, I. On-line preconcentration system for cadmium determination in environmental samples by flame atomic absorption spectrometry. *Chem Anal (Warsaw)* 2005, *50*, 705-715.

[637] Pyrzynska, K.; Jonca, Z. Multielement preconcentration and removal of trace metals by solid-phase extraction. *Anal Lett* 2000, *33*, 1441-1450.

[638] Horvath, Z.; Lasztity, A.; Szakacs; Bozsai, G. Iminodiacetic acid/ethylcellulose as a chelating ion exchanger: Part 1. Determination of trace metals by atomic absorption spectrometry and collection of uranium. *Anal Chim Acta* 1985, *173*, 273-280.

[639] Brajter, K.; Slonawka, K. The efficiency of Cellix-P for the preconcentration of lead and other trace metals from waters. *Anal Chim Acta* 1986, *185*, 271-277.

[640] Pérez-Serradilla, J.A.; Luque de Castro, M.D. Integrated sorption-energy-dispersive X-ray fluorescence detection for automatic determination of lead and cadmium in low-concentration solutions. *Anal Bioanal Chem* 2007, *389*, 1541-1547.

[641] Minamisawa, H.; Okunugi, R.; Minamisawa, M.; Tanaka, S.; Saitoh, K.; Arai, N.; Shibukawa, M. Preconcentration and determination of cadmium by GFAAS after solid phase extraction with synthetic zeolite. *Anal Sci* 2006, *22*, 709-713.

[642] Kenduzler, E. Determination of cadmium(II) in water and soil samples after preconcentration with a new solid phase extractor. *Sep Sci Technol* 2006, *41*, 1645-1659.

[643] Beck, N.G.; Franks, R.P.; Bruland, K.W. Analysis for Cd, Cu, Ni, Zn, and Mn in estuarine water by inductively coupled plasma mass spectrometry coupled with an automated flow injection system. *Anal Chim Acta* 2002, *455*, 11-22.

[644] Wan, C.C.; Chiang, S.; Corsini, A. Two-column method for preconcentration of trace metals in natural waters on acrylate resin. *Anal Chem* 1985, *57*, 719-723.

[645] Subramanian, K.S.; Meranger, J.C.; Wan, C.C.; Corsini, A. Preconcentration of cadmium, chromium, copper and lead in drinking-water on the polyacrylic ester resin, XAD-7. *Int J Environ Anal Chem* 1985, *19*, 261-272.

[646] Soylak, M.; Elci, L. Solid phase extraction of trace metal ions in drinking water samples from Kayseri-Turkey. *J Trace Microprobe Tech* 2000, *18*, 397-403.

[647] Hiraide, M.; Arima, Y.; Mizuike, A. Separation and determination of traces of heavy metals complexed with humic substances in river waters by sorption on indium-treated Amberlite XAD-2 resin. *Anal Chim Acta* 1987, *200*, 171-179.

[648] Tokalioglu, S.; Kartal, S.; Elci, L. Determination of trace metals in waters by FAAS after enrichment as metal-HMDTC complexes using solid phase extraction. *Bull Korean Chem Soc* 2002, *23*, 693-698.

[649] Nechar, M.; Molina, M.F.; Bosque-Sendra, J.M. Application of Doehlert optimization and factorial designs in developing and validating a solid-phase spectrophotometric determination of trace levels of cadmium *Anal Chim Acta* 1999, *382*, 117-130.

[650] Yaguchi, S.; Yanazaki, S.; Yamamoto, H.; Urayana, Y.; Hata, W.; Kasahara, Y.; Goto, K. Acid-soluble membrane filter for the pre-concentration and electrothermal atomisation atomic absorption spectrometric determination of trace levels of cadmium in water. *Analyst* 1988, *113*, 1695-1698.

[651] Leepipatpiboon, V. Trace enrichment by solid-phase extraction for the analysis of heavy metals in water. *J Chromatogr A* 1995, *697*, 137-43.

[652] Wells, M.L.; Bruland, K.W. An improved method for rapid preconcentration and determination of bioactive trace metals in seawater using solid phase extraction and high resolution inductively coupled plasma mass spectrometry. *Mar Chem* 1998, *63*, 145-153.

[653] Abbasse, G.; Ouddane, B.; Fischer, J.C. Determination of total and labile fraction of metals in seawater using solid phase extraction and inductively coupled plasma atomic emission spectrometry (ICP-AES). *J Anal At Spectrom* 2002, *17*, 1354-1358.

[654] Hu, Q.; Yang, G.; Zhao, Y.; Yin, J. Determination of copper, nickel, cobalt, silver, lead, cadmium, and mercury ions in water by solid-phase extraction and RP-HPLC with UV-Vis detection. *Anal Bioanal Chem* 2003, *375*, 831-835.

[655] Ouddane, B.; Abbasse, G.; Halwani, J.; Fischer, J.C. Determination of metal partitioning in porewater extracted from the Seine River Estuary sediment (France). *J Environ Monitor* 2004, *6*, 243-253.

[656] Liu, Z.S.; Huang, S.D. Determination of copper and cadmium in sea water by preconcentration and electrothermal atomic absorption spectrometry. *Anal Chim Acta* 1992, *267,* 31-37.

[657] Otero-Romaní, J.; Moreda-Piñeiro, A.; Bermejo-Barrera, A.; Bermejo-Barrera, P.. Evaluation of commercial C18 cartridges for trace elements solid phase extraction from seawater followed by inductively coupled plasma – optical emission spectrometry determination. *Anal Chim Acta* 2005, *536*, 213-218.

[658] Taher, M.A.; Puri, B.K.; Bansal, R.K. Simultaneous determination of cadmium and lead in real and environmental samples by differential pulse polarography after adsorption of their 2-nitroso-1-naphthol-4-sulfonic acid-tetradecyldimethyl benzy lammoniumion -associated complex on microcrystalline naphthalene. *Microchem J* 1998, *58*, 21-30.

[659] Costa, A.C.S.; Lopes, L.; Korn, M.G.A.; Portela, J.G. Separation and preconcentration of cadmium, copper, lead, nickel and zinc by solid-liquid extraction of their cocrystallized naphthalene dithizone chelate in saline matrices. *J Brazil Chem Soc* 2002, *13*, 674-678.

[660] Cesur, H.; Bati, B. Determination of cadmium by FAAS after solid-phase extraction of its 1-benzylpiperazinedithiocarbamate complex on microcrystalline naphthalene. *Turk J Chem* 2002, *26*, 29-35.

[661] Bhalotra, A.; Atamiyot; Puri, B.K. Simultaneous determination of cadmium and lead in a standard alloy, various biological and environmental samples by differential pulse

polarography after preconcentration by solid phase extraction of their 1-(2-thiazolylazo)-2-naphthol-tetraphenylborate complexes on microcrystalline naphthalene. *Quim Anal* 2002, *20*, 229-236.

[662] Mohammad Ali, T. Differential pulse polarographic determination of cadmium after solid liquid extraction and preconcentration using PAN. *J Chem* 2003, *27*, 529-537.

[663] Saracoglu, S.; Elci, L. Column solid-phase extraction with Chromosorb-102 resin and determination of trace elements in water and sediment samples by flame atomic absorption spectrometry. *Anal Chim Acta* 2002, *452*, 77-83.

[664] Saracoglu, S.; Soylak, M.; Elci, L.; Dogan, M. Determination of Cu, Fe, Ni, Co, Pb, Cd, Mn, and Cr in natural water samples after solid phase extraction on Chromosorb 102. *Anal Lett* 2002, *35*, 2603-2616.

[665] Tokman, N.; Akman, S. Determination of bismuth and cadmium after solid-phase extraction with Chromosorb-107 in a syringe. *Anal Chim Acta* 2004, *519*, 87-91.

[666] Tuzen, M.; Parlar, K.; Soylak, M. Enrichment/separation of cadmium(II) and lead(II) in environmental samples by solid phase extraction. *J. Hazard. Mater.* 2005, *121*, 79-87.

[667] Melek, E.; Tuzen, M.; Soylak, M. Flame atomic absorption spectrometric determination of cadmium(II) and lead(II) after their solid phase extraction as dibenzyldithiocarbamate chelates on Dowex Optipore V-493. *Anal Chim Acta* 2006, *578*, 213-219.

[668] Myasoedova, G.V.; Shcherbinina, N.I.; Svanidze, Z.S.; Varshal, G.M.; Myasoedova, B.F. Atomic-absorption determination of cadmium in mineral waters after its pre-concentration by sorption. *Zh Anal Khim* 1986, *41*, 477-480.

[669] Svanidze, Z.S.; Sedykh, E.M.; Bannykh, L.N.; Myasoedov, B.F. Atomic-absorption determination of cadmium after direct introduction of an organic sorbent into the atomizer. *Zh Anal Khim* 1987, *42*, 1989-1994.

[670] Ishmiyarova, G.R.; Shcherbinina, N.I.; Sedykh, E.M.; Myasoedova, G.V.; Vul'Fron, E.K.; Vernadskii, V.I. Sorption pre-concentration of copper, lead, cobalt, nickel and cadmium from sea-water and their electrothermal atomic-absorption determination in sorbent suspension. *Zh Anal Khim* 1988, *43*, 1981-1986.

[671] Sedykh, E.M.; Myasoedova, G.V.; Ishmiyarova, G.R.; Kasimova, O.G. Direct analysis of sorbent-concentrate in a graphite furnace. *Zh Anal Khim* 1990, *45*, 1895-1903.

[672] Sedykh, E.M.; Yu, G.; Ishmiyarova, G.R.; Ostronova, M.M. Methods for the analysis of sorbent concentrate in graphite furnace AAS. *At Spectrosc* 1994, *15*, 245-249.

[673] Samchuk, A.I.; Kazakevich; Yu. E.; Danilova, E.; Khabazova, T.A.; Emets, L.V.; Kokot, T.K. Atomic-absorption determination of heavy metals in natural waters. *Zh Anal Khim* 1988, *43*, 629-631.

[674] Blain, S.; Appriov, P.; Handel, H. Preconcentration of trace metals from sea water with the chelating resin Chelamine. *Anal Chim Acta* 1993, *272*, 91-97.

[675] Cesur, H.; Macit, M.; Bati, B. Determination of copper, nickel and cadmium by FAAS after preconcentration with zinc-piperazinedithiocarbamate loaded on activated carbon by solid-phase extraction *Anal Lett* 2000, *33*, 1991-2004.

[676] Daorattanachai, P.; Unob, F.; Imyim, A Multi-element preconcentration of heavy metal ions from aqueous solution by APDC impregnated activated carbon. *Talanta* 2005, *67*, 59-64.

[677] Kiran, K.; Janardhanam, K. Determination of trace metals in water samples by flame atomic absorption spectrometry using column solid-phase extraction. *Asian J Chem* 2007, *19*, 3468-3474.

[678] Di Nezio, M.S.; Palomeque, M.E.; Band, B.S.F. Automated flow injection method for cadmium determination with preconcentration and on-line reagent preparation. *Quim Nova* 2005, *28*, 145–148.

[679] Liu, Y.; Guo, Y.; Chang, X.; Meng, S.; Yang, D.; Din, B. Column solid-phase extraction with 2-acetylmercaptophenyldiazoaminoazobenzene (AMPDAA) impregnated Amberlite XAD-4 and determination of trace heavy metals in natural waters by flame atomic absorption spectrometry. *Microchim Acta* 2005, *149*, 95-101.

[680] dos Santos, W.N.L.; dos Santos, C.M.; Ferreira, S.L.C. Field sampling system for determination of cadmium and nickel in fresh water by flame atomic absorption spectrometry. *J Brazil Chem Soc* 2005, *16*, 727-732.

[681] Liu, Y.; Guo, Y.; Meng, S.; Feng, F.; Chang, X. Determination of trace heavy metals in waters by flame atomic absorption spectrometry after preconcentration with 2,4-dinitrophenyldiazoaminoazobenzene on Amberlite XAD-2. *Microchim Acta* 2007, *157*, 209-214.

[682] Nakashima, S.; Sturgeon, R.E.; Willie, S.N.; Bernan, S.S. Determination of trace metals in sea-water by graphite-furnace atomic-absorption spectrometry with pre-concentration on silica-immobilized quinolin-8-ol in a flow system. *Fresenius'Z Anal Chem* 1988, *330*, 592-595.

[683] Karadjova, I. Determination of Cd, Co, Cr, Cu, Fe, Mn, Ni and Pb in natural waters, alkali and alkaline earth salts by electrothermal atomic absorption spectrometry after preconcentration by column solid phase extraction. *Microchim Acta* 1999, *130*, 185-190.

[684] Abou-El-Sherbini, K.S.; Kenawy, I.M.M.; Hamed, M.A.; Issa, R.M.; Elmorsi, R. Separation and preconcentration in a batch mode of Cd(II), Cr(III, VI), Cu(II), Mn(II, VII) and Pb(II) by solid-phase extraction by using of silica modified with N-propylsalicylaldimine. *Talanta* 2002, *58*, 289-300.

[685] Kocjan, R.; Blazewicz, A.; Matosiuk, D. Properties and application of a chelating sorbent prepared by modification of LiChroprep-NH2 with calcon. *Microchim Acta* 2004, *144*, 221-226.

[686] Xie, Z.H.; Xie, F.Z.; Guo, L.Q.; Lin, X.C.; Chen, G.N. Thioacetamide chemically immobilized on silica gel as a solid phase extractant for the extraction and preconcentration of copper(II), lead(II), and cadmium(II). *J Sep Sci* 2005, *28*, 462-470.

[687] Cui, Y.; Chang, X.; Zhu, X.; Zou, X. Selective solid phase extraction of trace cadmium(II) and lead(II) from biological and natural water samples by ofloxacin-modified-silica gel. *Int J Environ Anal Chem* 2008, *88*, 857-868.

[688] Huang, X.; Chang, X.; He, Q.; Cui, Y.; Zhai, Y.; Jiang, N. Tris(2-aminoethyl) amine functionalized silica gel for solid-phase extraction and preconcentration of Cr(III), Cd(II) and Pb(II) from waters. *J Hazard Mat* 2008, *157*, 154-160.

[689] Tewari, P.K.; Singh, A.K. Thiosalicylic acid-immobilized Amberlite XAD-2: metal sorption behaviour and applications in estimation of metal ions by flame atomic absorption spectrometry. *Analyst* 2000, *125*, 2350-2355.

[690] Tuzen, M.; Narin, I.; Soylak, M.; Elci, L. XAD-4/PAN Solid phase extraction system for atomic absorption spectrometric determinations of some trace metals in environmental samples. *Anal Lett* 2004, *37*, 473-489.

[691] Lemos, V.A.; Baliza, P.X. Amberlite XAD-2 functionalized with 2- aminothiophenol as a new sorbent for on-line preconcentration of cadmium and copper. *Talanta* 2005, *67*, 564-570.

[692] Liu, Y.; Chang, X.; Guo, Y.; Ding, B.; Meng, S. Solid phase extraction and preconcentration of trace heavy metal ions in natural water with 2,2'-dithiobisaniline modified Amberlite XAD-2. *Solvent Extr Ion Exc* 2005, *23*, 725-740.

[693] Kasahara, Y.; Willie, S.N.; Sturgeon, R.E.; Bernan, S.S.; Taguchi, S.; Goto, K. Preparation of quinolin-8-ol immobilized adsorbents with minimum contamination for the pre-concentration of trace metals in water. *Bunseki Kagaku* 1993, *42*, 107-110.

[694] Gama, E.M.; Lima, A.S.; Lemos, V.A. Preconcentration system for cadmium and lead determination in environmental samples using polyurethane foam/Me-BTANC. *J Hazard Mater* 2006, *136*, 757-762.

[695] Magidi, V.; Holcombe, J.A. Pre-concentration of cadmium from environmental samples by an alga and analysis by graphite furnace atomic absorption spectrometry. *J Anal At Spectrom* 1989, *4*, 439-442.

[696] Bag, H.; Lale, M.; Türker, A.R. Determination of Cu, Zn and Cd in water by FAAS after preconcentration by baker's yeast (*Saccharomyces cerevisiae*) immobilized on sepiolite. *Fresenius J Anal Chem* 1999, *363*, 224-230.

[697] Baytak, S.; Türker, A.R.; Cevrimli, B.S. Application of silica gel 60 loaded with *Aspergillus niger* as a solid phase extractor for the separation/preconcentration of chromium(III), copper(II), zinc(II), and cadmium(II). *J Sep Sci* 2005, *28*, 2482-2488.

[698] Baytak, S.; Kenduzler, E.; Türker, A.R. Separation/preconcentration of Zn(II), Cu(II), and Cd(II) by *Saccharomyces carlsbergensis* immobilized on silica gel 60 in various samples. *Sep Sci Tech* 2006, *41*, 3449-3465.

[699] Pereira, M.G.; Arruda, M.A.Z. Preconcentration of Cd(II) and Pb(II) using humic substances and flow systems coupled to flame atomic absorption spectrometry. *Microchim Acta* 2004, *146*, 215-222.

[700] Martin-Esteban, A. Molecularly imprinted polymers: new molecular recognition materials for selective solid-phase extraction of organic compounds. *Fresenius J Anal Chem* 2001, *370*, 795-802.

[701] Iijima, S. Helical microtubules of graphitic carbon. *Nature*. 1991, *354*, 56-58.

[702] Iijima, S.; Ichihashi, T. Single-shell carbon nanotubes of 1-nm diameter. *Nature*. 1993, *363*, 603-605.

[703] Liang, P.; Liu, Y.; Guo, L.; Zeng, J.; Lu, H. Multiwalled carbon nanotubes as solid-phase extraction adsorbent for the preconcentration of trace metal ions and their determination by inductively coupled plasma atomic emission spectrometry. *J Anal At Spectrom* 2004, *19*, 1489-1492.

[704] El-Sheikh, A.H.; Sweileh, J.A.; Al-Degs, Y.S. Effect of dimensions of multi-walled carbon nanotubes on its enrichment efficiency of metal ions from environmental waters. *Anal Chim Acta* 2007, *604*, 119-126.

[705] Teixeira Tarley, C.R.; Barbosa, A.F.; Gava Segatelli, M.; Costa Figueiredo, E.; Orival Luccas, P. Highly improved sensitivity of TS-FF-AAS for Cd(II) determination at ng

L^{-1} levels using a simple flow injection minicolumn preconcentration system with multiwall carbon nanotubes. *J Anal At Spectrom* 2006, *21*, 1305-1313.

[706] Henglein, A. Small-particle research: Physicochemicalproperties of extremely small colloidal metal and semiconductor particles. *Chem Rev* 1989, *89*, 1861-1873

[707] Zheng, H.; Chang, X.J.; Lian, N.; Wang, S.; Cui, Y.M.; Zhai, Y.H. A pre-enrichment procedure using diethyldithiocarbamate-modified TiO_2 nanoparticles for the analysis of biological and natural water samples by ICP-AES. *Int J Environ Anal Chem* 2006, *86*, 431-441.

[708] Yin, J.; Jiang, Z.; Chang, G.; Hu, B. Simultaneous on-line preconcentration and determination of trace metals in environmental samples by flow injection combined with inductively coupled plasma mass spectrometry using a nanometer-sized alumina packed micro-column. *Anal Chim Acta* 2005, *540*, 333-339.

[709] Kresage, C.T.; Leonowicz, M.E.; Roth, W.J.; Vartuli, J.C.; Beck, J.S. Ordered mesoporous molecular sieves synthesized by a liquid-crystal template mechanism. *Nature* 1992, *359*, 710-712.

[710] Chen, D.H.; Hu, B.; Huang, C.Z. Chitosan modified ordered mesoporous silica as micro-column packing materials for on-line flow injection-inductively coupled plasma optical emission spectrometry determination of trace heavy metals in environmental water samples. *Talanta* 2009, *78*, 491-497.

[711] Arshady, R. Suspension, emulsion, and dispersion polymerization – A methodological survey. *Colloid Polym Sci* 1992, *270*, 717-732.

[712] Rao, T.P.; Daniel, S.; Gladis, J.M. Tailored materials for preconcentration or separation of metals by ion-imprinted polymers for solid-phase extraction (IIP-SPE). *Trends Anal Chem* 2004, *23*, 28-35.

[713] Rao, T.P.; Kala, R.; Daniel, S. Metal ion-imprinted polymers-Novel materials for selective recognition of inorganics. *Anal Chim Acta* 2006, *578*, 105-116.

[714] Zhai, Y.; Liu, Y.; Chang, X.; Chen, S.; Huang, X. Selective solid-phase extraction of trace cadmium(II) with an ionic imprinted polymer prepared from a dual-ligand monomer. *Anal Chim Acta* 2007, *593*, 123-128.

[715] Liu, Y.; Chang, X.; Wang, S.; Guo, Y.; Din, B.; Meng, S. Solid-phase extraction and preconcentration of cadmium(II) in aqueous solution with Cd(II)-imprinted resin (poly-Cd(II)-DAAB-VP) packed columns. *Anal Chim Acta* 2004, *519*, 173-179.

[716] Lu, Y.-K.; Yan. X.-P. An imprinted organic-inorganic hybrid sorbent for selective separation of cadmium from aqueous solution. *Anal. Chem.* 2004, *76*, 453-457.

[717] Olsen, S.; Pessenda, L.C.R; Růžička, J.; Hansen, E.H. Combination of flow injection analysis with flame atomic-absorption spectrophotometry: Determination of trace amounts of heavy metals in polluted seawater. *Analyst* 1983, 108, 905-917.

[718] Liljegren, G.; Pettersson, J.; Markides, K.E.; Nyholm, L. Electrochemical solid-phase microextraction of anions and cations using polypyrrole coatings and an integrated three-electrode device. *Analyst* 2002, *127*, 591-597.

[719] He, Y.; Tang, L.; Wu, X.; Hou, X.D.; Lee. Y.-I. Spectroscopy: The best way toward green analytical chemistry?. *Appl Spectrosc Rev* 2007, *42*, 119-138.

[720] Anastas, P.T.; Warner, J.C. *Green Chemistry: Theory and Practice;* Oxford University Press: Oxford, 2000.

[721] Koel, M.; Kalijurand, M. Application of the principles of green chemistry in analytical chemistry. *Pure Appl Chem* 2006, *78*, 1993-2002.

[722] Armenta, S.; Garrigues, S.; de la Guardia, M. Green analytical chemistry. *Trends Anal Chem* 2008,. *27*, 497-511.
[723] Moreda-Piñeiro, A.; Barciela-Alonso, M.C.; Domínguez-González, R.; Peña-Vázquez, E.; Herbello-Hermelo, P.; Bermejo-Barrera. P, Alternative solid sample pre-treatment methods in green analytical atomic spectrometry. *Spectrosc Lett* 2009, *42*, 394-417.
[724] Zlotorzynski, A. The application of microwave radiation to analytical and environmental chemistry. *Critical Rev Anal Chem* 1995, *25*, 43-76.
[725] Jin, Q.; Liang, F.; Zhang, H.; Zhao, L.; Huan, Y.; Song, D. Application of microwave techniques in analytical chemistry. *Trends Anal Chem* 1999, *18*, 479-484.
[726] Abu-Samra, A.; Morris, J.S.; Koirtyohann, S.R. Wet ashing of some biological samples in a microwave oven. *Anal Chem* 1975, *47*, 1475-1477.
[727] Agazzi, A.; Pirola, C. Fundamentals, methods and future trends of environmental microwave sample preparation. *Microchem J* 2000, *67*, 337-341.
[728] Srogi, K. A review: Application of microwave techniques for environmental analytical chemistry. *Anal Lett* 2006, *39*, 1261-1288.
[729] Kingston, H.M.; Jassie, L.B. *Introduction to Microwave Sample Preparation. Theory and Practice*; ACS Profesional Referente Book: Washington, 1988.
[730] Arain, M.B.; Kazi, T.G.; Jamali, M.K.; Jalbani, N.; Afridi, H.I.; Sarfraz, R.A.; Shah, A.Q. Determination of toxic elements in muscle tissues of five fish species using ultrasound-assisted pseudodigestion by electrothermal atomic absorption spectrophotometry: optimization study. *Spectrosc Lett* 2007, *40*, 861-878.
[731] Arain, M.B.; Kazi, T.G.; Jamali, M.K.; Afridi, H.I.; Jalbani, N.; Memon, A.R. Ultrasound-assisted pseudodigestion for toxic metals determination in fish muscles followed by electrothermal atomic absorption spectrophotometry: multivariate strategy. *J AOAC Int* 2007, *90*, 1118-1127.
[732] Brunori, C.; Ipolyi, I.; Macaluso, L.; Morabito, R. Evaluation of an ultrasonic digestion procedure for total metal determination in sediment reference materials. *Anal Chim Acta* 2004, *510*, 101-110.
[733] Bellotto, V.R.; Miekeley, N. Trace metals in mussel shells and corresponding soft tissue samples: A validation experiment for the use of *Perna perna* shells in pollution monitoring. *Anal Bioanal Chem* 2007, *389*, 769-776.
[734] Saavedra, Y.; González, A.; Fernández, P.; Blanco, J. A simple optimized microwave digestion method for multielement monitoring in mussel samples. *Spectrochim Acta B* 2004, *59*, 533-541.
[735] LaBrecque, J.J.; Benzo, Z.; Alfonso, J.A.; Cordoves, P.R.; Quintal, M.; Gómez, C.V.; Marcano, E. The concentrations of selected trace elements in clams, *Trivela mactroidea* along the Venezuelan coast in the state of Miranda. *Mar Pollut Bull* 2004, *49*, 664-667
[736] Yang, K.X.; Swami, K. Determination of metals in marine species by microwave digestion and inductively coupled plasma mass spectrometry analysis. *Spectrochim Acta B* 2007, *62*, 1177-1181.
[737] Hamilton, M.A.; Rode, P.W.; Merchant, M.E.; Sneddon, J. Determination and comparison of heavy metals in selected seafood, water, vegetation and sediments by inductively coupled plasma-optical emission spectrometry from an industrialized and pristine waterway in Southwest Louisiana. *Microchem J* 2008, *88*, 52-55.

[738] Cubadda, F.; Raggi, A.; Coni, E. Element fingerprinting of marine organisms by dynamic reaction cell inductively coupled plasma mass spectrometry. *Anal Bioanal Chem* 2006, *384*, 887-896.

[739] Uluozlu, O.D.; Tuzen, M.; Mendil, D.; Soylak, M. Trace metal content in nine species of fish from the Black and Aegean Seas, Turkey. *Food Chem* 2007, *104*, 835-840.

[740]] Meucci, V.; Laschi, S.; Minunni, M.; Pretti, C.; Intorre, L.; Soldani, G.; Mascini, M. An optimized digestion method coupled to electrochemical sensor for the determination of Cd, Cu, Pb and Hg in fish by square wave anodic stripping voltammetry. *Talanta* 2009, *77*, 1143-1148.

[741] Bocca, B.; Conti, M.E.; Pino, A.; Mattei, D.; Forte, G.; Alimonti, A. Simple, fast, and low-contamination microwave-assisted digestion procedures for the determination of chemical elements in biological and environmental matrices by sector field ICP-MS. *Int J Environ Anal Chem* 2007, *87*, 1111-1123.

[742] Sucharova, J.; Suchara, I. Determination of 36 elements in plant reference materials with different Si contents by inductively coupled plasma massspectrometry: Comparison of microwave digestions assisted by three types of digestion mixtures. *Anal Chim Acta* 2006, *576*, 163-176.

[743] Tuncel, S.G.; Yenisoy-Karakas, S.; Dogangun, A. Determination of metal concentrations in lichen samples by inductively coupled plasma atomic emission spectroscopy technique after applying different digestion procedures. *Talanta* 2004, *63*, 273-277.

[744] Pino, A.; Alimonti, A.; Botre, F.; Minoia, C.; Bocca, B.; Conti, M.E. Determination of twenty-five elements in lichens by sector field inductively coupled plasma mass spectrometry and microwave-assisted acid digestion. *Rapid Commun Mass Spectrom* 2007, *21*, 1900-1906.

[745] Engstroem, E.; Stenberg, A.; Baxter, D.C.; Malinovsky, D.; Maekinen, I.; Poenni, S.; Rodushkin, I. Effects of sample preparation and calibration strategy on accuracy and precision in the multi-elemental analysis of soil by sector-field ICP-MS. *J Anal At Spectrom* 2004, *19*, 858-866.

[746] Riebe, G.; Pritzkow, W.; Vogl, J.; Jochen. Determination of heavy metals in soil. A comparison of various digestion methods. *GIT Labor-Fachzeitschrift* 2006, *50*, 688-690.

[747] Marin, B.; Chopin, E.I.B.; Jupinet, B.; Gauthier, D. Comparison of microwave-assisted digestion procedures for total trace element content determination in calcareous soils. *Talanta* 2008, *77*, 282-288.

[748] Rodríguez, L.; Ruíz, E.; Alonso-Azcarate, J.; Rincón, J. Heavy metal distribution and chemical speciation in tailings and soils around a Pb-Zn mine in Spain. *J Environ Manag* 2009, *90*, 1106-1116.

[749] Al-Hefne, J.; Al-Dyel, O.; Chowdhury, D.A.; Al-Ajayan, T. Distribution and ICP-MS determination of heavy elements in the surfacial sand along the red sea coastline of Saudi Arabia. *At Spectros* 2005, *26*, 51-58.

[750] Melaku, S.; Dams, R.; Moens, L. Determination of trace elements in agricultural soil samples by inductively coupled plasma-mass spectrometry: Microwave acid digestion versus aqua regia extraction. *Anal Chim Acta* 2005, *543*, 117-123.

[751] Aneva, Z.; Chepanova, L. Assessment of heavy metals levels in urban dusts and soils by microwave-assisted aqua regia digestion. *J Environ Protec Ecol* 2005, *6*, 293-304.

[752] Landajo, A.; Arana, G.; de Diego, A.; Etxebarria, N.; Zuloaga, O.; Amouroux, D. Analysis of heavy metal distribution in superficial estuarine sediments (estuary of Bilbao, Basque Country) by open focused microwave-assisted extraction and ICP-OES. *Chemosphere* 2004, *56*, 1033-1041.

[753] Tuzen, M.; Sari, H.; Soylak, M. Microwave and wet digestion procedures for atomic absorption spectrometric determination of trace metals contents of sediment samples. *Anal Lett* 2004, *37*, 1925-1936.

[754] Gao, B.; Liu, Y.; Sun, K.; Liang, X.; Peng, P.; Sheng, G.; Fu, J. Precise determination of cadmium and lead isotopic compositions in river sediments. *Anal Chim Acta* 2008, *612*, 114-120.

[755] Bettiol, C.; Stievano, L.; Bertelle, M.; Delfino, F.; Argese, E. Evaluation of microwave-assisted acid extraction procedures for the determination of metal content and potential bioavailability in sediments. *Appl Geochem* 2008, *23*, 1140-1151.

[756] Karanasiou, A.A.; Thomaidis, N.S.; Eleftheriadis, K.; Siskos, P.A. Comparative study of pretreatment methods for the determination of metals in atmospheric aerosol by electrothermal atomic absorption spectrometry. *Talanta* 2005, *65*, 1196-1202.

[757] Pekney N.J.; Davidson, C.I. Determination of trace elements in ambient aerosol samples. *Anal Chim Acta* 2005, *540*, 269-277.

[758] Fernández Álvarez, F.; Ternero Rodríguez, M.; Fernández Espinosa, A.J.; Gutierrez Daban, A. Physical speciation of arsenic, mercury, lead, cadmium and nickel in inhalable atmospheric particles. *Anal Chim Acta* 2004, *524*, 33-40.

[759] Kulkarni, P.; Chellam, S.; Flanagan, J.B.; Jayanty, R.K.M. Microwave digestion-ICP-MS for elemental analysis in ambient airborne fine particulate matter: Rare earth elements and validation using a filter borne fine particle certified reference material. *Anal Chim Acta* 2007, *599*, 170-176.

[760] Karthikeyan, S.; Joshi, U.M.; Balasubramanian, R. Microwave assisted sample preparation for determining water-soluble fraction of trace elements in urban airborne particulate matter: Evaluation of bioavailability. *Anal Chim Acta* 2006, *576*, 23-30.

[761] Turkoglu, O.; Saracoglu, S.; Soylak, M.; Elci, L. Monitoring copper, nickel, cobalt, lead, cadmium, manganese and chromium levels in house dust samples from Kayseri, Turkey. *Trace Elem Electrol* 2004, *21*, 4-9.

[762] Nakatsuka, S.; Okamura, K.; Norisuye, K.; Sohrin, Y. Simultaneous determination of suspended particulate trace metals (Co, Ni, Cu, Zn, Cd and Pb) in seawater with small volume filtration assisted by microwave digestion and flow injection inductively coupled plasma mass spectrometer. *Anal Chim Acta* 2007, *594*, 52-60.

[763] Bermejo-Barrera, P.; Moreda-Piñeiro, A.; Bermejo-Barrera, A. Sample pre-treatment methods for the trace elements determination in seafood products by atomic absorption spectrometry. *Talanta* 2002, *57*, 969-984.

[764] Lorentzen, E.M.L.; Kingston, H.M. Comparison of microwave-assisted and conventional leaching using EPA method 3050B. *Anal Chem* 1996, *68*, 4316-4320

[765] Puchyr, R.F.; Shapiro, R. Determination of trace elements in foods by HCl-HNO_3 leaching and flame atomic absorption spectroscopy. *J Assoc Anal Chem* 1986, *69*, 868-870.

[766] Santos Júnior, D.; Krug, F.J.; de Godoy Pereira, M.; Korn.,M. Currents on ultrasound-assisted extraction for sample preparation and spectroscopic analytes determination. *Appl Spectrosc Rev* 2006, *41*, 305-321.

[767] Chmilenko, F.A.; Baklanov, A.N. Atomic absorption determination of toxic and biologically active trace elements in dairy products using ultrasound for acceleration of mineralization. *Zh Anal Khim* 1992, *47*, 1322-1327.

[768] Nakamura, Y.; Etoh, T.; Noto, Y.; Murai, Y. Determination of wear metals in lubricating oils by inductively coupled plasma emission spectrometry. *Bunseki Kagaku* 1985, *34*, T85-T88.

[769] Sánchez, J.; García, R.; Millán, E. Ultrasonic bath digestion procedures for analysis for heavy metals in several reference materials. *Analusis* 1994, *22*, 222-225.

[770] Sánchez, J.; Millán, M.E. Use of ultrasound in digestion of biological reference materials for determination of heavy metals. *Quím Anal* 1992, *11*, 3-10.

[771] Mason, T.J. *Sonochemistry*; Oxford University Press: Oxford, 1999.

[772] Luque-García, J.L.; Luque de Castro, M.D. Ultrasound: A powerful tool for leaching. *Trends Anal Chem* 2003, *22*, 41-46.

[773] Mason, T.J.; Lorimer, J.P. *Applied sonochemistry: Uses of Power Ultrasounds in Chemistry and Processing*; Wiley-VCH: Weinheim, 2002.

[774] Santos, H.M.; Capelo, J.L. Trends in ultrasonic-based equipment for analytical sample treatment. *Talanta* 2007, *73*, 795-802.

[775] Capelo, J.L.; Maduro, C.; Viena, C. Discussion of parameters associated with the ultrasonic solid-liquid extraction for elemental analysis (total content) by electrothermal atomic absorption spectrometry. An overview. *Ultrasounds Sonochem* 2005, *12*, 225-232.

[776] Kumina, D.M.; Gribovskaya, I.F.; Karyakin, A.V. Method for extraction of elements from plants into solution using ultrasound. *Zh Anal Khim* 1985, *40*, 1184-1187.

[777] Domínguez-González, R.; Moreda-Piñeiro, A.; Bermejo-Barrera, A.; Bermejo-Barrera, P. Application of ultrasound-assisted acid leaching procedure for major and trace elements determination in edible seaweed by inductively coupled plasma-optical emission spectrometry. *Talanta* 2005, *66*, 937-942.

[778] Brisbin, J.A.; Caruso, J.A. Comparison of extraction procedures for the determination of arsenic and other elements in lobster tissue by inductively coupled plasma mass spectrometry. *Analyst* 2002, *127*, 921-929.

[779] Borkowska-Burnecka, J.; Wisz, J.; Zyrnicki, W. Applicability of ultrasonic leaching by diluted acids for determination of total metal contents in plant materials. *Chem Anal* 2003, *48*, 115-126.

[780] Minami, H.; Honjyo, T.; Atsuya, I. A new solid-liquid extraction sampling technique for direct determination of trace elements in biological materials by graphite furnace atomic absorption spectrometry. *Spectrochim Acta B* 1996, *51*, 211-220.

[781] Borkowska-Burnecka, J.; Jankowiak, U.; Zyrnicki, W.; Wilk, K.A. Effect of surfactant addition on ultrasonic leaching of trace elements from plant samples in inductively coupled plasma-atomic emission spectrometry. *Spectrochim Acta B* 2004, *59*, 585-590.

[782] Filgueiras, A.V.; Lavilla, I.; Bendicho, C. Ultrasound-assisted solubilization of trace and minor metals from plant tissue using ethylenediaminetetraacetic acid in alkaline medium. *Fresenius' J Anal Chem* 2001, *369*, 451-456.

[783] García-Rey, R.M.; Quiles-Zafra, R.; Luque de Castro, M.D. New methods for acceleration of metal sample preparation prior to determination of the metal content by atomic absorption spectrometry. *Anal Bioanal Chem* 2003, *377*, 316-321.

[784] Maduro, C.; Vale, G.; Alves, S.; Galesio, M.; da Silva, M.D.R.G.; Fernández, C.; Catarino, S.; Rivas, M.G.; Mota, A.M.; Capelo, J.L. Determination of Cd and Pb in biological reference materials by electrothermal atomic absorption spectrometry: a comparison of three ultrasonic-based sample treatment procedures. *Talanta* 2006, *68*, 1156-1161.

[785] Lavilla, I.; Capelo, J.L.; Bendicho, C. Determination of cadmium and lead in mussels by electrothermal atomic absorption spectrometry using an ultrasound-assisted extraction method optimized by factorial design. *Fresenius' J Anal Chem* 1999, *363*, 283-288.

[786] El Azouzi, H.; Cervera, M.L.; de la Guardia, M. Multi-elemental analysis of mussel samples by atomic absorption spectrometry after room temperature sonication. *J Anal At Spectrom* 1998, *13*, 533-538.

[787] Moreda-Piñeiro, A.; Bermejo-Barrera, P.; Bermejo-Barrera, A. Chemometric investigation of systematic error in the analysis of biological materials by flame and electrothermal atomic absorption spectrometry. *Anal Chim Acta* 2006, *560*, 143-152.

[788] Manutsewee, N.; Aeungmaitrepirom, W.; Varanusupakul, P.; Imyim, A. Determination of Cd, Cu and Zn in fish and mussel by AAS after ultrasound-assisted acid leaching extraction. *Food Chem* 2007, *101*, 817-824.

[789] Väisänen, A.; Suontamo, R.J. Comparison of ultrasound-assisted extraction, microwave-assisted acid leaching and reflux for the determination of arsenic, cadmium and copper in contaminated soil samples by electrothermal atomic absorption spectrometry. *J Anal At Spectrom* 2002, *17*, 739-742.

[790] Shiowatana, J.; Tantidanai, N.; Nookabkaew, S.; Nacapricha, D. A novel continuous-flow sequential extraction procedure for metal speciation in solids. *J Environ Quality* 2001, *30*, 1195-1205.

[791] Moreno-Cid, A.; Yebra, M.C. Flow Injection determination of copper in mussels by flame atomic absorption spectrometry after on-line continuous ultrasound-assisted extraction. *Spectrochim Acta B* 2002, *57*, 967-974.

[792] Priego-Capote, F.; Luque de Castro, M.D. Dynamic ultrasound-assisted leaching of essential macro and micronutrient metal elements from animal feeds prior to flame atomic absorption spectrometry. *Anal Bioanal Chem* 2004, *378*, 1376-1381.

[793] Yebra-Biurrun, M.C.; Cancela-Pérez, S.; Moreno-Cid-Barinaga A. Coupling continuous ultrasound-assisted extraction, preconcentration and flame atomic absorption spectrometric detection for the determination of cadmium and lead in mussel samples. *Anal Chim Acta* 2005, *533*, 51-56.

[794] Cespón-Romero, R.M.; Yebra-Biurrun, M.C. Application of factorial designs for optimisation of on-line determination of cadmium, lead and nickel in welding fumes by atomic absorption spectrometry. *Int J Environ Anal Chem* 2008, *88*, 539-547.

[795] Seco-Gesto, E.V.; Moreda-Piñeiro, A.; Bermejo-Barrera, A.; Bermejo-Barrera, P. Multi-element determination in raft mussels by fast microwave-assisted acid leaching and inductively coupled plasma-optical emission spectrometry. *Talanta* 2007, *72*, 1178-1185.

[796] Borkowska-Burnecka, J. Microwave assisted extraction for trace element analysis of plant materials by ICP-AES. *Fresenius' J Anal Chem* 2000, *358*, 633-637.

[797] Herrera, M.C.; Luque de Castro, M.D. Dynamic approach based on iiterative change of the flow direction for microwave-assisted leaching of cadmium and lead from plant prior to GF-AAS. *J Anal At Spectrom* 2002, *17*, 1530-1533.

[798] Fisher, J.A.; Scarlett, M.J.; Stott, A.D. Accelerated solvent extraction: An evaluation for screening of soils for selected U.S. EPA semivolatile organic priority pollutants. *Environ Sci Technol* 1997, *31*, 1120-1127.

[799] Giergielewicz-Mozajska, H.; Dabrowski, L.; Namiesnik, J. Accelerated solvent extraction (ASE) in the analysis of environmental solid samples - Some aspects of theory and practice. *Critical Rev Anal Chem* 2001, *31*, 149-165.

[800] Alonso-Rodríguez, E.; Moreda-Piñeiro, J.; López-Mahía, P.; Prada-Rodríguez, D.; Fernández-Fernández, E.; Muniategui-Lorenzo, S.; Moreda-Piñeiro, A.; Bermejo-Barrera, A.; Bermejo-Barrera, P. Pressurized liquid extraction of organometals and its feasibility for total metal extraction. *Trends Anal Chem* 2006, *25*, 511-519.

[801] Kronholm, J.; Hartonen, K.; Riekkola, M.-L. Analytical extractions with water at elevated temperatures and pressures. *Trends Anal Chem* 2007, *26*, 396-412.

[802] Rico Varadé, C.M.; Luque de Castro, M.D. Determination of selenium in solid samples by continuous subcritical water extraction, flow injection derivatisation and atomic fluorescence detection. *J Anal At Spectrom* 1998, *13*, 787- 791.

[803] Fernández-Pérez, V.; Jiménez-Carmona, M.M.; Luque de Castro, M.D. Continuous extraction with acidified subcritical water of arsenic, selenium and mercury from coal prior to on-line derivatisation-atomic fluorescence detection. *J Anal At Spectrom* 1999, *14*, 1761-1765.

[804] Jiménez-Carmona, M.M.; Fernández-Pérez, V.; Gualda-Bueno, M.J.; Cabanás-Espejo, J.M.; Luque de Castro, M.D. Acidified subcritical water extraction of major ash-forming elements from coal. *Anal Chim Acta* 1999, *395*, 113-118.

[805] Salvador, F.; Merchán, M.D. USA Patent 5 400 642. 1995.

[806] Morales-Muñoz, S.; Luque-García, J.L.; Luque de Castro, M.D. Acidified pressurized hot water for the continuous extraction of cadmium and lead from plant materials prior to ETAAS. *Spectrochim Acta B* 2003, *58*, 159-165.

[807] Fernández-Pérez, V.; Jiménez-Carmona, M.M.; Luque de Castro, M.D. Continuous liquid–liquid extraction using modified subcritical water for the demetalisation of used industrial oils. *Anal Chim Acta* 2001, *433*, 47-52.

[808] Morales-Riffo, J.J.; Richter, P. Rapid determination of inorganic elements in airborne particulate matter by using acidified subcritical-water extraction and inductively-coupled plasma–optical-emission spectrometry. *Anal Bioanal Chem* 2004, *380*, 129-134.

[809] Tavakoli, O.; Yoshida, H. Effective recovery of harmful metal ions from squid wastes using subcritical and supercritical water treatments. *Environ Sci Technol* 2005, *39*, 2357-2363.

[810] Heltai, G.; Percsich, K.; Halász, G.; Jung, K.; Fekete, I. Estimation of ecotoxicological potential of contaminated sediments based on a sequential extraction procedure with supercritical CO_2 and subcritical H_2O solvents. *Microchem J* 2005, *79*, 231-237.

[811] Wanekaya, A.K.; Myung, S.; Sadik, O.A. Pressure assisted chelating extraction: a novel technique for digesting metals in solid matrices. *Analyst* 2002, *127*, 1272-1276.

[812] Alonso-Rodríguez, E.; Moreda-Piñeiro, J.; López-Mahía, P.; Muniategui-Lorenzo, S.; Prada-Rodríguez, D.; Moreda-Piñeiro, A.; Bermejo-Barrera, P. Use of chelating

solvent-based pressurized liquid extraction combined with inductively coupled plasma-optical emission spectrometry for trace element determination in atmospheric particulate matter. *J Anal At Spectrom* 2007, *22*, 1089-1096.

[813] Maurí-Aucejo, A.R.; Arnandis-Chover, T.; Marín-Sáez, R.; Llobat-Estellés, M. Application of pressurized fluid extraction to determine cadmium and zinc in plants. *Anal Chim Acta* 2007, *581*, 78-82.

[814] Moreda-Piñeiro, J.; Alonso-Rodríguez, E.; López-Mahía, P.; Muniategui-Lorenzo, S.; Fernández-Fernández, E.; Prada-Rodríguez, D.; Moreda-Piñeiro, A.; Bermejo-Barrera, A.; Bermejo-Barrera, P.. Pressurized liquid extraction as a novel sample pre-treatment for trace element leaching from biological material. *Anal Chim Acta* 2006, *572*, 172-179.

[815] Moreda-Piñeiro, J.; Alonso-Rodríguez, E.; López-Mahía, P.; Muniategui-Lorenzo, S.; Prada-Rodríguez, D.; Moreda-Piñeiro, A.; Bermejo-Barrera, A.; Bermejo-Barrera, P. As, Cd, Cr, Ni and Pb pressurized liquid extraction with acetic acid from marine sediment and soil samples. *Spectrochim Acta B* 2006, *61*, 1304-1309.

[816] Moreda-Piñeiro, J.; Alonso-Rodríguez, E.; López-Mahía, P.; Muniategui-Lorenzo, S.; Prada-Rodríguez, D.; Moreda-Piñeiro, A.; Bermejo-Barrera, P. Development of a new sample pre-treatment procedure based on pressurized liquid extraction for the determination of metals in edible seaweed. *Anal Chim Acta* 2007, *598*, 95-102.

[817] Peña-Farfal, C.; Moreda-Piñeiro, A.; Bermejo-Barrera, A.; Bermejo-Barrera, P.; Pinochet-Cancino, H.; de Gregori-Henríquez, I. Use of enzymatic hydrolysis for the multi-element determination in mussel soft tissue by inductively coupled plasma-atomic absorption spectrometry. *Talanta* 2004, *64*, 671-681.

[818] Bermejo, P.; Capelo, J.L.; Mota, A.; Madrid, Y.; Cámara, C. Enzymatic digestion and ultrasonication: A powerful combination in analytical chemistry. *Trends Anal Chem* 2004, *23*, 654-663.

[819] Carpenter, R.C. The Determination of Cadmium, copper, lead and thallium in human liver and kidney tissue by flame atomic absorption spectrometry after enzymatic digestion. *Anal Chim Acta* 1981, *125*, 209-213.

[820] Vale, G.; Rial-Otero, R.; Mota, A.; Fonseca, L.; Capelo, J.L. Ultrasonic-assisted enzymatic digestion (USAED) for total elemental determination and elemental speciation: A tutorial. *Talanta* 2008, *75*, 872-884.

[821] Peña-Farfal, C.; Moreda-Piñeiro, A.; Bermejo-Barrera, A.; Bermejo-Barrera, P.; Pinochet-Cancino, H.; de Gregori-Henríquez, I. Speeding up enzymatic hydrolysis procedures for the multi-element determination in edible seaweed. *Anal Chim Acta* 2005, *548*, 183-191.

[822] Moreda-Piñeiro, A.; Bermejo-Barrera, A.; Bermejo-Barrera, P.; Moreda-Piñeiro, J.; Alonso-Rodríguez, E.; Muniategui-Lorenzo, S.; López-Mahía, P.; Prada-Rodríguez, D. Feasibility of pressurization to speed up enzymatic hydrolysis of biological materials for multielement determinations. *Anal Chem* 2007, *79*, 1797-1805.

[823] Peachey, E.; McCarthy, N.; Goenaga-Infante, H. Acceleration of enzymatic hydrolysis of protein-bound selenium by focused microwave energy. *J Anal At Spectrom* 2008, *23*, 487-492.

[824] Peña-Farfal, C.; Moreda-Piñeiro, A.; Bermejo-Barrera, A.; Bermejo-Barrera, P.; Pinochet-Cancino, H.; de Gregori-Henríquez, I. Ultrasound bath-assisted enzymatic hydrolysis procedures as sample pretreatment for the multielement determination in

mussels by inductively coupled plasma atomic absorption spectrometry. *Anal Chem* 2004, *76*, 3541-3547.

[825] Kurfüst, V. *Solid Sample Analysis*; Springer: Berlin, 1998.

[826] Bendicho, C.; de Loos-Vollebregt, M.T.C. Solid sampling in electrothermal atomic absorption spectrometry using commercial atomizers. A review. *J Anal At Spectrom* 1991, *8*, 353-374.

[827] Cal-Prieto, M.J.; Felipe-Sotelo, M.; Carlosena, A.; Andrade, J.M.; López-Mahía, P.; Muniategui, S.; Prada, D. Slurry sampling for direct analysis of solid materials by electrothermal atomic absorption spectrometry (ETAAS). A literature review from 1990 to 2000. *Talanta* 2002 *56*, 1-51.

[828] Vale, M.G.R.; Oleszczuk, N.; dos Santos, W.N.L. Current status of direct solid sampling for electrothermal atomic absorption spectrometry. A critical review of the development between 1995 and 2005. *Appl Spectrosc Rev* 2006, *41*, 377-400.

[829] Carnrick, G.R.; Daley, G.; Fotinopoulos, A. Design and use of a new automated ultrasonic slurry sampler for graphite-furnace atomic absorption. *At Spectrosc* 1989, *10*, 170-174.

[830] Majidi, V.; Holcombe, J.A. Error analysis for sampling of slurries: Sedimentation errors. *Spectrochim Acta B* 1990, *45*, 753-761.

[831] Miller-Ihli, N.J. Ultrasonic slurry sampling graphite furnace atomic absorption spectrometry: Analytical considerations for analysis of high density glass material. *Spectrochim Acta B* 1997, *52*, 431-436.

[832] Miller-Ihli, N.J. influence of slurry preparation on the accuracy of ultrasonic slurry electrothermal atomic absorption spectrometry. *J Anal At Spectrom* 1994, *9*, 1129-1134.

[833] Almeida-Pereira, L.; Amorim, I.; Borba da Silva, J.B. Determination of cadmium, chromium and lead in marine sediment slurry samples by electrothermal atomic absorption spectrometry using permanent modifiers. *Talanta* 2006, *68*, 771-775.

[834] Ebdon, L.; Foulkes, M.; Sutton, K. Slurry nebulizations in plasmas. *J Anal At Spectrom* 1997, *12*, 213- 229.

[835] Santos, M.C.; Nóbrega, J.A. Slurry nebulization in plasmas for analysis of inorganic materials. *Appl Spectros Rev* 2006, *41*, 427-448.

[836] Matusiewicz, H. Chemical vapour generation with slurry sampling: A review of atomic absorption applications. *Appl Spectros Rev* 2003, *3*, 263-294.

[837] Moreda-Piñeiro, J.; Moscoso-Pérez, C.; López-Mahía, P.; Muniategui-Lorenzo, S.; Fernández-Fernández, E.; Prada-Rodríguez, D. Use of aqueous slurries of coal fly ash samples for the direct determination of As, Sb and Se by hydride generation-atomic fluorescence spectrometry. *At Spectros* 2006, *27*, 19-25.

[838] Moreda-Piñeiro, J.; Moscoso-Pérez, C., Piñeiro-Iglesias, M.; López-Mahía, P.; Muniategui-Lorenzo, S.; Fernández-Fernández, E., Prada-Rodríguez, D. Aqueous and acidified slurry sampling approaches in the As, Sb and Sn determination of urban dust samples by HG-ETAAS. *At Spectros* 2007, *28*, 137-143.

[839] Tepleton, D.M.; Ariese, F.; Cornelis, R.; Danielsson, L.G.; Muntau, H.; VanLeeuwen, H.P.; Łobinski, R. Guidelines for terms related to chemical speciation and fractionation of elements. Definitions, structural aspects, and methodological approaches (IUPAC Recommendations 2000). *Pure Appl Chem* 2000, *72*, 1453-1470.

[840] Kozelka, P.B.; Bruland, K.W. Chemical speciation of dissolved Cu, Zn, Cd, Pb in Narragansett Bay, Rhode Island. *Mar Chem* 1998, *60*, 267-282.

[841] Xue, H.; Sigg L. Cadmium speciation and complexation by natural organic ligands in fresh water. *Anal Chim Acta* 1998, *363*, 249-259.

[842] Pongratz, R.; Heumann, K.G. Determination of monomethylcadmium in the environment by differential pulse anodic stripping voltammetry. *Anal Chem* 1996, *68*, 1262-1266.

[843] Pongratz, R.; Heumann, K.G. Production of methylated mercury, lead, and cadmium by marine bacteria as a significant natural source for atmospheric heavy metals in polar regions. *Chemosphere* 1999, *39*, 89-102.

[844] Quevauviller, P. Operationally defined extraction procedures for soil and sediment analysis - I. Standardization. *Trends Anal Chem* 1998, *17*, 289-198.

[845] López-Sánchez, J.F.; Sahuquillo, A.; Fiedler, H.D.; Rubio, R.; Rauret, G.; Muntau, H.; Quevauviller, P. CRM 601, a stable material for its extractable content of heavy metals. *Analyst* 1998, *123*, 1675-1677.

[846] Ure, A.M.; Quevauviller, P.; Muntau, H.; Griepink, B. Speciation of heavy-metals in soils and sediments. An account of the improvement and harmonization of extraction techniques undertaken under the auspices of the BCR of the Commision of the European Communities. *Int J Environ Anal Chem* 1993, *51*, 135-151.

[847] Tessier, A.; Campbell, P.G.C.; Visón, M. Sequential extraction procedure for the speciation of particulate trace-metals. *Anal Chem*, 1979, *51*, 844-851.

[848] Filgueiras, A.V.; Lavilla, I.; Bendicho, C. Chemical sequential extraction for metal partitioning in environmental solid samples. *J Environ Monit*, 2002, *4*, 823-857.

[849] Rao, C.R.M.; Sahuquillo, A.; López-Sánchez, J.F. A review of the different methods applied in environmental geochemistry for single and sequential extraction of trace elements in soils and related materials. *Water Air Soil Pollut* 2008, *189*, 291–333.

[850] Davidson, C.M.; Delevoye, G. Effect of ultrasonic agitation on the release of copper, iron, manganese and zinc from soil and sediment using the BCR three-stage sequential extraction. *J Environ Monit* 2001, *3*, 398-403.

[851] Rauret, G.; López Sánchez, J. F.; Sahuquillo, A.; Barahona, E.; Lachica, M.; Ure, A.M.; Davidson, C.M.; Gómez, A.; Luck, D.; Bacon, J.; Yli-Halla, M.; Muntau, H.; Quevauviller, P. Application of a modified BCR sequential extraction (three-step) procedure for the determination of extractable trace metal contents in a sewage sludge amended soil reference material (CRM 483), complemented by a three-year stability study of acetic acid and EDTA extractable metal content. *J Environ Monitor* 2000, *2*, 228–233.

[852] Ho, M.D.; Evans, G.J. Sequential extraction of metal contaminated soils with radiochemical assessment of readsorption effects. *Environ Sci Technol* 2000, *34*, 1030-1035.

[853] Szakova, J.; Tlustos, P.; Balik, J.; Pavlikova, D.; Vanek, V. The sequential analytical procedure as a tool for evaluation of As, Cd, and Zn mobility in soil. *Fresenius' J Anal Chem* 1999, *363*, 594–595.

[854] Davidson, C.M.; Duncan, A.L.; Littlejohn, D.; Ure, A.M.; Garden, L.M. A critical evaluation of the three-stage BCR sequential extraction procedure to assess the potential mobility and toxicity of heavy metals in industrially-contaminated land. *Anal Chim Acta* 1998, *363*, 45-55.

[855] Ho, M.D.; Evans, G.J. Operational speciation of cadmium, copper, lead and zinc in the NIST standard reference materials 2710 and 2711 (Montana soil) by the BCR

sequential extraction procedure and flame atomic absorption spectrometry. *Anal Commun* 1997, *34*, 363-364.

[856] Tokalioglu, S.; Kartal, S.; Gueltekin, A. Investigation of heavy-metal uptake by vegetables growing in contaminated soils using the modified BCR sequential extraction method. *Int J Environ Anal Chem* 2006, *86*, 417–430.

[857] Bacon, J.R.; Hewitt, I.J.; Cooper, P. Reproducibility of the BCR sequential extraction procedure in a long-term study of the association of heavy metals with soil components in an upland catchment in Scotland. *Sci Total Environ* 2005, *337*, 191–205.

[858] Fernández, E.; Jiménez, R.; Lallena, A.M.; Aguilar, J. Evaluation of the BCR sequential extraction procedure applied for two unpolluted Spanish soils. *Environ Pollut* 2004, *131*, 355–364.

[859] Pueyo, M.; Sastre, J.; Hernández, E.; Vidal, M.; López-Sánchez, J.F.; Rauret, G. Prediction of trace element mobility in contaminated soils by sequential extraction. *J Environ Quality* 2003, *32*, 2054–2066.

[860] Larner, B.L.; Seen, A.J.; Townsend, A.T. Comparative study of optimised BCR sequential extraction scheme and acid leaching of elements in the certified reference material NIST 2711. *Anal Chim Acta* 2006, *556*, 444-449.

[861] Sahuquillo, A.; López-Sánchez, J.F.; Rubio, R.; Rauret, G.; Thomas, R.P.; Davidson, C. M.; Ure, A. M. Use of a certified reference material for extractable trace metals to assess sources of uncertainty in the BCR three-stage sequential extraction procedure. *Anal Chim Acta* 1999, *382*, 317-327.

[862] Rauret, G.; López-Sánchez, J.F.; Sahuquillo, A.; Rubio, R.; Davidson, C.; Ure, A.; Quevauviller, P.. Improvement of the BCR three step sequential extraction procedure prior to the certification of new sediment and soil reference materials. *J Environ Monit* 1999, *1*, 57-61.

[863] Davidson, C.M.; Wilson, L.E.; Ure, A.M. Effect of sample preparation on the operational speciation of cadmium and lead in a freshwater sediment. *Fresenius' J Anal Chem* 1999, *363*, 134-136.

[864] Quevauviller, P.; Rauret, G.; Muntau, H.; Ure, A. M.; Rubio, R.; López-Sánchez, J.F.; Fiedler, H.D.; Griepink, B. Evaluation of a sequential extraction procedure for the determination of extractable trace metal contents in sediments. *Fresenius' J Anal Chem* 1994, *349*, 808-814.

[865] Fiedler, H.D.; López-Sánchez, J.F.; Rauret, G.; Rubio, R.; Quevauviller, P.; Ure, A.M.; Muntau, H. Study of the stability of extractable trace metal contents in a river sediment using sequential extraction. *Analyst* 1994, *119*, 1109-1114.

[866] Heltai, G.; Percsich, K.; Fekete, I.; Barabas, B.; Jozsa, T. Speciation of waste water sediments. *Microchem J* 2000, *67*, 43-51.

[867] da Silva, C.L.; Masini, J.C. Determination of Cu, Pb, Cd, and Zn in river sediment extracts by sequential injection anodic stripping voltammetry with thin mercury film electrode. *Fresenius' J Anal Chem* 2000, *367*, 284-290.

[868] Marin, B.; Valladon, M.; Polve, M.; Monaco, A. Reproducibility testing of a sequential extraction scheme for the determination of trace metal speciation in a marine reference sediment by inductively coupled plasma-mass spectrometry. *Anal Chim Acta* 1997, *342*, 91-112.

[869] Thomas, R.P.; Ure, A.M.; Davidson, C.M.; Littlejohn, D.; Rauret, G.; Rubio, R.; López-Sánchez, J.F. 3-Stage sequential extraction procedure for the determination of metals in river sediments. *Anal Chim Acta* 1994, *286*, 423-429.

[870] Svete, P.; Milačič, R.; Pihlar, B. Partitioning of Zn, Pb and Cd in river sediments from a lead and zinc mining area using the BCR three-step sequential extraction procedure. *J Environ Monit* 2001, *3*, 586-590.

[871] Cuong, D.T.; Obbard, J.P. Metal speciation in coastal marine sediments from Singapore using a modified BCR-sequential extraction procedure. *Appl Geochem* 2006, *21*, 1335-1346.

[872] Filgueiras, A.V.; Lavilla, I.; Bendicho, C. Evaluation of distribution, mobility and binding behavior of heavy metals in surficial sediments of Louro River (Galicia, Spain) using chemometric analysis: a case study. *Sci Total Environ* 2004, *330*, 115-129.

[873] Scancar, J.; Milačič, R.; Strazar, M.; Burica, O. Total metal concentrations and partitioning of Cd, Cr, Cu, Fe, Ni and Zn in sewage sludge. *Sci Total Environ* 2000, *250*, 9-19.

[874] Greenway, G.M.; Song, Q.J. Heavy metal speciation in the composting process. *J Environ Monit* 2002, *4*, 300-305.

[875] Walter, I.; Martínez, F.; Cala, V. Heavy metal speciation and phytotoxic effects of three representative sewage sludges for agricultural uses. *Environ Pollut* 2006, *139*, 507–514.

[876] Kim, B.; McBride, M.B. A test of sequential extractions for determining metal speciation in sewage sludge-amended soils. *Environ Pollut* 2006, *144*, 475–482.

[877] Kazi, T.G.; Jamali, M.K.; Kazi, G.H.; Arain, M.B.; Afridi, H.I.; Siddiqui, A. Evaluating the mobility of toxic metals in untreated industrial wastewater sludge using a BCR sequential extraction procedure and a leaching testimate. *Anal Bioanal Chem* 2005, *383*, 297–304.

[878] Fuentes, A.; Llorens, M.; Saez, J.; Soler, A.; Aguilar, M.I.; Ortuno. J.F.; Meseguer, V.F. Simple and sequential extractions of heavy metals from different sewage sludges. *Chemosphere* 2004, *54*, 1039–1047.

[879] Margui, E.; Queralt, I.; Carvalho, M.L.; Hidalgo, M. Assessment of metal availability to vegetation (*Betula pendula*) in Pb–Zn ore concentrate residues with different features. *Environ Pollut* 2006, *145*, 179–184.

[880] Margui, E.; Salvado, V.; Queralt, I.; Hidalgo, M. Comparison of three-stage sequential extraction and toxicity characteristic leaching tests to evaluate metal mobility in mining wastes. *Anal Chim Acta* 2004, *524*, 151–159.

[881] Hlavay, J.; Polyák, K.; Weisz, M. Monitoring of the natural environment by chemical speciation of elements in aerosol and sediment samples. *J Environ Monit* 2001, *3*, 74–80.

[882] Sysalova, J.; Szakova, J. Mobility assessment and validation of toxic elements in tunnel dust samples- Subway and road using sequential chemical extraction and ICP-OES/GF AAS measurements. *Environ Res* 2006, *101*, 287–293.

[883] Soylak, M.; Uzek, U.; Narın, I.; Tuezen, M.; Turkoglu, O.; Elci, I. Application of the sequential extraction procedure for dust samples from Kayseri-Turkey. *Fresenius Environ Bull* 2004, *13*, 454–457.

[884] Petit, M.D.; Rucandio, M.I. Sequential extractions for determination of cadmium distribution in coal fly ash, soil and sediment samples. *Anal Chim Acta* 1999, *401*, 283–291.

[885] Arain, M.B.; Kazi, T.G.; Jamali, M.K.; Jalbani, N.; Afridi, H.I.; Baig, J.A. Speciation of heavy metals in sediment by conventional, ultrasound and microwave assisted single extraction methods: A comparison with modified sequential extraction procedure. *J Hazard Mat* 2008, *154*, 998-1006.

[886] Jamali, M.K.; Kazi, T.G.; Arain, M.B.; Afridi, H.I.; Jalbani, N.; Kandhro, G.A.; Shah, Q.A.; Baig, J.A. Speciation of heavy metals in untreated sewage sludge by using microwave assisted sequential extraction procedure. *J Hazard Mat* 2009, *163*, 1157-1164.

[887] Krasnodębska-Ostręga, B.; Kaczorowska, M.; Golimowski, J. Ultrasound-assisted extraction for the evaluation of element mobility in bottom sediment collected at mining and smelting Pb-Zn ores area in Poland. *Microchim Acta* 2006, *154*, 39-43.

[888] Domouhtsidou, G.P.; Dailianis, S.; Kaloyianni, M.; Dimitriadis, V.K. Lysosomal membrane stability and metallothionein content in *Mytilus galloprovincialis (L.)*, as biomarkers - Combination with trace metal concentrations. *Mar Pollut Bull* 2004, *48*, 572–586.

[889] Romeo, M.; Mourgaud, Y.; Geffard, A.; Gnassia-Barelli, M.; Amiard, J.C.; Budzinski, H. Multimarker approach in transplanted mussels for evaluating water quality in Charentes, France, coast areas exposed to different anthropogenic conditions. *Environ Toxicol* 2003, *18*, 295–305.

[890] Giguere, A.; Couillard, Y.; Campbell, P.G.C.; Perceval, O.; Hare, L.; Pinel–Alloul, B.; Pellerin, J. Steady-state distribution of metals among metallothionein and other cytosolic ligands and links to cytotoxicity in bivalves living along a polymetallic gradient. *Aquatic Toxicol* 2003, *64*, 185–200.

[891] Dragun, Z.; Erk, M.; Raspor, B.; Ivankovic, D.; Pavicic, J. Metal and metallothionein level in the heat-treated cytosol of gills of transplanted mussels *Mytilus galloprovincialis Lmk. Environ Int* 2004, *30*, 1019–1025.

[892] Mourgaud, Y.; Martínez, E.; Geffard, A.; Andral, B.; Stanisiere, J.-I.; Amiard, J.-C. Metallothionein concentration in the mussel *Mytilus galloprovincialis* as a biomarker of response to metal contamination: validation in the field. *Biomarkers* 2002, *7*, 479–490.

[893] Nordberg, M. Metallothioneins: historical review and state of knowledge. *Talanta* 1998, *46*, 243-254.

[894] Hunziker, P.E.; Kägi, J.H.R. Metallothionein; Harrison, P.; Ed.; *Metalloproteins Part 2. Metal proteins with non–redox roles*; Verlag Chemie: Weinheim, 1985.

[895] Vasak, M. Advances in metallothionein structure and functions. *J Trace Element Med Biol* 2005, *19*, 13-17.

[896] Suzuki, K.T. Preparation of metallothioneins; Stillman, M.J.; Shaw III, C.F.; Suzuki, K.T.; Eds; *Metallothioneins. Synthesis, structure and properties of metallothioneins, phytochelatins and metal–thiolate complexes*; VCH Publishers Inc.: Weinheim, 1992.

[897] Chassaigne, H.; Łobiński, R. Polymorphism and identification of metallothionein isoforms by reversed-phase HPLC with on-line ion spray mass spectrometric detection. *Anal Chem* 1998, *70*, 2536-2543.

[898] Chassaigne, H.; Łobiński, R. Characterization of metallothionein isoforms by reversed-phase high-performance liquid chromatography with on-line post-column acidification and electrospray mass spectrometric detection. *J Chromatogr A* 1998, *829*, 127-136.

[899] Mounicou, S.; Połeć, K.; Chassaigne, H.; Potin–Gautier, M.; Łobiński, R. Characterization of metal complexes with metallothioneins by capillary zone

electrophoresis (CZE) with ICP-MS and electrospray (ES)-MS detection. *J Anal At Spectrom* 2000, *15*, 635-642.

[900] Ferrarello, C.N.; Fernández de la Campa, M.R.; Goenaga–Infante, H.; Fernández–Sánchez, M.L.; Sanz–Medel, A. Improved separation of rabbit liver metallothioneins by FPLC-ICP-MS: a comparison with the conventional anion-exchange chromatography. *Analusis* 2000, *28*, 351-357.

[901] Polec, K.; Szpunar, J.; Palacios, O.; González-Duarte, P.; Atrian, S.; Łobinski, R. Investigation of metal binding by recombinant and native metallothioneins by capillary zone electrophoresis (CZE) coupled with inductively coupled plasma mass spectrometry (ICP-MS) via a self-aspirating total consumption micronebulizer. *J Anal At Spectrom* 2001, *16*, 567–574.

[902] Polec-Pawlak, K.; Schaumlöffel, D.; Szpunar, J.; Orange, A.; Łobinski, R. Analysis or metal complexes with metallothionein in rat liver by capillary zone electrophoresis using ICP double-focussing sector-field isotope dilution MS and electrospray MS detection. *J Anal At Spectrom* 2002, *17*, 908-912.

[903] Chassaigne, H.; Łobinski, R. Polymorphism and identification of metallothioneins isoforms by reversed-phase HPLC with on-line ion-spray and inductively coupled plasma mass spectrometric detection. *Fresenius J Anal Chem* 1998, *361*, 267–273.

[904] Polec, K.; Mounicou, S.; Chassaigne, H.; Łobinski, R. Probing metal-complexes with metallothioneins by reversed phase microbore chromatography and capillary zone electrophoresis coupled with inductively coupled plasma and electrospray mass spectrometry. *Cell Mol Biol* 2000, *46*, 221–235.

[905] Polec, K.; García-Arribas, O.; Peréz-Calvo, M.; Szpunar, J.; Ribas-Ozonas, N.; Łobinski, R. Identification of cadmium-bioinduced ligands in rat liver using parallel HPLC-ICP-MS and HPLC-electrospray MS. *J Anal At Spectrom* 2000, *15*, 1363–1368.

[906] Dal Corso, G.; Farinati, S.; Maistri, S.; Furini, A.. How plants cope with cadmium: Staking all on metabolism and gene expression. *J Integrative Plant Biol* 2008, *50*, 1268-1280.

[907] Domenech, J.; Mir, G.; Huguet, G.; Capdevila, M.; Molinas, M.; Atrian, S. Plant metallothionein domains: Functional insight into physiological metal binding and protein folding. *Biochim* 2006, *88*, 583-593.

[908] Gupta, R.K. Dobritsa, S.V. Stiles, C.A. Essington, M.E. Liu, Z.Y. Chen, C.-H. Serpersu, E.H.; Mullin, B.C. Metallohistins: A new class of plant metal-binding proteins. *J Protein Chem* 2002, *21*, 529-536.

[909] Roesijadi, G.; Fowler, B. A. Purification of invertebrate metallothioneins. *Meth Enzymol* 1991, *205*, 263-273.

[910] Wolf, C.; Rösick, U.; Brätter, P. Sampling and processing of biopsy samples for speciation studies of cytosolic metalloproteins. *Anal Bional Chem* 2002, *372*, 491-494.

[911] Rudolph, C.; Adam, G.; Simm, A. Determination of copy number of c-Myc protein per cell by quantitative Western blotting. *Anal Biochem* 1999, *269*, 66-71.

[912] Santiago-Rivas, S.; Moreda-Pineiro, A.; Bermejo-Barrera, P.; Moreda-Pineiro, J.; Alonso-Rodriguez, E., Muniategui-Lorenzo, S.; López-Mahía, P.; Prada-Rodríguez, D. Pressurized liquid extraction-assisted mussel cytosol preparation for the determination of metals bound to metallothnionein-like proteins. *Anal Chim Acta* 2007, *603*, 36-43.

[913] Pedersen, K.L.; Pedersen, S.N.; Knudsen, J.; Bjerregaard, P. Quantification of metallothionein by differential pulse polarography overestimates concentrations in crustaceans. *Environ Sci Technol* 2008, *42*, 8426-8432.

[914] López, M.J.; Arino, C.; Díaz-Cruz, S.; Díaz-Cruz, J.M.; Tauler, R.; Esteban, M. Voltammetry assisted by multivariate analysis as a tool for speciation of metallothioneins: Competitive complexation of α- and β-metallothionein domains with cadmium and zinc. *Environ Sci Technol* 2003, *37*, 5609-5616.

[915] Gómez-Ariza, J.L.; García-Barrera, T.; Lorenzo, F.; Bernal, V.; Villegas, M.J.; Oliveira, V. Use of mass spectrometry techniques for the characterization of metal bound to proteins (metallomics) in biological systems. *Anal Chim Acta* 2004, *524*, 15-22.

[916] Prange, A.; Schaumlöffel, D. Hyphenated techniques for the characterization and quantification of metallothionein isoforms. *Anal Bioanal Chem* 2002, *373*, 441-453.

[917] Ferrarello, C.N.; Fernández de la Campa, M.R.; Sanz-Medel, A. Multielement trace-element speciation in metal-biomolecules by chromatography coupled with ICP-MS. *Anal Bioanal Chem* 2002, *373*, 412-421.

[918] Minami, T.; Ichida, S.; Kubo, K. Study of metallothionein using capillary zone electrophoresis. *J Chromatogr B* 2002, *781*, 303-311.

[919] Ferrarello, C.N.; Fernández de la Campa, M.; Muñiz, C.S.; Sanz–Medel, A. Metal distribution patterns in the mussel *Mytilus edulis* cytosols using size-exclusion chromatography and double focusing ICP-MS detection. *Analyst* 2000, *125*, 2223–2229.

[920] Ferrarello, C.N.; Fernández de la Campa, M.; Montes-Bayón, M.; Sanz–Medel, A. Multi-elemental speciation studies of trace elements associated with metallothionein-like proteins in mussels by liquid chromatography with inductively coupled plasma time of fligh mass spectrometric detection. *J Anal At Spectrom* 2000, *15*, 1558-1563.

[921] Montes-Bayón, M.; Proefrock, D.; Sanz-Medel, A.; Prange, A. Direct comparison of capillary electrophoresis and capillary liquid chromatography hyphenated to collision-cell inductively coupled plasma mass spectrometry for the investigation of Cd-, Cu- and Zn-containing metalloproteins. *J Chromatogr A* 2006, *1114*, 138-144.

[922] Digilio, G.; Bracco, C.; Vergani, L.; Botta, M.; Osella, D.; Viarengo, A. The cadmium binding domains in the metallothionein isoform Cd7-MT10 from *Mytilus galloprovincialis* revealed by NMR spectroscopy. *J Biol Inorg Chem* 2009, *14*, 167-178.

[923] Wang, J.; Dreessen, D.; Wiederin, D.R.; Houk, R.S. Measurement of trace elements in proteins extracted from liver by size exclusion chromatography-inductively coupled plasma-mass spectrometry with a magnetic sector mass spectrometer. *Anal Biochem* 2001, *288*, 89–96.

[924] Ogra, Y.; Suzuki, K.T. Biological significance of non-acetylated metallothionein. *J Chromatogr B* 1999, *735*, 17–24.

[925] Nischwitz, V.; Michalke, B.; Kettrup, A. Extraction and characterization of trace element species from porcine liver samples using online HPLC-ICP-MS and off line HPLC-ESI-MS. *J Anal At Spectrom* 2003, *18*, 444-451.

[926] Wolf. C.; Rösick, U.; Brätter, P. Quantification of the metal distribution in metallothioneins of the human liver by HPLC coupled with ICP-AES. *Fresenius J Anal Chem* 2000, *368*, 839–843.

[927] Boulyga, S.F.; Loreti, V.; Bettmer, J.; Heumann, K.G. Application of SEC-ICP-MS for comparative analyses of metal-containing species in cancerous and healthy human thyroid samples. *Anal Bioanal Chem* 2004, *380*, 198-203.

[928] Richarz, A.-N.; Wolf, C.; Brätter, P. Determination of protein-bound trace elements in human cell cytosols of different organs and different pathological states. *Analyst* 2003, *128*, 640-645.

[929] Richarz, P.; Brätter, A.-N. Speciation analysis of trace elements in the brains of individuals with Alzheimer's disease with special emphasis on metallothioneins. *Anal Bioanal Chem* 2002, *372*, 412-417.

[930] Van Campenhout, K.; Goenaga Infante, H.; Goemans, G.; Belpaire, C.; Adams, F.; Blust, R.; Bervoets, L. A field survey of metal binding to metallothionein and other cytosolic ligands in liver of eels using an on-line isotope dilution method in combination with size exclusion (SE) high pressure liquid chromatography (HPLC) coupled to inductively coupled plasma time-of-flight mass spectrometry (ICP-TOFMS). *Sci Total Environ* 2008, *394*, 379-389.

[931] Goenaga Infante, H.; Van Campenhout, K.; Schaumlöffel, D.; Blust, R.; Adams, F.C. Multi-element speciation of metalloproteins in fish tissue using size-exclusion chromatography coupled "on-line" with ICP-isotope dilution-time-of-flight-mass spectrometry. *Analyst* 2003, *128*, 651-657.

[932] Rodríguez-Cea, A.; Linde Arias, A.R.; Fernández de la Campa, M.R.; Costa Moreira, J.; Sanz-Medel, A. Metal speciation of metallothionein in white sea catfish, *Netuma barba*, and pearl cichlid, *Geophagus brasiliensis*, by orthogonal liquid chromatography coupled to ICP-MS detection. *Talanta* 2006, *69*, 963-969.

[933] Van Campenhout, K.; Goenaga Infante, H.; Adams, F.; Blust, R. Induction and binding of Cd, Cu, and Zn to metallothionein in carp (*Cyprinus carpio*) using HPLC-ICP-TOFMS. *Toxicol Sci* 2004, *80*, 276-287.

[934] Goenaga Infante, H.; Van Campenhout, K.; Blust, R.; Adams, F.C. Inductively coupled plasma time-of-flight mass spectrometry coupled to high-performance liquid chromatography for multi-elemental speciation analysis of metalloproteins in carp cytosols. *J Anal At Spectrom* 2002, *17*, 79-87.

[935] Goenaga Infante, H.; Cuyckens, F.; Van Campenhout, K.; Blust, R.; Claeys, M.; Van Vaeck, L.; Adams, F.C. Characterization of metal complexes with metallothioneins in the liver of the carp *Cyprinus carpio* by reversed-phase HPLC with ICP-MS and electrospray ionization (ESI)-MS. *J Anal At Spectrom* 2004, *19*, 159-166.

[936] Ferrarello, C.N.; Fernández de la Campa, M.R.; Carrasco, J.F.; Sanz-Medel, A. Speciation of metallothionein-like proteins of the mussel *Mytilus edulis* by orthogonal separation mechanisms with inductively coupled plasma-mass spectrometry detection: effect of selenium administration. *Spectrochim Acta B* 2002, *57*, 439-449.

[937] Vacchina, V.; Łobinski, R.; Oven, M.; Zenk, M.H. Signal identification in size-exclusion HPLC-ICP-MS chromatograms of plant extracts by electrospray tandem mass spectrometry (ES MS/MS). *J Anal At Spectrom* 2000, *15*, 529–534.

[938] Goenaga Infante, H.; Fernández Sánchez, M.L.; Sanz-Medel, A. Cadmium-bound species in human urine using high-performance liquid chromatography-vesicular hydride generation-inductively coupled plasma mass spectrometry. *J Anal At Spectrom* 1999, *14*, 1343–1348.

[939] Nöstelbacher, K.; Kirchgessner, M.; Stangl, G.I. Separation and quantitation of metallothionein isoforms from liver of untreated rats by ion-exchange high-performance liquid chromatography and atomic absorption spectrometry. *J Chromatogr B* 2000, *744*, 273–282.

[940] Goenaga Infante, H.; Van Campenhout, K.; Blust, R.; Adams, F.C. Anion-exchange high performance liquid chromatography hyphenated to inductively coupled plasma-isotope dilution-time-of-flight mass spectrometry for speciation analysis of metal complexes with metallothionein isoforms in gibel carp (*Carassius auratus gibelio*) exposed to environmental metal pollution. *J Chromatogr A* 2006, *1121*, 184-190.

[941] Rodríguez-Cea, A.; Fernández de la Campa, M.R.; Blanco González, E.; Andon Fernández, B.; Sanz-Medel, A. Metal speciation analysis in eel (*Anguilla anguilla*) metallothioneins by anionic exchange-FPLC-isotope dilution-ICP-MS. *J Anal At Spectrom* 2003, *18*, 1357-1364.

[942] Chassaigne, H.; Łobinski, R. Speciation of metal complexes with biomolecules by reversed-phase HPLC with ion-spray and inductively coupled plasma mass spectrometric detection. *Fresenius J Anal Chem* 1998, *361*, 267–273.

[943] Goenaga Infante, H.; Fernández Sánchez, M.L.; Sanz-Medel, A. Vesicle-mediated high performance liquid chromatography coupled to hydride generation inductively coupled plasma mass spectrometry for cadmium speciation in fish cytosols. *J Anal At Spectrom* 2000, *15*, 519–524.

[944] Ferrarello, C.N.; Ruiz Encinar, J.; Centineo, G.; García Alonso, J.I.; Fernández de la Campa, M.R.; Sanz-Medel, A. Comparison of three different ICP-MS instruments in the study of cadmium speciation in rabbit liver metallothionein-1 using reversed-phase HPLC and post-column isotope dilution analysis. *J Anal At Spectrom* 2002, *17*, 1024-1029.

[945] Lu, Q.; Bird, S.M.; Barnes, R.M. Interface for capillary electrophoresis and inductively-coupled plasma-mass spectrometry. *Anal Chem* 1995, *67*, 2949–2956.

[946] Lu, Q.; Barnes, R.M. Evaluation of an ultrasonic nebulizer interface for capillary electrophoresis and inductively coupled plasma mass spectrometry. *Microchem J* 1996, *54*, 129–143.

[947] B'Hymer, C.; Day, J.A.; Caruso, J.A. Evaluation of a microconcentric nebulizer and its suction effect in a capillary electrophoresis interface with inductively coupled plasma mass spectrometry. *Appl Spectr* 2000, *54*, 1040–1046.

[948] Taylor, K.A.; Sharp, B.L.; Lewis, D.J.; Crews, H.M. Design and characterisation of a microconcentric nebuliser interface for capillary electrophoresis-inductively coupled plasma mass spectrometry. *J Anal At Spectrom* 1998, *13*, 1095–1100.

[949] Majidi, V.; Miller-Ihli, N.J. Two simple interface designs for capillary electrophoresis inductively coupled plasma mass spectrometry. *Analyst* 1998, *123*, 803–808.

[950] Majidi. V.; Miller-Ihli, N.J. Potential sources of error in capillary electrophoresis inductively coupled plasma mass spectrometry for chemical speciation. *Analyst* 1998, *123*, 809–813.

[951] Prange, A.; Schaumlöffel, D.; Brätter, P.; Richarz, A.-N.; Wolf, C. Species analysis of metallothionein isoforms in human brain cytosols by use of capillary electrophoresis hyphenated to inductively coupled plasma-sector field mass spectrometry. *Fresenius J Anal Chem* 2001, *371*, 764–774.

[952] Schaumlöffel, D.; Prange, A.; Marx, G.; Heumann, K.G.; Brätter, P. Characterization and quantification of metallothionein isoforms by capillary electrophoresis-inductively coupled plasma-isotope-dilution mass spectrometry. *Anal Bioanal Chem* 2002, *372*, 155–163.

[953] Wang, Z.; Prange, A. Use of surface-modified capillaries in the separation and characterization of metallothionein isoforms by capillary electrophoresis inductively coupled plasma mass spectrometry. *Anal Chem* 2002, *74*, 626–631.

[954] Álvarez-Llamas, G.; Fernández de la Campa, M.R.; Sanz-Medel, A. An alternative interface for CE-ICP-MS cadmium speciation in metallothioneins based on volatile species generation. *Anal Chim Acta* 2005, *546*, 236-243.

[955] Álvarez-Llamas, G.; Fernández de la Campa, M.R.; Fernández Sánchez, M.L.; Sanz-Medel A. Comparison of two CE-ICP-MS interfaces based on microflow nebulizers: application to cadmium speciation in metallothioneins using quadrupole and double focusing mass analyzers. *J Anal At Spectrom* 2002, *17*, 655-661.

[956] Schaumlöffel, D.; Prange, A.; Marx, G.; Heumann, K.G.; Brätter, P. Characterization and quantification of metallothionein isoforms by capillary electrophoresis-inductively coupled plasma-isotope-dilution mass spectrometry. *Anal Bioanal Chem* 2002, *372*, 155-163.

[957] Baker, S.A.; Miller-Ihli, N.J. Comparison of a cross-flow and microconcentric nebulizer for chemical speciation measurements using CZE-ICP-MS. *Appl Spectrosc* 1999, *53*, 471–478.

[958] Wolf, C.; Schaumlöffel, D.; Richarz, A.-N.; Prange, A.; Brätter, P. CZE-ICP-MS separation of metallothioneins in human brain cytosols: comparability of electropherograms obtained from different sample matrices. *Analyst* 2003, *128*, 576-580.

[959] Proefrock, D.; Leonhard, P.; Prange, A. Determination of sulfur and selected trace elements in metallothionein-like proteins using capillary electrophoresis hyphenated to inductively coupled plasma mass spectrometry with an octopole reaction cell. *Anal Bioanal Chem* 2003, *377*, 132-139.

[960] Profrock, D.; Prange, A.; Schaumlöffel, D.; Ruck, W. Optimization of capillary electrophoresis-inductively coupled plasma mass spectrometry for species analysis of metallothionein-like proteins extracted from liver tissues of Elbe-bream and Roe deer. *Spectrochim Acta B* 2003, *58*, 1403-1415.

[961] Álvarez-Llamas, G.; Fernández de la Campa, M.R.; Sanz-Medel, A. Sample stacking capillary electrophoresis with ICP-(Q)MS detection for Cd, Cu and Zn speciation in fish liver metallothioneins. *J Anal At Spectrom* 2003, *18*, 460-466.

[962] Álvarez-Llamas, G.; Rodríguez-Cea, A.; Fernández de la Campa, M.R.; Sanz-Medel, A. Metallothionein isoforms separation and cadmium speciation by capillary electrophoresis with ultraviolet and quadrupole-inductively coupled plasma mass spectrometric detection. *Anal Chim Acta* 2003, *448*, 105-119.

[963] International Standard Organization. 1994. Water quality. Determination of cadmium by atomic absorption spectrometry. ISO 5961. International Organization for Standardization, Case Postale 56, CH-1211, Geneva 20 Switzerland.

[964] International Standard Organization. 1986. Water quality. Determination of cobalt, nickel, copper, zinc, cadmium and lead. Flame atomic absorption spectrometric

methods. ISO 8288. International Organization for Standardization, Case Postale 56, CH-1211, Geneva 20 Switzerland.

[965] International Standard Organization. 2003. Water quality. Determination of trace elements using atomic absorption spectrometry with graphite furnace. ISO 15586. International Organization for Standardization, Case Postale 56, CH-1211, Geneva 20 Switzerland.

[966] International Standard Organization. 2002. Water quality. Digestion for the determination of selected elements in water. Part 1. Aqua regia digestion. ISO 15587-1. International Organization for Standardization, Case Postale 56, CH-1211, Geneva 20 Switzerland.

[967] International Standard Organization. 2002. Water quality. Digestion for the determination of selected elements in water. Part 2. Nitric acid digestion. ISO 15587-1. International Organization for Standardization, Case Postale 56, CH-1211, Geneva 20 Switzerland.

[968] International Standard Organization. 2007. Water quality. Determination of selected elements by inductively coupled plasma optical emission spectroscopy (ICOP-OES). ISO 11885. International Organization for Standardization, Case Postale 56, CH-1211, Geneva 20 Switzerland.

[969] International Standard Organization. 2004. Water quality. Application of inductively coupled plasma mass spectrometry (ICP-MS). Part 1: General guidelines. ISO 17294-1. International Organization for Standardization, Case Postale 56, CH-1211, Geneva 20 Switzerland.

[970] International Standard Organization. 2003. Water quality. Application of inductively coupled plasma mass spectrometry (ICP-MS). Part 2: Determination of 62 elements. ISO 17294-2:2003. International Organization for Standardization, Case Postale 56, CH-1211, Geneva 20 Switzerland.

[971] Official methods of analysis of The Association of Official Analytical Chemists, Methods Manual, 18th edition, 2005. Cadmium, chromium, copper, iron, lead, magnesium, manganese, silver, and zinc in water. Atomic absorption spectrometric method. AOAC Official method 974.27.

[972] APHA. 1999. Standard Methods for the Examination of Water and Wastewater. 3111 Metals by Flame Atomic Absorption Spectrometry. American Public Health Association (APHA), American Water Works Association (AWWA), Water Environment Federation publication (WPCF). APHA, Washington, DC.

[973] APHA. 2004. Standard Methods for the Examination of Water and Wastewater. 3113 Metals by Electrothermal Atomic Absorption Spectrometry. American Public Health Association (APHA), American Water Works Association (AWWA), Water Environment Federation publication (WPCF). APHA, Washington, DC.

[974] APHA. 1999. Standard Methods for the Examination of Water and Wastewater. 3120 Metals by Plasma Emission Spectroscopy. American Public Health Association (APHA), American Water Works Association (AWWA), Water Environment Federation publication (WPCF). APHA, Washington, DC.

[975] APHA. Standard Methods for the Examination of Water and Wastewater. 3125 Metals by Inductively Coupled Plasma/Mass Spectrometry. American Public Health Association (APHA), American Water Works Association (AWWA), Water Environment Federation publication (WPCF). APHA, Washington, DC.

[976] APHA.2004. Standard Methods for the Examination of Water and Wastewater. 3130 Metals by Anodic Stripping Voltammetry. American Public Health Association (APHA), American Water Works Association (AWWA), Water Environment Federation publication (WPCF). APHA, Washington, DC.

[977] International Standard Organization. 1995. Water quality. Soil quality. Extraction of trace elements soluble in aqua regia. ISO 11466. International Oganization for Standardization, Case Postale 56, CH-1211, Geneva 20 Switzerland.

[978] International Standard Organization. 2001. Soil quality. Dissolution for the determination of total element content. Part 1: Dissolution with hydrofluoric and perchloric acids. ISO 14869-1. International Organization for Standardization, Case Postale 56, CH-1211, Geneva 20 Switzerland.

[979] International Standard Organization. 1998. Soil quality. Determination of cadmium, chromium, cobalt, copper, lead, manganese, nickel and zinc in aqua regia extracts of soils. Flame and electrothermal atomic absorption spectrometric methods. ISO 11047. International Organization for Standardization, Case Postale 56, CH-1211, Geneva 20 Switzerland.

[980] International Standard Organization. 2008. Soil quality. Determination of trace elements in extracts of soil by inductively coupled plasma - atomic emission spectrometry (ICP – AES). ISO 22036. International Organization for Standardization, Case Postale 56, CH-1211, Geneva 20 Switzerland.

[981] European Standard 2004. Stationary sources emissions. Determination of the total emission of As, Cd, Cr, Co, Cu, Mn, Ni, Pb, Sb, Tl and V. EN 14385. European Committee for Standardization, Brussels, Belgium.

[982] European Standard 2005. Ambient air quality. Standard method for measurement of Pb, Cd, As and Ni in the PM10 fraction of suspended particulate matter. EN 14902. European Committee for Standardization, Brussels, Belgium.

INDEX